Trace Elements in Waterlogged Soils and Sediments

Trace Elements in Waterlogged Soils and Sediments

Editor

Dalmiro Martell

scitus
academics

Trace Elements in Waterlogged Soils and Sediments
Edited by **Dalmiro Martell**

ISBN: 978-1-68117-703-8
Library of Congress Control Number: 2016958748

SCITUS Academics LLC,
616, Corporate Way, Suite 2, 4766,
Valley Cottage, NY 10989

www.scitusacademics.com

Notice

Printed in United States of America on Acid Free Paper

Preface

Elements occur naturally in soils and some are nutrients that, in addition to animal and human health, are essential to plant growth. However, at elevated levels, all trace elements become potentially toxic. Therefore, the anthropogenic input of trace elements into the natural environment presents a variety of environmental and health problems. The development of monitoring systems is of primary importance in environmental sciences. Increasingly stringent environmental regulations call for the development of new analytical techniques and for meaningful and simple tools to obtain information in solid phases on metal fractions of different mobility and bioavailability. The monitoring objectives are to evaluate the effects of pollution on humans and the environment, to identify possible sources and to establish relationships between pollutant concentrations and environmental or health effects. Fifteen or more elements, normally in small quantities, in soils and rocks, are important for plant and/or animal nutrition. They are known as trace elements, minor elements or possibly micronutrients because of their great abundance in natural uncontaminated earth materials or perhaps plants. Animals require zinc, vanadium, silicone, molybdenum, manganese, iron, copper and boron; copper, molybdenum, manganese, iron, iodine, cobalt, zinc and selenium. In order to understand their chemical cycles in nature, it is therefore important to investigate and understand the transport mechanisms of trace elements and their complexities.

This book analyzes state-of-the-art research and trends in the field, including chemistry, analysis, regulation and destination of trace elements in soils, as well as contaminated soil remediation strategies. Factors that control the dynamics are discussed and kinetics of trace elements and their main biogeochemical processes are released. It also examines current technologies for the remediation of trace metal contaminated sites and the role of bioavailability in regulatory decision-making and risk assessment. It will be a valuable tool for professionals worldwide in aquatic organisms, contaminant destination and transport, remediation technologies and risk assessment of aquatic and wetland ecosystems.

Table of Contents

Chapter 1

Screening for Barley Waterlogging Tolerance in Nordic Barley Cultivars (*Hordeum vulgare* L.) Using Chlorophyll Fluorescence on Hydroponically-Grown Plants

Nils-Ove Bertholdsson

Department of Plant Breeding, Swedish University of Agricultural Sciences, P.O. Box 101, 230 53 Alnarp, Sweden

ABSTRACT

Waterlogging can reduce crop yield by 20%–50% or more, and lack of efficient selection methods is an obstacle in plant breeding. The methods currently used are mainly indices based on germination ability in Petri dishes and leaf chlorosis in plants grown in waterlogged soils. Cultivation in oxygen-depleted nutrient solution is the ultimate waterlogging system. Therefore methods based on root growth inhibition and on fluorescence in plant material hydroponically grown in oxygen-depleted solution were evaluated against data on biomass accumulation in waterlogged soils. Both traits were correlated with waterlogging tolerance in soil, but since it was easier to measure fluorescence, this method was further evaluated. A selection of F_2 plants with high and low fluorescence revealed a small but significant screening effect in F_3 plants. A test of 175 Nordic cultivars showed large variations in chlorophyll fluorescence in leaves from oxygen-stressed seedlings, indicating that adaptation to waterlogging has gradually improved over the past 40–50 years with the introduction of new cultivars onto the market. However, precipitation also increased during the period and new cultivars may have inadvertently been adapted to this while breeding barley for grain yield. The results suggest that the hydroponic method can be used for screening barley populations, breeding lines or phenotyping of populations in developing markers for quantitative trait loci.

Keywords: biomass accumulation; chlorophyll fluorescence; Nordic barley; breeding; climate change; root growth; waterlogging

1. INTRODUCTION

Barley (*Hordeum vulgare* L.) is more susceptible to waterlogging stress than other cereals [1] and waterlogging is a limiting factor in barley production worldwide. In China and Japan, improved tolerance to waterlogging is one of the major objectives in barley breeding programmes [2], and with expected climate change its importance will also increase in northern Europe [3]. Today waterlogging is reported to reduce grain yield by 20%–25% or more, depending on stage of development [1]. As with other abiotic stresses, barley adaptation to waterlogging is a complex trait. However, there is currently no useful and efficient selection method of screening for this trait, which is an obstacle in breeding programmes. There are several recent reviews of the area [1,2,4,5] and thus only some basic information of importance for the present study is mentioned here.

The timing and duration of waterlogging vary with soil type, frequency of heavy rainfall and risk of flooding. Waterlogging is a state in which excess water in the root zone affects the oxygen concentration in the soil, resulting in an energy crisis in roots [4] and reducing plant growth at any growth stage [1]. During short-term waterlogging, differences in constitutive root porosity (*i.e.*, more aerenchyma) are probably the main reason why cereals differ in tolerance, but anoxia tolerance could also have an impact [1,4,6]. The plant may also escape stress by resorting to low metabolic activity during the stress period and showing rapid recovery ability when the soil is drained. During long-term waterlogging, ethanolic fermentation and various recovery mechanisms may also take place [1,7,8]. One important feature among wetland species is a barrier to radial oxygen losses (ROL), but this seems to be of less importance in wheat and barley [6,8]. Even short-term waterlogging in wheat has long-term effects on growth and physiology of the plant due to death of root apices and impaired root and shoot growth [9].

Adaptation of barley to waterlogging at the germination stage has been studied in Japan [10] and Australia [2], and extensive screening of barley germplasm for this trait has been carried out in China and Japan [1]. The main methods used are indices based on leaf chlorosis [11] and germination ability [10]. Pang *et al.* [7] suggest the use of photosynthetic rate or total chlorophyll for selection of parental genotypes and chlorophyll fluorescence when screening large numbers of breeding lines. However, there are no reports on the heritability of these traits and their use in practical breeding. Li *et al.* [12] used indices of leaf chlorosis, plant survival and plant biomass reduction to identify quantitative trait loci (QTL) associated with waterlogging in barley and found seven such QTL. Some of these QTL affected multiple adaptation to waterlogging-related traits, e.g. by reducing leaf chlorosis and increasing plant biomass under waterlogging stress. In a recent study, Zhou [2] concluded that accurate phenotyping is crucial for identifying better QTL.

Most studies to date have been performed using waterlogged soils, but the use of hydroponics could be an interesting alternative. Cultivation in oxygen-depleted nutrient solution is the ultimate waterlogging system and can be carefully controlled, unlike the soil system. However, there are few studies using hydroponically grown plants or plants grown in stagnant water (gels) or with sand as substrate [13]. The aim of the present study was therefore to use hydroponically-grown material to identify useful selection criteria at the seedling stage for adaptation of barley to waterlogging. Two traits of particular interest are root growth and chlorophyll fluorescence, both of which can easily be observed in hydroponically-grown seedlings. The root system is the first plant part that is directly exposed to the stress and the plant responds to waterlogging with reduced root growth [14]. An even faster response to waterlogging is apparent as changes in leaf chlorophyll fluorescence caused by abiotic stresses [15,16]. Measurement of leaf chlorophyll fluorescence also offers the possibility of non-destructive selection of plants.

Specific objectives of the study were: (1) to develop a screening method for barley adaptation to waterlogging based on hydroponically-grown seedlings; (2) to evaluate the method by screening a segregating population; and (3) to characterise barley cultivars released in the Nordic and Baltic countries over more than 100 years in order to identify useful genetic resources for breeding waterlogging-tolerant accessions.

2. EXPERIMENTAL SECTION

A preliminary test conducted with 10 barley cultivars submitted to official variety testing in Sweden between 1996 and 2001 revealed 40% differences in adaptation to waterlogging [17]. These cultivars and two more recent cultivars were therefore selected as reference material in different studies to develop a screening method for waterlogging tolerance. For the evaluation of selection efficiency, a F_2 population from the cross cv. Henni × cv. Gustav was used. For the screening of possible sources of adaptation to waterlogging, a collection of 175 Nordic and Baltic cultivars released in the last 100+ years was used [18], complemented with some new cultivars. Initially the reference material was studied in waterlogged soil and in stagnant nutrient solution to determine which of the two traits (root growth or chlorophyll fluorescence) was most suitable for use in screening of waterlogging tolerance.

2.1. Soil Studies

The reference material was sown in cylindrical pots (diameter 10 cm, height 20 cm) filled with fertilised peat mould with Leca (NPK 180/90/195 g m^{3} + Micro©) (Weibulls Horto, Hammenhög, Sweden). The drainage holes in the base of half the pots were sealed with adhesive tape to allow waterlogging. Five seeds were planted in each of four replicate pots and thinned out to four after germination. At sowing, 150 mL water were added to the pots, which were randomly positioned with the four replicate pots on separate trolleys in a growth chamber with 16 h artificial light (150 µmol $m^{-2}s^{-1}$ in the range 400–700 nm) for 12 days at 18/15 °C day/night temperature. After 12 days, all pots were given 150 mL nutrient solution (Rika-S NPK 84/12/56 g L^{-1} +Micro©, diluted 1:400 (Weibulls Horto, Hammenhög, Sweden) and pots with sealed holes were given additional water until the soil was fully saturated (55%–57% water content; Theta Probe type ML1, Delta T Devices, Cambridge, UK). Control pots were kept at 25%–33% water content. After 18 days of stress, the control and oxygen-stressed plants were cut and dried at 70 °C for one week. The study was repeated twice and a waterlogging tolerance index was calculated as the mean seedling weight of oxygen-stressed plants relative to mean seedling weight of control plants × 100. The values obtained were used as reference values for waterlogging tolerance (WLT) when optimising the hydroponic methods.

2.2. Hydroponics

A previously developed hydroponic system [18] was used to study seedling root and shoot growth of the reference material. The seeds were sown on strips of corrugated paper suspended in frames in 25-litre containers with a balanced phosphate-buffered complete 2 mM nitrogen nutrient solution (pH = 6.5) [19]. Each strip was 3.5 cm wide and the ridges and furrows were flattened out at the lower two-thirds with a clothes iron to prevent the seeds falling down and to facilitate mounting in the frames. The ends of the flattened part were cut so that the strip could hang down between plastic rulers lying on the frame and fixed in place by masking tape. Similar strips of filter paper, but 10 cm wide, were suspended alongside the corrugated strips as wicks. After sowing, the strips were sprayed with a detergent (Lissapol, 0.5%, Syngenta, Copenhagen, Denmark) and covered with filter paper strips partly hanging down into the solution until the seeds had germinated. The cultivars were arranged in a randomised complete block design, with four replicates of six plants, and repeated twice. The plants in the first strip at each end of the frame and the first plant in each strip were border plants. The containers were placed in a growth chamber with 16 h artificial light (150 μmol $m^{-2}s^{-1}$ in the range 400–700 nm) at 18/15 °C day/night temperature. Containers with fully aerated (control) and non-aerated (oxygen-stressed/waterlogged) nutrient solution were used. Aerated solution was obtained by bubbling air through the nutrient solution. Both the control and oxygen-stressed containers were aerated from day 4 to 7, but after day 7 only the controls were aerated. The nutrient solution was renewed after one week and then at 4-day intervals until the end of the experiment. After 12 days of stress, the length of the longest seminal root (Root length) in all plants, as well as the pooled dry weight of roots (Root weight), shoots (Shoot weight) and whole plants (Plant weight), were analysed.

2.3. Chlorophyll Fluorescence

Measurements were made with a pulsed amplitude modulation (PAM) portable MINI-PAM fluorometer with Leaf-Clip Holder 2030-B attached (WALZ, GmbH, Germany). Six plants each of two replicates per treatment, were measured for quantum yield (QY) as (M-F)/M)), where F is fluorescence yield at various light intensities and M is maximum fluorescence yield during a saturation light pulse. To facilitate the measurements fully light-illuminated leaves were used, which was possible since only QY and not quenching coefficients were of interest [20]. In all measurements cv Henni was used as an internal standard and QY is presented relative to Henni when not all cultivars could be analysed in the same container. During the development of the method, F was evaluated at five light intensities: ambient light giving a PAR reading of 89 μmol $m^{-2}s^{-1}$ with the built-in sensor, ambient light closer to the light source giving a PAR reading of 125 μmol $m^{-2}s^{-1}$, light from an external LED lamp giving 400 μmol $m^{-2}s^{-1}$, 5 sec of actinic light from the internal lamp giving a PAR reading of 275 μmol $m^{-2}s^{-1}$ and 5 sec of actinic light giving a PAR reading of 376 μmol $m^{-2}s^{-1}$. The last three intensities are a combination of ambient and external or internal lamp. The material was grown in the same hydroponic system as the plants for studies of root and shoot growth. The plants were subjected to stress by removing the aerating device after 10 days. The PAM measurements were made after 0 and until 9 days of stress. The seedling was removed from the corrugated paper prior the measurement and the first leaf of the plant was clipped on to the leaf holder at the middle of the leaf and upper side up. The clip holder was fixed to a table to give standard conditions during measurements. For the measurement at 275 and 378 µmol $m^{-2}s^{-1}$, the first leaf was cut off and placed horizontally in ambient light of 155 µmol $m^{-2}s^{-1}$ for 5 min prior to measurements. When the measurements were made immediately after removing the plant from the hydroponic container, QY was lower, with higher standard deviation. After only one minute with the leaves lying on the table, QY stabilised at a higher value and remained there for at least 20 min, and the standard error was also reduced. In the final method, leaves were cut off and placed horizontally in ambient light of 155 µmol $m^{-2}s^{-1}$ for 5 min for light-adaptation and then measured with 5 sec of extra light of 378 ± 20 µmol $m^{-2}s^{-1}$. It was important that the part of the leaf being measured did not curl away from the light but stayed fully illuminated.

During the development of the method, 2 × 6 plants were measured in each test run. In the selection of plants from the segregation population cv. Henni × cv. Gustav, two hydroponic frames were filled with 500 F_2 seeds from the population. Twenty plants with high and low QY were selected and planted in pots and tested again in the next generation as breeding lines with 4 × 6 plants per line. In the study of Nordic cultivars, the cultivars were divided into sub-sets of 11 cultivars + cv. Henni as an internal check to calculate relative QY (QYrel). Relative values were used since in repeated tests of the reference material, QYrel varied less than QY. Each sub-set was tested in a single hydroponic tank with a randomised block design with four blocks and six plants per block. Some cultivars were tested two or three times if they showed a particularly high or low QY.

2.4. Statistical Analysis

The soil and hydroponic experiments were arranged as randomised split-plot designs with treatment as main plot and varieties as subplots. Data for growth parameters in pots and hydroponics with oxygen-stressed and non-stressed treatments were analysed by analysis of variance (ANOVA) with the general linear model (GLM) in the Statgraphics Plus (Manugistics, Inc, Rockville, MD, USA) software. Differences between treatments for all measurements were compared using the *t*-test ($P < 0.05$) and 2 × standard errors of means are presented in the diagrams. The number of observations varied with parameter studied and is shown in each table or diagram. Regression and correlation analyses were performed between mean values of parameters with physiological relevance. Although the measurements of fluorescence were carried out in standardised environment, some very low values deviated more than others and were excluded if the studentised residual was greater than 2.5. Box-and-whisker plots were used to illustrate how QYrel changed by introducing new cultivars grouped by decades or half-decades in the Nordic countries.

3. RESULTS

3.1. Growth Reduction in Waterlogged Soil

The seedling shoot weight of the reference material was reduced during 18 days of waterlogging by on average 29% ($P < 0.001$). However, the growth reduction varied with cultivar ($P < 0.01$ for treatment × cultivar). Cultivars Cecilia and Pongo were most affected, with a growth reduction of 37.6% and 37.1%, respectively ($P < 0.001$), and cvs. Henni and Otira were least affected, with a non-significant reduction of 16.3% and 14.8%, respectively (Table 1). Cv. Pongo and particularly cv. Cecilia also showed yellowing leaves, while cvs. Henni and Otira were still green. The relative growth reduction in soil (WLT) is used from here on as the reference for adaptation to waterlogging and to identify possible other traits that could be used for screening of waterlogging tolerance.

3.2. Seedling Root Growth in Oxygen-Depleted Nutrient Solution

Shoot and root weight of hydroponically grown plants were analysed, but since seminal root length was strongly correlated with WLT and QY, only the results for this parameter are presented here. The length of the longest seminal root was decreased by between 17% and 49% by non-aeration ($P < 0.001$) (Table 1). As found for shoot growth in waterlogged soil, cvs. Henni and Otira were less affected (22% and 17% decrease, respectively). Most affected was cv. Scarlet (49% decrease) followed by cv. Meltan (47%). In absolute figures, cvs. Henni and Otira had the longest seminal roots under hypoxia and cvs. Barke and Scarlett the shortest. Seminal root length was highly correlated with WLT ($r = 0.81$, $P < 0.01$) under stress, but not correlated at all when aerated ($r = 0.01$, non-significant) (Table 2).

3.3. Quantum Yield of Seedlings Grown in Oxygen-Depleted Solution

The correlation between WLT and QY at ambient light intensity of hypoxia-stressed plants increased with number of days without aeration, but remained non-significant. With light intensities above ambient, the correlations improved and reached significance already after 3–5 days. After a longer stress period, the correlation coefficient decreased again (Figure 1).

Table 1. Shoot dry weight of 12 barley cultivars grown in pots with (Stress) and without (Control) 19 days of waterlogging, seminal root length of hydroponically grown plants with (Control) and without aerated (Stress) nutrient solution for 14 days, and fluorescence (Quantum yield measured at 376 (µmol $m^{-2}s^{-1}$) of the first leaf of hydroponically-grown plants with (Control) and without aerated (Stress) nutrient solution during four days prior to fluorescence measurements. Each shoot weight value is the mean of five replicate plants, root length is the mean of 8 plants and QY is the mean of 4 × 6 plants. Rel = Stress/Control×100. Rel for shoot weight is the reference measure of waterlogging tolerance (WLT).

Cultivar	Shoot weight (mg)			Root length (mm)			Quantum yield		
	Control	Stress	Rel(%)	Control	Stress	Rel(%)	Control	Stress	Rel(%)
Alexis	327	212	65.8	155	97	63	0.418	0.364	87.2
Baronesse	349	246	72.0	175	107	61	0.471	0.432	91.8
Cecilia	356	222	62.4	170	99	59	0.408	0.372	91.4
Barke	356	244	68.4	155	89	58	0.388	0.383	98.8
Henni	294	245	83.7	164	127	78	0.462	0.440	95.2
Scarlett	297	205	69.8	180	92	51	0.439	0.394	89.8
Meltan	288	211	75.3	180	96	53	0.456	0.413	90.6
Otira	318	260	85.2	150	124	83	0.455	0.439	96.3
Pongo	286	173	62.9	138	98	72	0.418	0.385	92.2
Optic	326	228	70.7	184	101	55	0.450	0.391	86.8
Gustav	262	182	70.6	173	101	58	0.482	0.416	86.2
Quench	326	215	69.1	191	102	54	0.459	0.421	91.8
SEM	10.8	13.2	4.5	3.9	2.0	2.0	0.006	0.006	1.56
Treatment	***			***			***		
Cultivar	ns		***	*		***	***		***
T x C	**			***			***		

ns, *, ** and *** indicate non-significant ($P > 0.05$) or significant at $P < 0.05$, $P < 0.01$ and $P < 0.001$, respectively.

Several ways to increase light intensity above ambient were tested and the best results were obtained when the internal lamp was used. To avoid overheating, only 5 sec of light were given, to increase the light intensity to 275 µmol $m^{-2}s^{-1}$ and 376 µmol $m^{-2}s^{-1}$. In the study of the reference material, leaves from 12 plants were measured. However, when 24 leaves adapted to the same light intensity prior to measurement were used the correlation with WLT was stronger ($r = 0.82$, $P < 0.01$) (Table 2). The values obtained in these ways were also positively correlated with RLs ($r = 0.84$, $P < 0.01$) (Table

2) and grain yield from official variety testing in Sweden ($r = 0.83$, $P < 0.01$) (Figure 2). The correlation between grain yield and WLT was $r = 0.73$, $P < 0.05$ (not shown in any table or diagram).

Table 2. Correlations between characteristics of 12 cultivar seedlings (reference material) grown in aerated (Control) and non-aerated (Stress) hydroponic nutrient solution and a waterlogging tolerance (WLT) index, calculated as the shoot weight of waterlogged plants relative to that of control plants.

	WLT	SWc	SWs	RLc	RLs	RLs/RLc
Shoot weight control (SWc)	−0.24	1.00				
Shoot weight stress (SWs)	0.61*	0.61*	1.00			
Root length control (RLc)	0.01	0.03	0.02	1.00		
Root length stress (RLs)	0.81**	−0.13	0.53	−0.16	1.00	
RLs/RLc	0.55	−0.11	0.34	−0.72**	0.80**	1.00
QY control (QYc)	0.56	−0.51	0.01	0.49	0.51	0.05
QYstress (QYs)	0.82***	−0.28	0.41	0.21	0.75**	0.40
QYs/QYc	0.34	0.36	0.60*	−0.43	0.32	0.49

*, **, *** indicate correlations significantly different from zero at $P < 0.05$, $P < 0.01$ and $P < 0.001$, respectively. Degrees of freedom = 10. QY was measured at 376 ($\mu mol\ m^{-2} s^{-1}$).

3.4. Selection Efficiency of Quantum Yield

Of 20 selected plants with high and low QY from a cv. Henni (QYrel = 100) × cv. Gustav (QYrel = 95) cross, only 16 of each selection produced enough F_3 seeds for testing chlorophyll fluorescence. The mean QYrel of the plus and minus selection were 104.7 and 100.9, respectively, which were significantly different from each other ($P < 0.001$). A mean selection effect of 3.8 % is not much, but the parents only differed by 5%. Of the 16 lines selected to be high only two lines showed low QY and among those selected to be low thirteen showed low QY (Figure 3). However, since there is no absolute relationship between QY and WLT the study should be followed up with water logging studies in soils to confirm these results.

3.5. Waterlogging Tolerance of Nordic Barley Cultivars

With the introduction of new, in most cases six-row, barley cultivars in Norway and Finland, there has been an increasing trend in QY, although the fluctuation is large (Figure 4). Most landraces (introduced before 1900) showed lower QYrel than the standard cv. Henni and the very first cultivars were also less tolerant than more recent cultivars in general. A sub-set of 11 Norwegian cultivars (every second) was also studied in both well-aerated nutrient solution and oxygen-depleted nutrient solution. While the cultivars under stress showed a significantly increasing trend in QY, the trend was non-significant in well-aerated solution ($r = 0.74$, $P < 0.01$ compared with $r = 0.31$, non-significant) (Figure 5). Swedish and Danish cultivars had a negative trend until the 1960s, after which the trend was similar to that of cultivars from Norway and Finland (Figure 6). Among the landraces, Norwegian and Finnish types showed all low QY, while there were both low and high QY types among Swedish and Danish landraces. The changes in QYrel with time of introduction of cultivars from Estonia and Latvia are not presented here, but followed similar trends as in Sweden and Denmark.

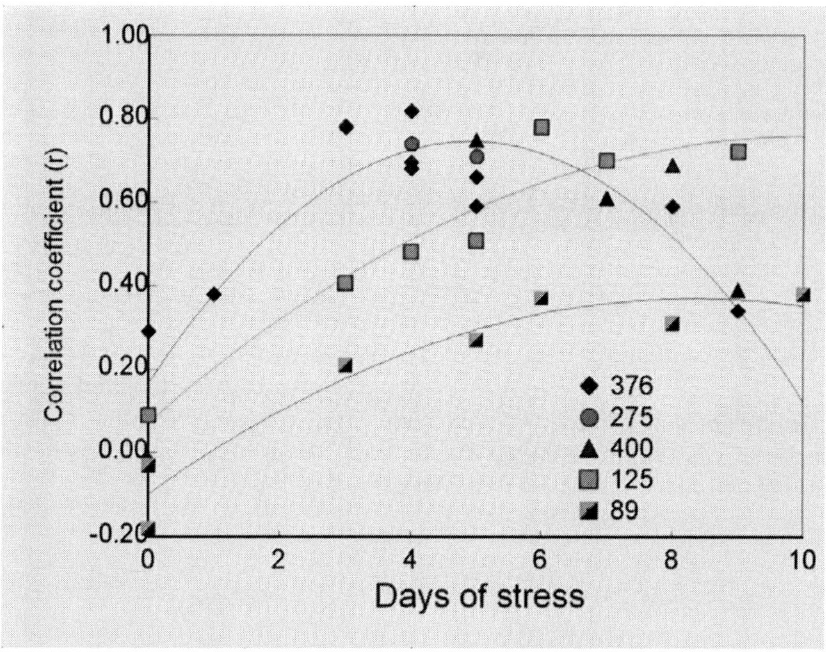

Figure 1. Correlation coefficients after increasing days with oxygen stress between quantum yield and a waterlogging tolerance index (WLT) of the reference material calculated as the growth reduction in pots with and without waterlogging. The different curves show the coefficients obtained with 376, 125 and 89 µmol $m^{-2}s^{-1}$ during the fluorescence measurements. The data are fitted to second order polynomials (p-values of models were

0.0046, 0.0005 and 0.0060, respectively). The plants were 10 days old at day 0. N = 12 for the daily regression and light intensity analyses.

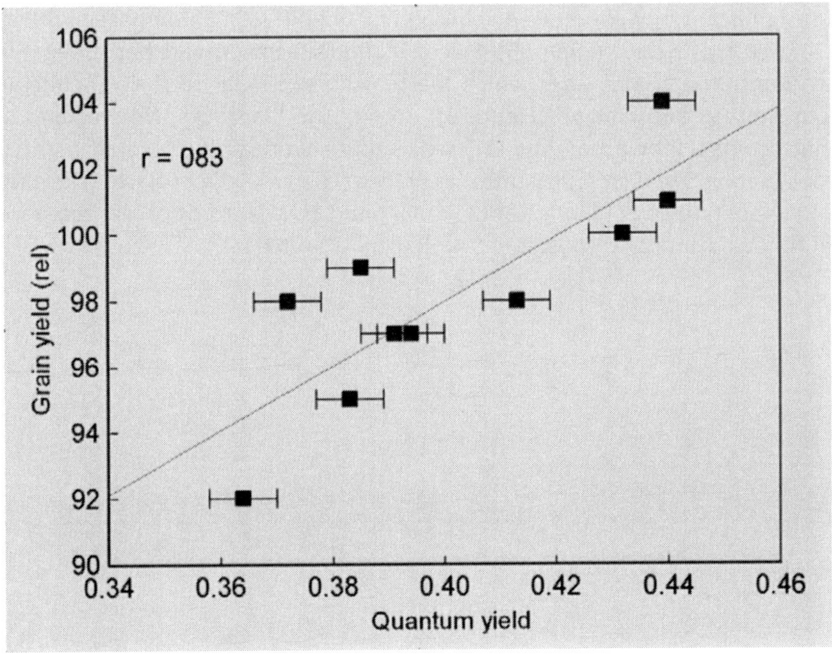

Figure 2. Relationship between mean relative quantum yield at 376 (μmol $m^{-2}s^{-1}$) (QYrel, Henni = 100) of plants grown in oxygen-depleted nutrient solution for 4 to 5 days and grain yield of 10 cultivars from the reference material with comparable yield data from Swedish official variety testing during five years in Sweden. Bars indicate 2 × standard error (SE).

4. DISCUSSION

There is a little, if any, direct screening for abiotic stresses in general, and waterlogging tolerance (WLT) in particular, in plant breeding of barley cultivars for temperate and maritime areas of Europe. Stress tolerance is obtained indirectly by testing, at best, advanced breeding lines in multi-environmental trials and finally by approval of cultivars with the highest mean grain yield in variety testing. A trait-based selection method, either for root growth or for chlorophyll fluorescence, to improve WLT would probably be more successful than just breeding for yield. The risk of discarding interesting genotypes on the way would be reduced and it would be possible to introgress special alleles for high WLT. The method used must also be cost-effective and the use of seedlings and hydroponics would allow testing of many plants in a limited space under controlled conditions.

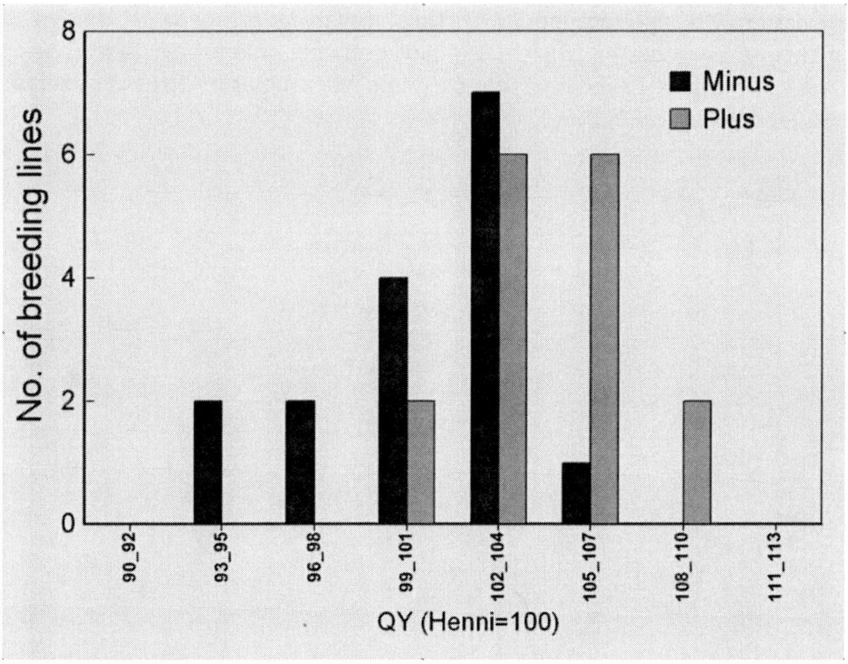

Figure 3. Frequency distribution of quantum yield (QY, Henni = 100) of F_3breeding lines from a plus and minus selection of single F_2 plants from a cross between cv. Henni (QYrel = 100) and cv. Gustav (QYrel = 95). Standard error (SE) = 0.006, *n* = 24.

In this study, both the seminal root length and chlorophyll fluorescence of waterlogged plants were positively correlated with WLT of the reference material and could hence be used for phenotyping and screening purposes without making comparisons with unstressed plants. Both would allow screening of single plants in segregating populations. One disadvantage of using the root growth of seedlings as a method is that seed size and seed vitality could have an influence and it is thus important that the seeds used are carefully selected [18]. The use of chlorophyll fluorescence is therefore a more interesting alternative, even if the correlation with WLT was similar for root growth. Fluorescence measurements are very quick and easy, but must follow a certain protocol. In this study no dark adaptation was used and instead all measurements were standardised by using detached leaves subjected to similar ambient light intensity prior to measurements, which were made at higher light intensities than ambient. With a limited time of acclimatisation and probably other within-plant variations, the standard deviation was high (around 5.5%). Thus, it was necessary to use at least 12 plants, or even better 24 plants, to get a reliable quantum yield (QY) value. This sample size is similar to that used in other studies in citrus [21], tomato [16] and lucerne [22].

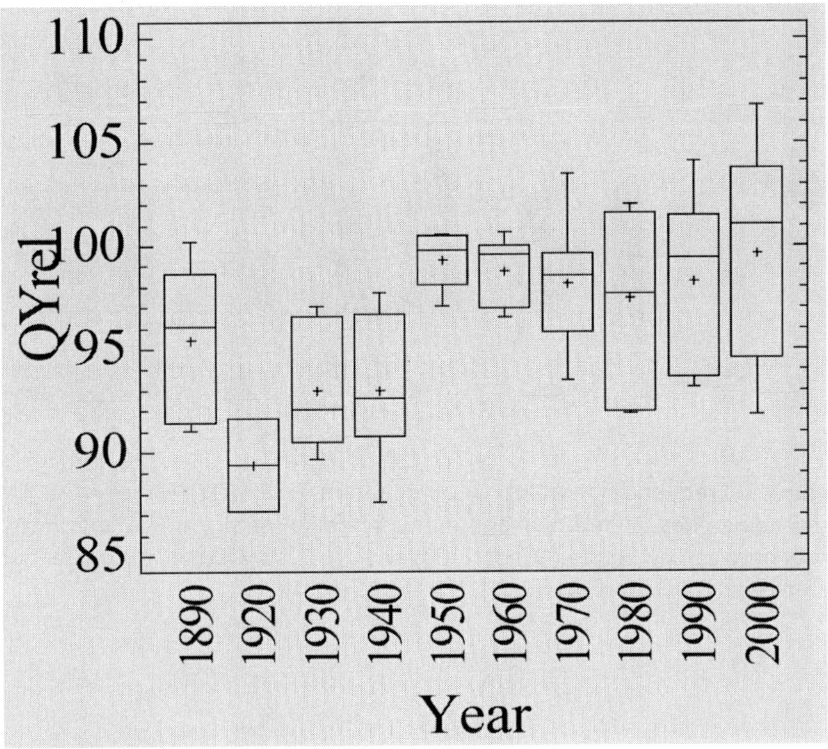

Figure 4. Relative quantum yield (cv. Henni = 100) of oxygen-stressed plants of 15 two-row and 49 six-row old and new cultivars from the Norwegian and Finish barley breeding programmes. The cultivars are grouped in decades and presented as box-and-whisker plots. The vertical line inside the box is the median value and + the mean value. The year of introduction of landraces was set to 1890.

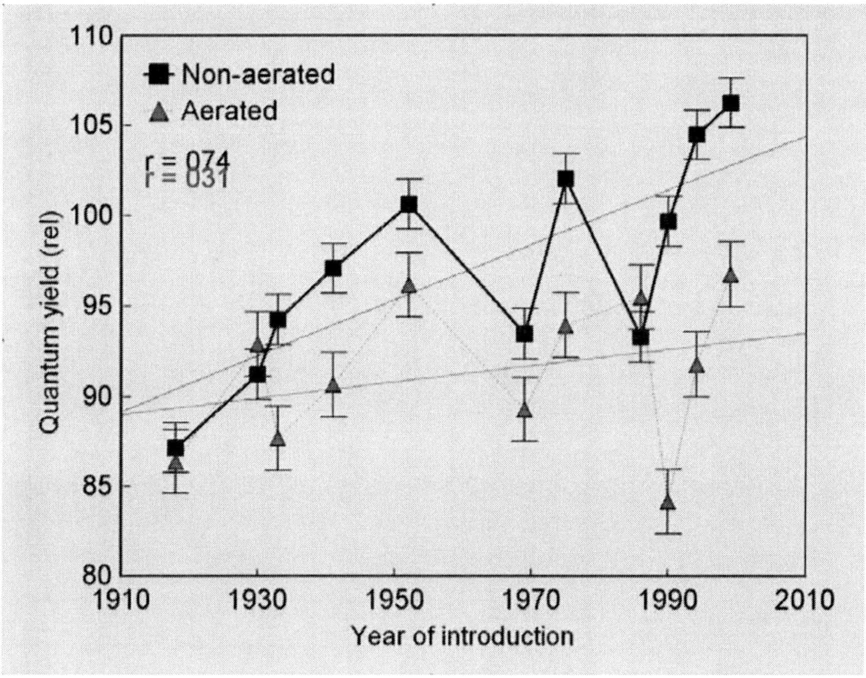

Figure 5. Relative quantum yield (cv. Henni = 100) of oxygen-stressed plants and non-oxygen-stressed plants six-row Norwegian cultivars (every second six-row Norwegian cultivar presented in Figure 4). Bars= 2 × SE.

The QY of unstressed plants of the reference material was initially not significantly correlated with WLT, but 3–5 days after interrupted aeration the QY changed and correlated significantly with WLT. Such quick responses are reported for aerenchyma development in tomato [16] and wheat [23], but changes in metabolic pathways may also play a role [24]. In this study there was a maximum effect after 4–5 days, whereas Pang *et al.* [7] reported maximum effects on QY in dark-adapted leaves after 2–3 weeks of waterlogging. They used different soils and this might have increased the response time. In our study, longer periods of waterlogging tended to reduce the correlation with WLT, especially if higher light intensity than ambient was used. The simple hydroponic system used did not permit studies over several weeks, but after three or more weeks there is a risk of competition, nutrient deficiency and other factors affecting the results.

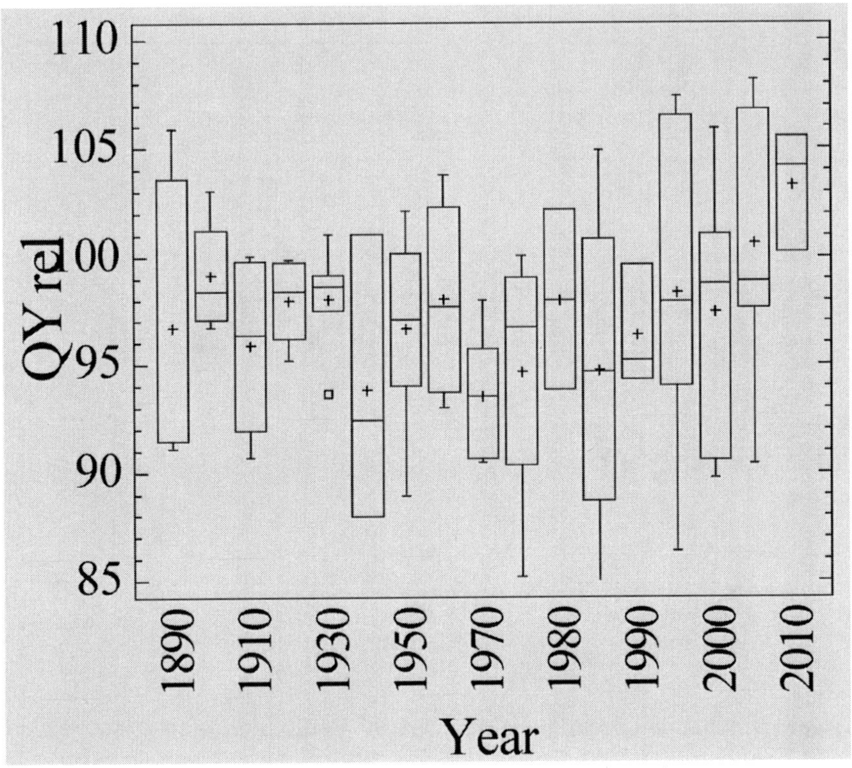

Figure 6. Relative quantum yield (cv. Henni = 100) of oxygen-stressed plants from 70 two-row and 10 six-row old and new cultivars of Swedish and Danish barley. Some newer cultivars are from other parts of Europe but approved for cultivation in Sweden and Denmark at the time indicated by year of introduction. The cultivars are grouped in decades (after 1970 as half-decades) and presented as box-and-whisker plots. The vertical line inside the box is the median value and + the mean value. The year of introduction of landraces was set to 1890, since the first cultivars were introduced in 1900.

While individual fluorescence measurements showed rather high fluctuations, the results of the +/– selection indicate that fluorescence can be used for screening single plants. With the use of hydroponics, a large number of plants can be screened and selected plants can be transferred to a soil medium. Zhou *et al.* [2] showed that WLT was mainly controlled by additive genes and had high heritability, and hence is suitable for screening in early generations. In their study leaf chlorosis was used, a symptom observed later than chlorophyll fluorescence changes.

The correlation between WLT and QY was higher for waterlogged plants than the correlation between WLT and QY of stressed/non-stressed plants. This is positive, since it is possible to select single plants from a segregating population or, if breeding lines are evaluated, test twice the number of genotypes. Of course, there is a risk that QY is high for reasons other than genotype and the selection may then have nothing to do with WLT. It is therefore important to follow up selected genotypes with studies in waterlogged soil and with non-waterlogged controls. If dark-adapted plants had been used there would not have been this risk, since QY of unstressed plants is fairly constant [16,22]. In this study, QY of non-stressed plants was lower than that of dark-adapted plants and also showed a positive but non-significant correlation with WLT in some cases. The reason for this could be that the plants were stressed by other factors in the hydroponic system, even if the solution was aerated. However, the results for the Norwegian cultivars with and without aeration confirm that the increased fluorescence of plants growth in oxygen-depleted solution is related to oxygen stress and not only to an improved genotype in general.

Both WLT and QY are positively correlated with grain yield. Thus, there is no conflict between WLT, high QY and grain yield, as also indicated by an inadvertent improvement in WLT by breeding for yield in the Nordic countries. Hence, trait-based selection could be done without any negative effects on grain yield. In fact grain yield would be improved, probably more efficiently, since a wide variety of cultivars could be evaluated and further used as parents [25].

The fluorescence method could also be used in the search for QTL. Some of the QTL involved in adaptation to waterlogging have been identified [12,26], but their efficiency may depend on the method used [27]. The use of fluorescence and hydroponics could hence improve the accuracy of searches for new QTL for introgressing into barley with the aim of creating new cultivars with enhanced adaptation to waterlogging.

5. CONCLUSIONS

The use of hydroponics is a simple and efficient way to simulate waterlogging stress of seedlings by interrupting the aeration of the nutrient solution. After only four days in oxygen-depleted nutrient solution, leaf chlorophyll fluorescence was affected and was correlated with a waterlogging tolerance index calculated from the weight of oxygen-stressed and unstressed pot-grown plants. A study of Nordic cultivars revealed a general trend for improved waterlogging tolerance during the last 50 years of barley breeding.

ACKNOWLEDGMENTS

The author thanks the Swedish Farmers' Foundation for Agricultural Research (SLF) for funding this study, Lantmännen AB for providing seeds, and Ingegerd Nilsson for technical assistance.

REFERENCES AND NOTES

1. Setter, T.L.; Waters, I. Review of prospects for germplasm improvement for waterlogging tolerance in wheat, barley and oats. Plant Soil 2003, 253, 1–34.

2. Zhou, M.X.; Li, H.B.; Mendham, N.J. Combining ability of waterlogging tolerance in barley. Crop Sci. 2007,47, 278–284.

3. Kjellstrom, E.; Nikulin, G.; Hansson, U.; Strandberg, G.; Ullerstig, A. 21st century changes in the European climate: Uncertainties derived from an ensemble of regional climate model simulations. Tellus Ser. A Dyn. Meteorol. Oceanogr. 2011, 63, 24–40.

4. Colmer, T.D.; Voesenek, L.A.C.J. Flooding tolerance: Suites of plant traits in variable environments. Funct. Plant Biol. 2009, 36, 665–681.

5. Ahmed, F.; Rafii, M.Y.; Ismail, M.R.; Juraimi, A.S.; Rahim, H.A.; Asfaliza, R.A.; Latif, M.A. Waterlogging tolerance of crops: Breeding, mechanism of tolerance, molecular approaches, and future prospects. BioMed. Res. Int. 2013.

6. Garthwaite, A.J.; von Bothmer, R.; Colmer, T.D. Diversity in root aeration traits associated with waterlogging tolerance in the genus Hordeum. Funct. Plant Biol. 2003, 30, 875–889.

7. Pang, J.Y.; Zhou, M.X.; Mendham, N.; Shabala, S. Growth and physiological responses of six barley genotypes to waterlogging and subsequent recovery. Aust. J. Agric. Res. 2004, 55, 895–906.

8. Malik, A.I.; Islam, A.K.M. R.; Colmer, T.D. Transfer of the barrier to radial oxygen loss in roots of Hordeum marinum to wheat (Triticum aestivum): Evaluation of four H-marinum-wheat amphiploids. New Phytol. 2011,190, 499–508.

9. Malik, A.I.; Colmer, T.D.; Lambers, T.L.; Setter, H.T.L.; Schortemeyer, M. Short-term waterlogging has long-term effects on the growth and physiology of wheat. New Phytol. 2002, 153, 225–236.

10. Takeda, K.; Fukuyama, T. Tolerance to pre-germination flooding in the world collection of barley varieties.Barley Genet. 1987, V, 735–740.

11. Hamachi, Y.; Yoshino, M.; Furusho, M.; Yoshida, T. Index of screening for wet endurance in malting barley.Jpn J. Breed. 1990, 40, 361–366.

12. Li, H.; Vaillancourt, R.; Mendham, N.; Zhou, M. Comparative mapping of quantitative trait loci associated with waterlogging tolerance in barley (Hordeum vulgare L.). BMC Genomics 2008, 9, 401.

13. Striker, G. Visiting the methodological aspects of flooding experiments: Quantitative evidence from agricultural and ecophysiological studies. J. Agron. Crop Sci. 2008, 194, 249–255.

14. Qiu, F.Z.; Zheng, Z.L.; Xu, S.Z. Mapping of QTL associated with waterlogging tolerance during the seedling stage in maize. Ann. Bot. 2007, 99, 1067–108.

15. Maxwell, K.; Johnson, G.N. Chlorophyll fluorescence—A practical guide. J. Exp. Bot. 2000, 51, 659–668.

16. Else, M.A.; JJanowiak, F.; Atkinson, C.J.; Jackson, M.B. Root signals and stomatal closure in relation to photosynthesis, chlorophyll a fluorescence and adventitious rooting of flooded tomato plants. Ann. Bot.2009, 103, 313–323.

17. Bertholdsson, N.-O. Pre-studies of water- and drought tolerance in Barley. unpublished.

18. Bertholdsson, N.-O.; Kolodinska Brantestam, A. Breeding for improved yield in Nordic barley germplasms and its effects on early vigour, straw length and harvest index. Eur. J. Agron. 2009, 30, 266–274.

19. Larsson, S. A simple, rapid and non-destructive method useful for drought resistance breeding in oats (Avena sativa L.). Zeitschrift für Pflanzenzüchtung 1982, 89, 206–221.

20. Walz, H. Photosynthesis yield analyzer MINI-PAM. In Handbook of Operation, 2nd ed; Heinz Walz GmbH: Effeltrich, Germany, 1999.

21. Arbona, V.; López-Climent, M.F.; Pérez-Clemente, R.M.; Gómez-Cadenas, A. Maintenance of a high photosynthetic performance is linked to flooding tolerance in citrus. Environ. Exp. Bot. 2009, 66, 135–142.

22. Smethurst, C.F.; Shabala, S. Screening methods for waterlogging tolerance in lucerne: Comparative analysis of waterlogging effects on chlorophyll fluorescence, photosynthesis, biomass and chlorophyll content. Funct. Plant Biol. 2003, 30, 335–343.

23. Hague, M.E.; Oyanagi, A.; Kawaguchi, K. Aerenchyma formation in the seminal roots of Japanese wheat cultivars in relatin to growth under waterlogged conditions. Plant Product. Sci. 2012, 15, 164–173.

24. Vartapetian, B. Plant anaerobic stress as a novel trend in ecological physiology, biochemistry, and molecular biology: 2. Further development of the problem. Russian J. Plant Physiol. 2006, 53, 711–738.

25. Lynch, J.P. Roots of the second green revolution. Aust. J. Bot. 2007, 55, 493–512.

26. Xue, D.; Zhou, M.; Zhang, X.; Chen, S.; Wie, K.; Chen, S.; Wie, K.; Zeng, F.; Mao, Y.; Wu, F.; Zhang, G. Identification of QTLs for yield and yield components of barley under different growth conditions. J. Zhejiang Univ. Sci. B. 2011, 11, 169–176.

27. Zhou, M. Accurate phenotyping reveals better QTL for waterlogging tolerance in barley. Plant Breed. 2011,130, 203–208.

Chapter 2

Geochemical Speciation and Risk Assessment of Heavy Metals in Soils and Sediments

Santosh Kumar Sarkar[1], Paulo J.C. Favas[2, 3], Dibyendu Rakshit[1] and K.K. Satpathy[4]

[1] *Department of Marine Science, University of Calcutta, Calcutta, West Bengal, India*

[2] *Department of Geology, School of Life Sciences and the Environment, University of Trás-os-Montes e Alto Douro, Vila Real, Portugal*

[3] *IMAR-CMA Marine and Environmental Research Centre, Faculty of Sciences and Technology, University of Coimbra, Coimbra, Portugal*

[4] *Indira Gandhi Centre for Atomic Research, Environment and Safety Division, Kalpakkam, Tamil Nadu, India*

1. INTRODUCTION

Heavy metal pollution is a serious and widely environmental problem due to the persistent and non-biodegradable properties of these contaminants. Sediments serve as the ultimate sink of heavy metals in the marine environment and they play an important role in the transport and storage of potentially hazardous metals. They are introduced into the aquatic system as a result of weathering of soil and rocks, from volcanic eruptions and from a variety of human activities involving mining, dredging, processing and use of metals and/or substances containing metal contaminants. Heavy metals entering natural water become part of the water-sediment system and their distribution processes are controlled by a dynamic set of physicochemical interactions and equilibria. The properties of metals in soils and sediments depend on the physiochemical form in which they occur [1]. Heavy metals are distributed throughout soil and sediment components and associated with them in various ways, including adsorption, ion exchange, precipitation and complexation and so on [2]. Changes in environmental conditions, such as temperature, pH, redox potential and organic ligand concentrations, can cause metals to be released from solid to liquid phase and sometimes cause contamination of surrounding waters in aquatic systems [3]. They are not permanently fixed by soil or sediment. Therefore, it cannot provide sufficient information about mobility, bioavailability and toxicity of metals if their total contents are studied alone.

Natural and anthropogenic activities have the capacity to cause changes in environment conditions, such as acidification, redox potential, or organic ligand concentrations, which can remobilize contaminated soils and sediments releasing the elements from soils and sediments and pore water to the water column resulting contamination of surrounding waters. Daily tidal currents, wind energies, and storms in coastal and estuarine systems can cause periodical remobilization of surface sediments [4]. More turbulent conditions, such as seasonal flooding or storms, or bioturbation, due to feeding and movement of benthic organisms, can expose anoxic sediments to oxidant conditions. In addition, activities such as dredging result in major sediment disturbances, leading to changes in chemical properties of sediment [5].

The remediation of heavy metal pollution is often problematic due to their persistence and non-degradability in the environment. As a sink and source, soils and sediments constitute a reservoir of bioavailable heavy metals and play a significant role in the remobilization of contaminants in the aquatic systems under favorable conditions. Such potential of sediment for being a sink as well as a source of contaminant can make sediment chemistry and toxicity key components of the quality of aquatic system. Much concern has been focused on the investigation of the total element contents in soils and sediments. However, it cannot provide sufficient information about mobility, bioavailability and toxicity of elements and thus may not be able to provide information about the exact dimension of pollution. The data on total contents of metals are quite insufficient to estimate the possible risk of remobilization of total metals under changing environmental conditions and potential uptake of liberated metals by biota and thus the determination of different fractions assume great importance. This has been described as “speciation” [6]. Since each form have different bioavailability and toxicity, the environmentalists are rightly concerned about the exact forms of metal present in the aquatic environment.

The concept of speciation dates back to 1954 when Goldberg introduced the concept of speciation to improve the understanding of the biogeochemical cycling of trace elements in seawater. Kinetic and thermodynamic information together with the analytical data made it possible to differentiate between oxidized versus reduced, complexed or chelated versus free metal ions in solution and dissolved between particulate species. Florence [7] has defined the term speciation analysis as the determination of the individual physicochemical forms of the element, which together make up its total concentration in a sample. According to Lung [8], speciation analysis involves the use of analytical methods that can provide information about the physicochemical forms of the elements. Schroeder [9] distinguishes physical speciation, which involves differentiation of the physical size or the physical properties of the metal, and chemical speciation, which entails differentiation among the various chemical forms. The main objective of measuring metal species relates to their relative toxicities to aquatic biota. The second and long term aim of speciation studies is to advance an understanding of metal interactions between water and bed sediments in an aquatic ecosystem. In the last decade researchers have followed different sequential extraction techniques for the fractionation of metals in sediments of different river systems. Rauret et al. [10] studied the speciation of copper and lead in the sediments of River Tenes (Spain) while Pardo et al. [11] studies the speciation of zinc, cadmium, lead, copper, nickel and cobalt in the sediments of Pisuerga River, Spain, in order to establish the extent to which these are polluted and their capacity to remobilization. Jardo and Nickless [12] investigated the chemical association of zinc, cadmium, lead and copper in soils and sediments of England and Wales. In most samples, these four metals were associated with all the chemical fractions. Tessier et al. [13] studied speciation of cadmium, cobalt, copper, nickel, lead, zinc, iron and manganese in water and sediments of St. Fransois River, Quebec, Canada. Elsokkary and Muller [14] studied speciation of chromium, nickel, lead and cadmium in the sediments of Nile River, Egypt, reporting that a high proportion of chromium, nickel and lead is bound to organic material and sulphides, while cadmium is bound to carbonate fraction. Ure [15] and Rauret [16] have reviewed the chemical extraction procedures used for heavy metal determinations in contaminated soils and sediments. Owing to the need for validation of extraction schemes, the EC Measurement and Testing Programme (formerly BCR) has organized a project for improving the quality of determinations of extractable heavy metals, where development and validation of extraction procedures has been discussed [17, 18].

The present article aims to summarize the potentials of sequential extraction technique adopting different analytical protocols for gaining information on the mobility and dynamics of operationally determined chemical forms of heavy metals in soils and sediments. The BCR (Community Bureau of Reference, now superseded by the Standards, Measurement and Testing Programme of the European Community) procedure has been illustrated considering the case study of Ganges (Hugli) River Estuary and adjacent Indian Sundarban mangrove wetland (a UNESCO World Heritage Site), northeastern part of the Bay of

Bengal. In addition, the authors also evaluate the modified BCR sequential extraction technique as devised by various scientists, the risk assessment code (RAC) as well as assessment of toxicity comparing with sediment quality guidelines. The RAC classification is based on the strength of the bond between the metals and the different geochemical fractions in sediments or soils and the ability of metals to be released and enter into the food chain.

2. SEQUENTIAL EXTRACTION: MERITS AND DEMERITS

The sequential extraction provides more or less detailed information concerning the origin, mode of occurrence, biological and physicochemical availabilities, mobilization and transport of heavy metals. The procedure stimulates the mobilization and retention of these species in the natural environment using changes in environmental condition such as pH, redox potential and degradation of organic matter [16]. A series of reagents is applied to the sample, increasing the strength of the extraction at each step, in order to dissolve the trace metal present in different sediment phases. The extractants are inert electrolytes, weak acids, reducing agent, oxidizing agents and strong mineral acids [19].

The 3-stage sequential extraction procedure proposed by the European Community Bureau of Reference (BCR) was developed in an attempt to standardize the various schemes described in the literature [2, 20, 21], since the use of different procedures, varying in the number of steps, types of reagents and extraction condition. Hindered comparison of results obtained in the many studies of heavy metals chemical fractionation in environmental samples [22].

The BCR methods has been widely adapted by various authors, and applied to a range of type of solid sample including fresh water sediment [23-25], salt water sediment [26-28], sewage sludge and particulate matter [29-31]. This scheme enables us to associate the meals with one of the following four geochemical phases:

I. Acid-soluble phase: This phase is made up of exchangeable metals and others bound to carbonates that are able to pass easily into the water column, for example, when the pH drops. It is the fraction with the most labile bond to the soil/sediment and, therefore, the most dangerous for the environment.

II. Reducible phase: This phase consists of metals bound to iron and manganese oxides that can be released if the sediment changes from the oxic to the anoxic state, which could be caused, for example, by the activity of microorganisms present in the soils/sediments.

III. Oxidisable phase: This shows the amount of metal bound to organic matter and sulphides, which can be released under oxidizing conditions. Such conditions can occur, for example, if the sediment is resuspended (by dredging, currents, flooding, tides, etc.) and the sediment participles come into contact with oxygen-rich water.

IV. Residual phase : Lithogenous and inert (Non-bioavailable).

The heavy metals in the soils and sediments are bound to different fractions with different strengths, the value can, therefore, give a clear indication of soil and sediment reactivity, which in turn assess the risk connected with the presence of heavy metals in a terrestrial or aquatic environment. The rationale of the sequential extraction procedure is that each successive reagent dissolves a different component, which can content heavy metals within their crystalline structures. Under natural conditions, metals in minerals are unlikely to experience significant release over the time frames of interest [32, 33].

3. ANALYTICAL PROTOCOLS FOR SEQUENTIAL EXTRACTION

In recent years a great number of papers have been published on various analytical techniques proposed for the fractionation analysis of trace elements in various environmental samples (soils, sediments, etc.). An approach that has been found to be preferable is the fractionation of heavy metal into operationally defined forms under the sequential action of different extractants [2]. Selective extractants, used in sequential extraction procedures, are aimed at the simulation of natural conditions whereby metals associated with certain soil (sediment) components can be released. For example, changes in the ionic composition affecting adsorption–desorption reactions or a decrease in pH may lead to the release of metals, retained on a matrix by weak electrostatic interactions or co-precipitated with carbonates ("exchangeable" and "acid soluble" forms). Decreasing the redox potential can result in dissolution of oxides, unstable under reducing conditions, and liberation of scavenged metals ("reducible" forms). Changes in oxidizing conditions may cause the degradation of organic matter and release of complexed metals ("oxidizable" forms). Finally, the destruction of primary and secondary mineral lattice releases heavy metal retained within the crystal structure, e.g., due to isomorphous substitution ("residual" forms) [2]. The nominal "forms" determined by operational fractionation can help to estimate the amounts of total metals in different reservoirs which could be mobilized under changes in the chemical properties of the soil [34]. Since the 1970s a considerable number of extraction procedures have been proposed for determining the forms of heavy metal [2, 35-39]. Most of these procedures are based on the scheme of

Tessier et al. [2]. Although most of the extracting reagents were originally used in the chemical analysis of soils, the procedures proposed have been tested on a wide variety of contaminated environmental samples—sediments, road dust, sewage sludge, etc.

Sequential extraction can be useful to have an operational classification of metals in different geochemical fractions [2] which is the most reliable criteria to quantify the potential effect of soil/sediment contamination by heavy metals. This can provide information about the identification of the main binding sites, the strength of element binding to the particulates and the phase associations of trace elements in soil/sediment. Following this basic scheme, some modified procedures with different sequences of reagents or operational conditions have been developed [40-43]. Considering the diversity of procedures and the lack of uniformity in different protocols, a European Community Bureau of Reference (BCR, now the European Community Standards Measurement and Testing Program) method was proposed [6] and was applied by a large group of researchers [31, 44-47]. In this study, we followed the sequential extraction procedure proposed by the European Union's Standards, Measurements and Testing program [3].

4. MODIFIED BCR SEQUENTIAL EXTRACTION PROCESS

As discussed above it is evident that sequential extraction provides valuable information regarding identification of main binding site, the strength of the element binding to the particulates and the phase associations of heavy metals in sediments. However, various complicated sequential extraction procedures were experimented to provide more detailed information regarding different metal phase associations [2, 48, 49]. A wide range of techniques is available whereby various extraction reagents and experimental conditions are used. These techniques involve a 5-step [2], 4-step (BCR, Bureau Commune de Reference of the European Commission), 6-step [50] and 7-step [51, 52] extraction, and are thus becoming popular methods to be used for sequential extraction [53, 54]. Following this basic scheme, some modified procedures with different sequences of reagents or operational conditions have been developed [40-43].

Several sophisticated instruments have been used for the determination of heavy metals contents in marine environments. These include; flame AAS [55, 56], atomic fluorescence spectrometry [57], anodic stripping voltametry [58, 59], ICP-AES [60] and ICP-MS [61, 62].

Heavy metal mobility and bioavailability depend strongly on their chemical and mineralogical forms in which they occur [63]. Several speciation studies have been conducted to determine study different forms of heavy metals rather their total metal content. These studies reveal the level of bioavailability of metals in harbour sediments and also confirm that sediments are indicators of heavy metal pollution in marine environment [64-67].

Since the early 1980s and 1990s sequential extraction methodology has been developed to determine speciation of metals in sediments [2, 68] due to the fact that the total concentration of metals often does not accurately represent their characteristics and toxicity. In order to overcome the above mentioned obstacles it is helpful to evaluate the individual fractions of the metals to fully understand their actual and potential environmental effects [2]. To date, strong acid digestion is used often for the determination of total heavy metals in the sediments. However, this method can be misleading when assessing environmental effects due to the potential for an overestimation of exposure risk. Moreover, in order to determine the mobility of heavy metals in sediments, various sequential extraction procedures have been developed [69-71].

Among a range of available techniques using various extraction reagents and experimental conditions to investigate the distribution of heavy metals in sediments and soils, the 5-step Tessier et al. [2] and the 6-step extraction method, Kersten and Fronstier [50] were mostly widely used. Following these two basic schemes, some modified procedures with different sequences of reagents or experimental conditions have been developed [40-43]. Considering the diversity of procedures and lack of uniformity in different protocols, a BCR, Bureau Commun de Recherche (now called the European Community (EC) Standards Measurement and Testing Programme) method was proposed [6]. It harmonized differential extraction schemes for sediment analysis. The method has been validated using a sediment certified reference material BCR-701 with certified and indicative extractable concentration of Cd, Cr, Cu, Ni, Pb and Zn [72]. This method was applied and accepted by a large group of specialists [31, 44, 45, 47, 73, 74] despite some shortcoming in the sequential extraction steps [75, 76].

Wang et al. [77] used a modified Tessier sequential extraction method to investigate the distribution and speciation of Cd, Cu, Pb, Fe, and Mn in the shallow sediments of Jinzhou Bay, Northeast China. This site was heavily contaminated by nonferrous smelting activities. They found out that the concentrations of Cd, Cu and Pb in sediments was to be 100, 73, 13 and 7 times, respectively, higher than the National guidelines (GB 18668-2002). The sequential extraction tests revealed that 39%-61% of Cd was found in exchangeable fractions. This shows that Cd in the sediments posed a high risk to the local environment. Copper and Pb were found to be at moderate risk levels. According to the relationships between percentage of metal speciation and total metal concentration, it was concluded that the distributions of Cd, Cu and Pb in some geochemical fractions were dynamic in the process of pollutants migration and stability of metals in marine sediments from Jinzhor Bay decrease in the order Pb>Cu>Cd.

Yuan et al. [78] applied BCR-sequential extraction protocol to obtain metal distribution patterns in marine sediments from the East China Sea. The results showed that both the total contents and the most dangerous non-residual fractions of Cd and Pb were extremely high. More than 90% of the total concentration of V, Cr, Mo and Sn existed in the residual fraction while more than 60% of Fe, Co, Ni, Cu, and Zn were mainly present in the residual fraction. Manganese, Pb, and Cd were dominantly present in the non-residual fractions in the top sediments.

Jones and Turki [79] worked on distribution and speciation of heavy metals in surface sediments from the Tees estuary, North East England. Tessier et al. [2] metal speciation scheme modified by Ajay and van Loon [80] was used for the study. They observed out that the sediments were largely organic-rich clayey silts in which metal concentrations exceed background levels, and which attain peak values in the upper and middle reaches of the estuary. Chromium, Pb and Zn were associated with the reducible, residual, and oxidizable fractions. Cobalt and Ni were not highly enriched while Cu is associated with the oxidizable and residual fractions. Cadmium is associated with the exchangeable fractions.

Pempkowlak et al. [81] investigated the speciation of heavy metals in sediments and their bioaccumulation by mussels. They used a 4-step sequential extraction procedure adapted from Forstner and Watmann [82]. Their investigation which was characterized by varying metal bioavailability was aimed at revealing differences in the accumulation pattern of heavy metals in mussel inhabiting that inhabit in sediments. The bioavailabilities of metals were measured using the contents of metals adsorbed to sediments and associated with Fe and Mn hydroxides. The biovailable fraction of heavy metals contents in sediments collected from Spitsbergen represented a small proportion (0.37% adsorbed metals and 0.11%, are associated with metals hydroxides). It was also revealed that the percentages of metals adsorbed and bound to hydroxides of the sediments ranged from 1 to 46% and 1 to 13%, respectively.

Wepener and Vermeulen [66] investigated on the concentration and bioavailability of selected metals in sediments of Richards Bay harbor, South Africa. Sequential extraction of sediments was carried out according to Tessier et al. [2] method. The following metals were investigated: Al, Cr, Fe, Mn, and Zn. Their studies revealed that metals concentrations in sediments samples varied only slightly between seasons, but showed significant spatial variation, which was significantly correlated to sediment particle size composition. Highest metal concentration was recorded in sites with substrates dominated by fine mud. Manganese and Zn had more than 50% of this concentration in reducible fraction while more than 70% of the Cr was associated with the inert fractions and the concentration recorded at some sites were still above action levels when considering only the bioavailable fractions. They also concluded that the concentration of Zn recorded was not elevated their results were compared with the historic data.

Coung and Obbard [54] used a modified 3-step sequential extraction procedure to investigate metal speciation in coastal marine sediments from Singapore as described by the European Community Bureau of Reference (ECBR). Highest percentages of Cr, Ni, and Pb were found in residual fractions in both Kranji (78.9%, 54.7% and 55.9% respectively) and Pulang Tokong (82.8%, 77.3% and 62.2% respectively). This means that these metals were strongly bound to sediments. In sediments from Kranji, the mobility order of heavy metals studied were Cd>Ni>Zn>Cu>Pb>Cr while sediments from Pulan Tekong showed the same order for Cd, Ni, Pb and Cr, but had a reverse order for Cu and Zn (Cu>Zn). The sum of the 4-steps (acid soluble + reducible + oxidizable + residual) was in good agreement with the total metal content, which confirmed the accuracy of the microwave extraction procedure in conjunction with the GFASS analytical method.

Fedotov et al. [83] applied a modified technique for accelerated fractionation of heavy metals in contaminated soils and sediments using rotating coiled columns. Rotating coiled columns (RCC) is valuable for the continuous-flow sequential extraction and can be successfully applied to the dynamic leaching of heavy metals from soil and sediments. This is a fluoroplastic or steel coil wound around a rigid cylindrical drum, which revolves about its axis and, at the same time, revolves around the central axis of the device called planet centrifuge. The stationary (liquid, solid, or heterogeneous) phase is retained in the column because of the centrifugal force field, and the mobile liquid phase is continuously pumped through the column. A solid sample was retrieved in the rotating column as the stationary phase under the action of centrifugal forces while different elements (aqueous solution of complexing reagents, mineral salts and acids) were continuously pumped through. This procedure developed is time saving and requires only 4-5 hr instead of the several days needed for individual sequential extraction. Losses of solid sample are minimal. Further studies are needed to better estimate the reproducibility of the technique.

Nemati et al. [84] used a modified BCR sequential extraction procedure (SEP) in combination with ICP-MS to obtain the metal distribution patterns in different depths of sediments from Sungai Buloh, Selangor, Malaysia. The results showed that heavy metal contaminations at Sungai Buloh River sediments were more severe than at other sampling sites, especially for Zn, Cu, Ni and Pb. Nevertheless, the element concentrations from top to bottom layers decreased predominantly.

Mossop et al. [85] compared of original and modified BCR sequential extraction procedures for the fractionation of Cu, Fe, Pb, Mn and Zn in soils and sediments. The procedures were applied to five soil and sediment substrates: a sewage sludge-amended soil, two different industrially contaminated soils, river sediment and intertidal sediment. Extractable Fe and Mn concentrations were measured to assess the effects of the procedural modifications on dissolution of the reducible matrix components. Statistical analyses (two-tailed t-tests at 95% confidence interval) indicated that recovery of Fe in step 2 was not markedly enhanced when the intermediate protocol was used. However, significantly greater amounts were isolated with the revised BCR scheme than with the original procedure. Copper behaved similarly to Fe. Lead recoveries were increased by use of both modified protocols, with the greatest effect occurring for the revised BCR extraction. In contrast, Mn and Zn extraction did not vary markedly between procedures. The work indicates that the revised BCR sequential extraction proves better attack on the Fe-based components of the reducible matrix for a wide range of soils and sediments.

5. SEQUENTIAL EXTRACTION OF METALS IN SEDIMENTS OF THE HUGLI RIVER ESTUARY AND INDIAN SUNDARBAN WETLAND: A CASE STUDY

5.1. Materials And Methods

5.1.1. Sample collection and sediment quality analysis

The delta region formed by Hugli (Ganges) River Estuary (HRE) and is famous for its luxuriant mangrove vegetation, known as Sundarban wetland, acclaimed as UNESCO World Heritage Site for its capacity of sustaining an excellent biodiversity. The wetland is characterized by a complex network of tidal creeks, which surrounds hundreds of tidal islands exposed to different elevations at high and low semi-diurnal tides. This is one of the most sensitive and vulnerable ecosystems in the world and suffers from environmental degradation due to rapid human settlement, tourism and port activities, operation of mechanized boats, deforestation, and increasing agricultural and aquaculture practices. The ongoing degradation is also related to huge siltation, flooding, storm runoff, atmospheric deposition, and other stresses resulting in changes in water quality, depletion of fishery resources, choking of river mouth and inlets, and overall loss of biodiversity. Moreover, the rapid economic development in this deltaic region has caused highly dense areas of human activity and led to serious contamination including heavy metals and persistent organic pollutants (POPs).

Nine sampling sites, namely Barrackpur (S_1), Dakshineswar (S_2), Babughat (S_3), Budge budge (S_4), Ulubaria (S_5), Diamond Harbor (S_6), Frezergunge (S_7), Gangasagar (S_8), and Haribhanga (S_9) were selected considering the existence of typical sediment dispersal patterns along the drainage network systems (as shown in Figure 1) and their position was fixed by a global positioning system (GPS). The stations are representative of the variable environmental and energy regimes that cover a wide range of substrate behavior, wave–tide climate, and intensity of bioturbation (animal–sediment interaction), geomorphological–hydrodynamic regimes and distances from the sea (Bay of Bengal). The sites are exposed to a variable level of heavy metal contamination mainly from anthropogenic sources as mentioned earlier. Six sampling sites (S_1 to S_6) have been chosen along the lower stretch of Hugli River Estuary, while residual three sites (S_7 to S_9) were taken into account in the coastal regions of Sundarban wetland. All sampling sites together with the main stresses to which they are subjected are presented in Table 1.

Table 1. Details of the nine sampling sites and the main stresses to which they are subjected.

Station number	Site	Main stresses
S_1	Barrackpur	Industrial effluents, domestic sewage disposal, boating, bathing.
S_2	Dakshineswar	Industrial and domestic effluents, boating, bathing, idol immersion site.
S_3	Babughat	Power plant discharges, domestic sewage, boating, idol immersion site.
S_4	Budge budge	Domestic and industrial effluents, bathing, boating.
S_5	Ulubaria	Domestic and industrial effluents
S_6	Diamond Harbour	Boating, recreational activities, bathing, fishing, jetties for fishing trawlers
S_7	Frezergunge	Tourist activities, ferry services, fishing
S_8	Gangasagar	Boating, tourist activities, dredging, fishing, agricultural, domestic and aquaculture practices
S_9	Haribhanga	Boating, fishing and ferrying

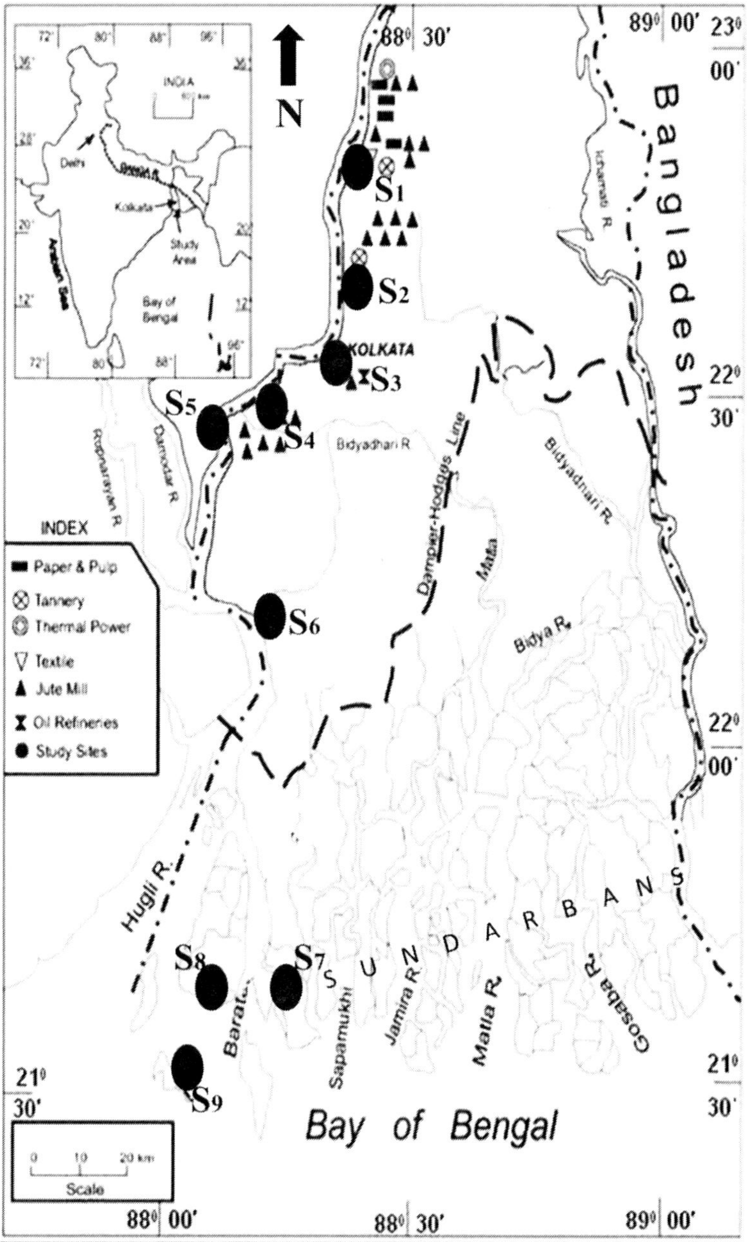

Figure 1. Map showing the location of the monitoring sites (S_1 to S_9) covering Hugli River Estuary and Sundarban mangrove wetland along with the location of the major industries.

During winter months (January–March 2009) surface sediment samples weighting 10 g were randomly collected in triplicate from the top 3–5 cm of the surface at each sampling site during low tide using a grab sampler, pooled and thoroughly mixed. Immediately after collection, the samples were placed in sterilized plastic bags in the ice box and transported to the laboratory. Samples were oven dried at 50°C, most gently disaggregated, transferred into precleaned inert polypropylene bags and stored in deep freeze prior to analyses. Each sample was divided into two aliquots: one unsieved (for the determination of sediment quality parameters) and the other sieved through 63 μm metallic sieves (for elemental analyses). Organic carbon content was determined following a rapid titration method [86] and pH with the help of a deluxe pH meter (model no. 101E) using combination glass electrode manufactured by M.S. Electronics Pvt. Ltd. (India). Mechanical analyses of sediment were done by sieving in a Ro-Tap Shaker manufactured by W.S. Tyler Company, Cleveland, Ohio.

5.1.2. Analytical Procedure

To determine the total element concentration, sediment samples were digested in polytetrafluoroethylene vessels with aqua regia (HCl/HNO_3, 3:1) and HF neutralized with H_3BO_3 in a 650 W microwave oven (CEM MDS 2000) with a program consisting of a 20-min ramp and a 30-min hold at 100% power in pressure and temperature controlled conditions (150 psi and 180°C). The digested samples were filtered, transferred to polyethylene containers and stored at +4°C until analysis. All reagents were Suprapur® grade (Merck). Reagent blank was processed with the samples and did not show any significant contamination. Accuracy of the procedure was checked using two different certified reference materials (CRM): MESS-2 and PACS-2, which are both marine sediments certified by the National Research Council of Canada for the element content. The MESS-2 recovery ranged between 91% and 116% for all the elements (Table 2). Precision, calculated as relative standard deviation (RSD%), resulted always lower than 5%.

In this study, we followed the sequential extraction procedure proposed by the European Union's Standards, Measurements and Testing program [3]. Selective extraction is based on the procedure used by Tessier et al. [2] with improvements made according to the BCR, which examined and finally eliminated irreproducibility sources. It is made up of three steps, which dissolve the following phases, respectively: exchangeable and bound to carbonate, bound to Fe and Mn oxides and hydroxides, bound to organic matter and sulphides. Exchangeable and bound to carbonate phase (phase 1) is extracted

Table 2. Results of certified reference materials MESS-2 and PACS-2 as well as the observed values. All the values are expressed in µg/g of dry weight. MESS-2 and PACS-2 recovery rates are also reported.

	Al	As	Cd	Co	Cr	Cu	Fe	Mn	Ni	Pb	Zn
Found MESS-2	86,613± 17,773	24.0± 2.4	0.230± 0.010	15.6± 1.5	112± 12	38.4± 6.1	47,385± 3,668	372± 42	54.1± 3.8	20.0± 1.4	170±12
Found PACS-2	70,190± 3,784	29.2± 1.3	1.59± 0.80	12.8± 0.6	94.9± 4.4	307± 22	46,630± 1,411	465± 23	44.3± 2.5	184± 10	398±16
Certified MESS-2	85,698± 2,600	20.7± 0.8	0.240± 0.010	13.8± 1.4	106± 8	39.3± 2.0	43,504± 2,266	365± 21	49.3± 1.8	21.9± 1.2	172±16
Certified PACS-2	66,125± 3,184	26.2± 1.5	2.11± 0.15	11.5± 0.3	90.7± 4.6	310± 12	43,738± 585	440± 19	39.5± 2.3	183±8	364±23
Recovery MESS-2	101%	116%	95.8%	113%	106%	97.8%	109%	102%	110%	91.2%	99.0%
Recovery PACS-2	106%	111%	75.4%	111%	105%	99.1%	107%	106%	112%	100%	109%

with 0.11 M acetic acid, while the fraction bound to Fe–Mn oxides (phase 2) with 0.5 M hydroxylamine hydrochloride, adjusted to pH 2 with nitric acid (65%). The phase bound to organic and sulphides (phase 3) is extracted with 8.8 M hydrogen peroxide (stabilized at a pH included between 2 and 3), treated at 80°C in a microwave oven using a program consisting of a 30-min ramp and a 60-min hold at 50% power in pressure and temperature controlled conditions (80 psi and 85°C), and 2 M ammonium acetate adjusted to pH 2 with nitric acid (65%). Each extraction was carried out overnight (16 h) at room temperature. All the reagents employed were Tracepur® grade (Merck Eurolab, Italy). After each extraction, the samples were separated from the aqueous phase by centrifuging at 4, 000 rpm for 15 min. The sediments were washed with Milli-Q water and centrifuged again. The wash water was added to supernatants. The element content of the residual phase was obtained from the difference between the total content and the sum of phases 1, 2 and 3, according to Ianni et al. [37, 38], Ramirez et al. [39], and Mester et al. [27]. Sequential extraction reagent blanks showed no detectable contamination. Accuracy of the procedure was checked with BCR-701 (SM&T). The recovery rates for trace elements in the standard reference material ranged between 77% and 118% (Table 3). Precision, calculated as RSD%, resulted generally lower than 5%, except As and Cr in the phase 1 (~20%).

Table 3. Results of certified reference materials BCR-701 as well as the observed values (expressed in µg/g of dry weight) together with recovery rates for each step. n.a.= not available.

	Al	As	Cd	Co	Cr	Cu	Fe	Mn	Ni	Pb	Zn
Found BCR-701 step 1	198±1	2.57± 0.28	6.09± 0.09	2.06± 0.08	2.41± 0.51	47.7± 1.7	43.8± 5.8	180±1	14.5± 0.3	3.38± 0.35	185±4
Found BCR-701 step 2	3,451± 46	16.5± 0.3	3.37± 0.08	3.22± 0.03	39.2± 0.4	100±2	7,042± 106	128±3	24.5± 0.4	111±2	102±1
Found BCR-701 step 3	1,912± 74	3.09± 0.20	0.28± 0.01	1.86± 0.17	169±4	64.8± 1.5	1,147± 56	31.9± 2.6	17.4± 1.7	7.15± 0.12	58.4± 5.0
Certified BCR-701 step 1	n.a.	n.a.	7.34± 0.35	n.a.	2.26± 0.16	49.3± 1.7	n.a.	n.a.	15.4± 0.9	3.18± 0.21	205±6
Certified BCR-701 step 2	n.a.	n.a.	3.77± 0.28	n.a.	45.7±2	124±3	n.a.	n.a.	26.6± 1.3	126±3	114±5
Certified BCR-701 step 3	n.a.	n.a.	0.27± 0.06	n.a.	143±7	55.2± 4.0	n.a.	n.a.	15.3± 0.9	9.3±2.0	54.2± 2.0
Recovery step 1	n.a.	n.a.	83.0%	n.a.	107%	96.8%	n.a.	n.a.	94.4%	106%	90.3%
Recovery step 2	n.a.	n.a.	89.4%	n.a.	85.9%	80.6%	n.a.	n.a.	92.3%	87.9%	89.4%
Recovery step 3	n.a.	n.a.	104%	n.a.	118%	117%	n.a.	n.a.	114%	76.9%	108%

The elemental concentrations were determined with an inductively coupled plasma atomic emission spectrometer Vista Pro (Varian), with the internal standard method. Cadmium was determined by electrothermal atomization atomic absorption spectrometry. A Varian Spectra A300 spectrometer with Zeeman effect background correction and autosampler Varian Model 96 was used employing the standard addition method for calibration. All the metal analyses were performed at the Department of Chemistry and Industrial Chemistry of the University of Genoa (Genoa, Italy).

5.1.3. Statistical analyses

Principal component analysis (PCA) was used to characterize the metal composition in sediments, and cluster analysis was used for grouping the sampling stations. Principal component analysis (PCA) is a multivariate statistical technique used for data reduction and for deciphering patterns within large sets of data. With PCA, a large data matrix is reduced to two smaller ones that consist of principal component (PC) scores and loadings. PC loadings are eigenvectors of the correlation or covariance matrix depending on which is used for the analysis. The PC scores contain information on all of the variables combined into a single number, with the loadings indicating the relative contribution of each variable to that score [87]. Hierarchical cluster analysis (HCA) characterizes similarities among samples by examining interpoint distances representing all possible sample pairs in high-dimensional space. The sample similarities are represented on two dimensional diagrams call dendrograms [88]. All statistical analyses were performed using the computer software STATISTICA (StatSoft, Inc. 2001).

5.2. Results And Discussion

5.2.1. Sediment Geochemistry

Table 4 shows values of pH; organic carbon (%); and percentage of sand, silt, and clay in sediments of the nine sampling sites. Organic carbon values, ranging from 0.22% (in station S_8) to 1.02% (in station S_2), are low in comparison with values found in sediments from other Indian coastal areas, such as Gulf of Mannar [89], Cochin [90], and Muthupet mangroves [91]. The low organic carbon values might be related with the poor absorbability of organics on negatively charged quartz grains, which predominate in sediments in this estuarine environment [92]. In addition, the constant flushing activity by tides along with the impact of waves can support the low percentage of organic carbon in the sediments. The sediments of the studied stations are characterized by slightly basic pH (7.50–8.36) with maximum values recorded in the stations closest to the sea (stations S_6, S_8, and S_9) and minimum in station S_7.

These were different from the low pH values in most of the mangrove swamps in Hong Kong [93], where sediments were not frequently flooded by the tide and become acidic in reducible conditions. With respect to texture, the sediment samples show a variable admixture of sand, silt, and clay. Clay fractions dominate in low-energy areas of suspensional deposits. On the contrary, silt, and sand dominates where the energy level is high. Sediments from station S_7 contain higher percentage of sand (98%) compared to the others, while sediments from S_1, S_2, S_5, and S_8 contains higher percentage of silt (more than 50%) compared to the others. A variable mixture of sand, silt, and clay is present in the other stations and reflect a variable amount of erosion and deposition.

Table 4. Geographical position, physicochemical and textural properties of sediment samples of 9 sampling sites.

Stations	Latitude and Longitude	Salinity	pH	Organic carbon (%)	Sand (%)	Silt (%)	Clay (%)
S1	22°43′ 16″ N 88°21′ 20″ E	0	7.86	0.35	4	87.1	8.9
S2	22°39′ 17″ N 88°12′ 25″ E	0	7.80	1.02	1	76.5	22.5
S3	22°33′ 53″ N 88°20′ 19″ E	0	7.90	0.52	2.24	41.97	55.79
S4	22°30′ 10″ N 88°11′ 48″ E	0–2.5	7.60	0.74	18.25	47.42	34.33
S5	22°28′ 06″ N 88°06′ 54″ E	0–1	7.90	0.91	16.7	69.6	13.7
S6	22°11′ 14″ N 88°11′ 15″ E	0–5.6	8.36	0.56	3.15	41.13	55.71
S7	21°34′ 44″ N 88°15′ 03″ E	30–34.3	7.50	0.36	98.02	0.18	0
S8	21°38′ 15″ N 88°03′ 53″ E	32–35	8.14	0.22	32.85	58.45	8.7
S9	21°34′ 20″ N 88°01′ 25″ E	35	8.10	0.46	39.3	44.25	16.45

5.2.2. Total Element Concentrations

Total element concentrations in the investigated stations varied in a narrow range of values (Table 5)and were comparable with data obtained for other Indian coastal areas [94, 95]. Datta and Subramanian [96] found very similar trace element concentrations throughout the Bengal Basin, where anthropogenic perturbation is low and river channel may receive a several centimeter-thick sediment layer in a single event during peak flow, preventing to bear the signature of an accumulation of trace elements. The highest concentrations for As, Cu, Fe, Mn, and Ni were measured at station S_9 while for Cd and Pb at station S_3, close to Calcutta city (about 4.5 million residents, but about 14.2 million including suburbs). An anthropogenic input from vehicular traffic and in-dustrial activities may cause high Cd and Pb con-centrations measured in samples collected in the Calcutta urban area. The lowest element concentrations were found at station S_5. Very low (close to the detection limit) Cd concentration was found in the coastal stations (S_7, S_8, and S_9).

Table 5. Total element concentrations (µg/g) in sediments of 9 sampling sites (instrumental precision, calculated as RSD%, resulted lower than 5% for each element in all samples).

Stations	Al	As	Cd	Co	Cr	Cu	Fe	Mn	Ni	Pb	Zn
S1	70,289	8.81	0.165	13.0	67.6	27.8	37,737	591	31.9	20.4	86.6
S2	70,879	8.44	0.452	14.0	74.8	36.8	39,405	625	34.2	22.3	90.7
S3	72,134	8.65	1.79	14.5	73.5	32.3	40,070	712	35.0	33.2	83.1
S4	72,613	8.49	0.492	14.9	76.8	27.9	40,303	726	35.0	19.6	80.4
S5	62,044	6.41	0.220	12.1	58.2	21.1	33,428	597	27.5	17.0	64.1
S6	64,325	6.79	0.106	12.0	64.8	32.0	34,273	613	31.3	17.9	69.6
S7	77,529	7.77	0.044	14.0	75.1	28.4	40,084	389	38.1	20.5	74.4
S8	68,146	8.08	0.027	12.7	62.5	22.2	36,786	511	34.3	19.7	61.4
S9	72,666	9.40	0.044	14.1	74.2	36.6	40,838	785	40.1	22.9	74.9

The geoaccumulation index (I_{geo}) of Muller [97] has been calculated for the analyzed elements, by comparing current concentrations with pre-industrial levels, in order to estimate the metal contamination in sediments. The equation used for the calculation of I_{geo} is: $\log_2 (C_n/1.5\ B_n)$, where C_n is the measured content of element "n" and B_n the element's content in "average shale" [98]. Factor 1.5 is used because of possible variations in background values for a given element in the environment, as well as very small anthropogenic influences [99]. As shown in Figure 2, all sediments fall in class 0 for Al, Co, Cr, Cu, Fe, Mn, Ni, Pb, and Zn, therefore the area is not contaminated for these elements. Unlike the Hugli river, in other rivers of the Bengal Basin, such as Meghna and Brahmaputra, Cr exhibits higher I_{geo} values respect to the other elements [96]. For Cd, two stations fall in class 1 and three in class 2 exhibiting a moderate contamination for this element. In all stations, As falls in class 2 (moderate pollution). In this area, As contamination was already observed in previous studies and it is probably due to groundwater contamination [100]. This contamination can have natural origin, such as coal seams in Rajmahal basin and arsenic mineral in mineral rocks in the upper reaches of the Ganges river system. The highly reducing nature of groundwater would reduce As, causing the possible desorption of As [101].

5.2.3. Speciation Patterns

The potential environmental risk of trace elements in sediments is associated with both their total content and their speciation. The chemical partitioning of the considered elements (Al, As, Cd, Co, Cr, Cu, Fe, Mn, Ni, Pb, and Zn) from each extraction step has been described. Aluminum, Cr, and Fe are present mainly in the residual phase, representing 95.8–96.8%, 88.9–91%, and 83.0–94.7% of the total concentration, respectively, which implies that these elements are strongly linked to the inert fraction of the sediments. This result was in good agreement with data reported by several studies carried out worldwide in marine coastal areas [45, 46, 78, 102]. The high percentage of Fe in the residual phase indicates that most of the Fe exists as crystalline Fe peroxides (goethite, limonite, magnetite, etc.). The remaining Fe is associated with the reducible phase (mean, 11.25%). Large amounts of Fe accumulate in the residual phase probably because it is basically of natural origin (it is the most common element in the earth's crust).

Concentrations of Al, Fe, and Cr are very low in exchangeable phase (0.08%, 0.26%, and 1.72% as mean values, respectively), limiting their potential toxicity as pollutants. It should be noted that sediments always act as reservoir for elements; therefore, their potential risk of pollution to environment has always to be considered.

Arsenic, Co, Ni, and to some extent Zn, are found mainly in the residue (~50% of the total concentration). Nickel and Co are associated to the residue respectively for 56% and 74% of the total concentration, with a speciation similar in all the samples. A mean of 23% of Co is present in the phase 2. The highest percentage of labile Co (~13%) was measured in S_6 (Diamond Harbour) and S_8(Gangasagar) and can be due to a recent input of this element. The dominant proportion of Ni in the residual phase is in agreement with the results of other studies [27, 46]. Nickel is present, apart from the residue, in phases 2 and 3 (about 10% in each phase). Arsenic is distributed mainly between the residual (mean 47%), the reducible, and the oxidizable phases (mean 19% and 22%, respectively). Acharyya et al. [101] observed that As is adsorbed to iron-hydroxide-coated sand grains and to clay minerals in the sediments of the Ganges delta from West Bengal. Among the studied elements, As is found with the greatest proportion in the oxidizable phase coinciding with organic and sulfur compounds. Arsenic is present in the phase 1 for about 10% of the total content, in station S_7 phase 1 percentage rises up to 16%. The lower land alluvial basin of the Ganges River is recognized as an arsenic-affected area. Arsenic in solution probably is easily entrapped in the fine grained organic-rich sediments deposited in the Ganges delta [101]. The percentage of silt (lower than 70% except in $S_1$and S_2) may have contributed to a low retention of dissolved As since coarse sediments are less efficient at retaining As.

Cadmium was mainly present in the labile phase (more than 60%) in all the stations with the exception of station S_7, where the Cd labile percentage represents only 25% of the total concentration. Cadmium concentrations were negligible in phases 2 and 3. The highest labile Cd concentration was measured at station S_3, the closest to the city of Calcutta. Datta and Subramanian [96] found that the concentrations of elements in the non-detrital phases were higher in stations sampled in the Hugli river around Calcutta than in samples collected along Brahmaputra and Meghna rivers. The petroleum refinery, industrial, and mining effluents carried by the Hugli river may be responsible for this higher concentrations of non detrital fractions.

About 40% of the total Cu concentration is associated to the residue, while 33% of Cu is bound to Fe-Mn oxide and hydroxide (phase 2). The high percentage of Cu in the residue is likely due to the fact that Cu is easily chemisorbed on or incorporated in clay minerals [103]. All the samples showed lower Cu concentrations in exchangeable phase, with percentage ranging from 7% (S_7) to 22% (S_5), with a mean of 15%. Copper is characterized by high complex constant with organic matter thus it can be hypothesized that Cu is bound to labile organic matter such as lipids, proteins, and carbohydrates. On the other hand, high-element concentration in labile phase could be related to recent coastal input [39].

Manganese was found in all the four sediment phases, as observed by other researchers [45, 104]. Manganese is the most mobile element since it is present with the highest percentage (a mean of 42%) in the labile phase. This is probably because of the known close association of Mn with carbonates [105] as endorsed by other workers [69, 106]. In this phase, weakly sorbed Mn retained on sediment surface by relatively weak electrostatic interactions may be released by ion exchange processes and dissociation of Mn-carbonate phase [2]. The result indicates that considerable amount of Mn may be released into environment if conditions become more acidic [107]. The highest Mn labile percentage was measured in S_6 (57%). Differently, in S_7, Mn in the residue represents 65% of the total concentration, while the labile Mn is only 15%. A substantial Mn percentage was also found in the residue (mean 37.8%), followed by the reducible phase (14.7%), in which Mn exists as oxides and may be released if the sediment is subjected to more reducing conditions [108].

The major geochemical phase for Pb in these sediments was the Fe-Mn oxides phase (mean 55.7%) followed by the residual phase (mean 30.2%) while lower percentage of the total Pb are bound to exchangeable-labile (mean 5.3%) and oxidizable phases (mean 6.8%). At S_3 (Babughat), the reducible part is as high as 65% and only 19.9% of the total is associated with the residue. Atmospheric input as fallout from vehicular emission can be probably the major input of Pb for this station. The relatively high percentage of Pb in reducible phase is in agreement with the known ability of amorphous Fe–Mn oxides to scavenge Pb from solution [109, 110]. Caille et al. [111] observed that resuspension of anoxic sediment results in a rapid desorption of Pb and Cu adsorbed to sulphides. Thus, a high element percentage in the reducible fraction is a hazard for the aquatic environment because Fe and Mn species can be reduced into the porewaters during early diagenesis by microbially mediated redox reactions

[112]. Dissolution will also release Pb associated with oxide phases to the porewater possibly to the overlying water column [113] and to benthic biota [79]. The major sources of Pb are from intensive human activities, including agriculture in the drainage basin [114], auto exhaust emission together with atmospheric deposition [115]. In addition, a substantial contribution from the factories located in the upstream of the Hugli river dealing with Pb producing lead ingots and lead alloys play a vital role as referred by Sarkar et al. [116].

The percentage of Zn in residue is highly variable (38.5–70%) and the distribution pattern in each fraction showed the following order: residual>reducible>oxidizable> exchangeable and bound to carbonates. There was some difference in Zn speciation among the sampling sites: in stations S_1, S_2, and S_3 about 40% of Zn is present in the residue, while in the other stations this percentage increases to more than 60%. In station S_1, the exchangeable and oxidisable phases shared over 22% of the total Zn, whereas labile Zn was as low as 4.6% at S_7. A major part of Zn (16.3%) is associated with Fe–Mn oxide phase, because of the high stability constants of Zn oxides. Iron oxides adsorb considerable quantities of Zn and these oxides may also occlude Zn in the lattice structures [117].

The BCR procedure as discussed above showed satisfactory recoveries, detection limits, and standard deviations for determinations of heavy metals/metalloid in the sediments. It is evident from the present results of the fractionation studies that the metals/metalloids in the sediments are bound to different fractions with different strengths leading to variations in mobility and availability and some of them show significant spatial variations subject to diverse environmental stresses. This type of association between metals and the sediments can be understood in detail by sequential extraction techniques. Hence the application of sequential extraction is fully justified as the quantification of different forms of metal is more meaningful than the estimation of its total metal concentrations. The strength values can, therefore, give a clear indication of sediment reactivity, which in turn assess the risk connected with the presence of metals in this wetland environment. The results obtained suggest the need for corrective remediation measures due to the higher accumulation of potentially dangerous metals/metalloids, which in most cases exceed the limits established by certain legislation.

5.2.4. Comparison With Sediment Quality Guidelines

Results obtained after total and sequential extraction are compared with Sediment Quality Guidelines (SQGs). Table 6 reports consensus-based values, such as TEC (concentration below which harmful effects on sediment-dwelling organisms were not expected) and PEC (concentration above which harmful effects on sediment-dwelling organisms were expected to occur frequently), and effect range-low and range-medium, such as ERL (concentrations below which adverse biological effects were observed in less than 10% of studies) and ERM (concentrations above which effects were more frequently observed in more than 75% of studies).

Table 6. Sediment Quality Guidelines concentrations with respect to total and labile element concentrations found in the analyzed samples (expressed as μg/g of dry weight).

Element	Phase	Si<TEC	TEC	Si<TEC<PEC	PEC	Si<ERL	ERL	ERL<Si<ERM	ERM
As	Total	All	9.79	None	33	S5,S6,S7,S8	8.2	S1,S2,S3,S4,S9	70
	Labile	All		None		All		None	
Cd	Total	S1,S2,S4,S5, S6,S7,S8,S9	0.99	S3	4.98	S1,S2,S4,S5, S6,S7,S8,S9	1.2	S3	9.6
	Labile	S1,S2,S4,S5, S6,S7,S8,S9		S3		S1,S2,S4,S5, S6,S7,S8,S9		S3	
Cr	Total	None	43.4	All	111	All	81	None	370
	Labile	All		None		All		None	
Cu	Total	S1,S4,S5,S7,S8	31.6	S2,S3,S6,S9	149	S1,S3,S4,S5, S6,S7,S8	34	S2,S9	270
	Labile	All		None		All		None	
Ni	Total	None	22.7	All	48.6	None	20.9	All	51.6
	Labile	All		None		All		None	
Pb	Total	All	35.8	None	128	All	46.7	None	218
	Labile	All		None		All		None	
Zn	Total	All	121	None	459	All	150	None	410
	Labile	All		None		All		None	

Comparing our results with the SQGs, it is revealed that for Pb and Zn in all the stations the measured concentrations are lower than both TEC and ERL. As regards Cd, concentration measured in station $S_3$is higher than TEC and ERL but lower than PEC and ERM both in term of total and labile concentration. For this station, some possible toxic effect on benthic organism can be hypothesized, in particular because of the large amount of element bound to the most labile phase of the sediment. Considering Cu, some stations (S_2, S_3, S_6, and S_9) exhibit total concentrations higher than TEC but lower than PEC. Concentrations of Cu are higher than ERL but lower than ERM only in stations $S_2$and S_9. Since only 7–22% of total Cu is bound to the labile phase, in all stations Cu labile concentrations are lower than TEC and ERL. Total As concentrations in stations S_1, S_2, S_3, S_4, and $S_9$are higher than ERL value but lower than TEC value. Since more than 50% of total As is not found in the residue, attention should be paid to a change in the environment conditions which could induce a release of As from the sediments. Total Ni and Cr concentrations are higher than TEC (Ni is also higher than ERL) but lower than PEC (and ERM in the case of Ni) in all the stations. Nevertheless, more than 70% of Ni as well as 90% of Cr are present in the residual fractions, therefore adverse impacts on organisms is very much negligible.

Mean sediment quality guidelines quotients (mSQGQ) have been developed for assessing the potential effects of contaminant mixtures in sediments [118]: they are determined by calculating the arithmetic mean of the quotients derived by dividing the concentrations of chemicals in sediments by their respective SQGs. The probability of observing sediment toxicity can be estimated by comparing the mSQGQ in a sample to previously published probability tables. It is important to keep in mind that mSQGQs cannot be used to accurately predict the uptake and bioaccumulation of sediment-bound chemicals by fish, wildlife, and humans, even if there is considerable evidence that this assessment tool can be predictive of the presence or absence of toxic effects [118].

SQGQs are calculated for seven elements considering ERM as sediment quality guidelines (Table 7). The mean quotient values ranges from 0.16 in station S_5 to 0.24 in station S_3. Using PEC values instead of ERM, the mean SQGQ ranges from 0.25 in station S_5 to 0.38 in station S_3 (Table 7).

Table 7. Mean Sediment Quality Guidelines Quotients calculated for the nine stations using PEC and ERM as SQGs.

Stations	$SQGQ_{PEC}$	$SQGQ_{ERM}$
S1	0.30	0.19
S2	0.33	0.21
S3	0.38	0.24
S4	0.33	0.21
S5	0.25	0.16
S6	0.28	0.18
S7	0.32	0.21
S8	0.28	0.18
S9	0.34	0.22

Compiled data from multiple data sets reporting 10-day toxicity test conducted on amphipod species in saltwater showed that the incidence of toxicity for a range of SQGQ of 0.25–0.5 is ~35%, while for a mean SQGQ range from 0.1 to 0.25, the incidence of toxicity lowers to ~20%. Measures recorded in a survey of Biscayne Bay (port of Miami and the adjoining saltwater reaches of the lower Miami River, FL, USA) showed that the average amphipod survival (*Ampelisca abdita*) decreased slightly from the least contaminated (ERMQ <0.03) to the intermediate category, (ERMQ included in 0.03–0.2 range) then decreased greatly in the most contaminated sediments (ERMQ included in 0.2–2 range). Therefore, we can presume a low toxicity of sediments sampled in the nine stations for benthic organisms. It is important to note that the benthic response to contaminants covaried among stations with both the mean ERM quotients and the effect of natural factors, such as the sediment texture, TOC, and salinity [118].

5.2.5. Statistical Analyses

The relationships between variables and the differences between stations were evaluated by PCA. The analysis was performed on 36 objects (four sediment phases for nine stations) and 11 variables (Al, As, Cd, Co, Cr, Cu, Fe, Mn, Ni, Pb, Zn). Two significant components were identified explaining 68.3% and 14.5% of the total variance, respectively. By studying the loadings of the variables (Figure 2a) on the components it can be seen that all the elements except Cd, Mn, and Pb are significantly correlated.

Unlike the other elements, most of Cd and Mn is present in the first phase: labile Cd and Mn represent more than 60% and 40% of the total concentration, respectively, except in station S_7. Cadmium and Mn speciation can be ascribed to their considerable affinity for carbonates. Lead is the only element which is bound to the reducible phase for more than 50%. Lead is a very reactive element in water column and, having scavenging type behavior, is easily bound to hydroxy- and oxyligands. Copper is positively and significantly correlated with all elements except Cd and Mn, but with lower correlation coefficients (0.66–0.81).

In the score plot (Figure 2b) phases 1, 2, 3, and 4 (corresponding to labile, reducible, oxidizable, and residual phases respectively) are identifiable by 1, 2, 3, and 4 suffix, respectively. In all stations, residue concentrations were characterized by negative values of PC1 and consequently by high concentrations of Al, As, Co, Cr, Fe, Ni, and Zn. Conversely, in the positive PC1 semi-axis labile and oxidizable metal concentrations, which represent a small percentage of the total elements, are distributed. For all stations, reducible concentrations are distributed along the positive PC2 semi-axis, i.e., high Pb concentrations, with a maximum for station S_3 and a minimum for S_5. The group formed by elements bound to organic matter and sulphides (phase 3) is

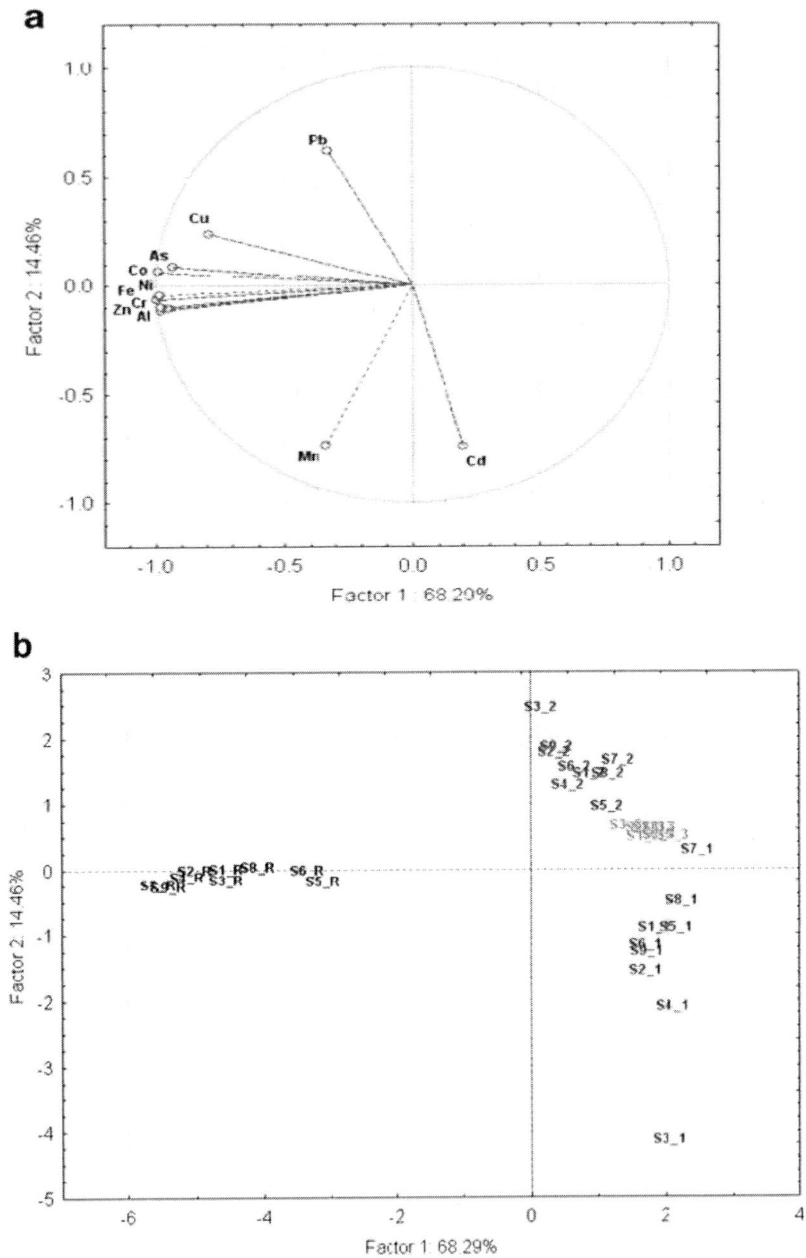

Figure 2. Principal component analysis: a) variable plot; b) score plot (phase 1, 2, 3 and 4, corresponding to labile, reducible, oxidizable and residue, are identifiable by 1, 2, 3 and 4 suffix and different colors in the score plot).

characterized by low values of both PC1 and PC2. Therefore, a low percentage of elements (higher than 20% exclusively for all As data and for Zn in stations S_1 and S_3) is bound to the oxidizable phase, suggesting the presence of an oxidant environment. High Mn and Cd concentrations are associated with negative values of PC2, therefore a relatively high concentration of labile Mn and Cd is present in all samples (in particular in S_3), except station S_7. Samples are prevalently grouped in relation to the sediment geochemical phase, suggesting a similar element speciation among the stations. Station S_7 represents an exception, in fact the labile fraction is closely associated to the oxidizable phase group.

A HCA was carried out by applying Euclidean distances to quantitatively identify specific groups of similar stations. In the dendrogram of the sampling stations (Figure 3), we can note two main clusters: the first represented by station S_7, characterized by the highest element percentage bound to residue, and the second constituted by all the remainder stations. In the second group, a subgroup formed by station S_5 and S_6 can be individuated.

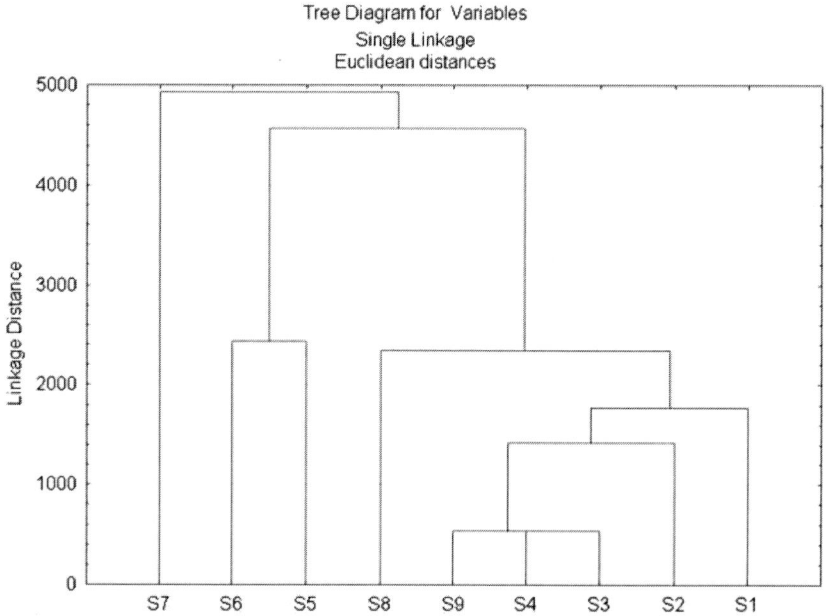

Figure 3. Dendrogram indicating linkage of sites on the basis of element concentrations.

Station S_7 was sampled in a marine coastal environment; it is characterized by a peculiar grain size percentage respect to the other stations, being the sand percentage as high as 98.6%. In general, the concentrations of elements are much higher in fine than in coarse fraction because the fine fraction larger specific surface facilitates absorption processes. As previously noted by Ramirez et al. [39], this pattern is particularly evident for Cd. It is interesting to note that the marine coastal stations S_8 and $S_9$are more similar to river stations than to station S_7. Both the stations S_8 and S_9 are, in fact, located in front of the Hugli river runoff, while station S_7 is located easternmost and probably is less influenced by the Hugli river discharge.

5.3. Conclusion And Recommendation

The study provides valuable information on the potential mobility of trace elements in sediments collected along the stretch of Hugli River and in the Sundarban mangrove wetland (northeastern part of the Bay of Bengal). The results obtained adopting BCR sequential extraction method provided the following important information: (i) Al, Cr, and Fe were found mostly in the residual phase while the other elements were found in the four phases sharing different proportions; (ii) the dominant Cd, Mn, and Pb proportion was found in the non-residual fractions and (iii) Mn had the highest percentage in the labile phase. This is worthwhile to mention that coastal environment of West Bengal is considerably constrained due to implementation of dredging, construction of port/ harbor and other industrial activities. The authors strongly recommend for periodical monitoring on the bioavailability and mobility of trace elements, control the mixing of effluent of the concentration of heavy metals in the region, environmental remediation, treatment of industrial effluent and municipal wastewater for effective management of this estuarine system. It is wisely suggested that an environmental recovery framework should be urgently implemented to avoid extension of heavy metal contamination (especially As).

6. CASE STUDIES FROM OTHER COASTAL REGIONS IN INDIA

Although the importance of metal speciation and fractionation has been realized in developed countries, the subject has not really taken off in India and only few references are available on the speciation of metals in Indian rivers. Speciation of selected heavy metals geochemistry in surface sediments (n=10 was studied by Venkatramanan et. al. [119] from Tirumalairajan river estuary, east coast of India. The results obtained from sequential extraction showed that a larger portion of the metals were associated with the residual phase, although they are available in other fractions too.

Trace metal fractionation in the Pichavaram mangrove–estuarine sediments in southeast coast of India was studied by Ranjan et. al. [120] considering the pronounced changes due to occurrence of tsunami (2004). A 5-step sequential extraction procedure was applied to assess the effects of tsunami on mobility and redistribution of selected elements (Cd, Cr, Cu, Fe, Mn, Ni, Pb, and Zn) in coastal sediments revealed that metals in the residual fraction (lattice bound) had the highest concentration suggesting their non-availability and limited biological uptake in the system. Majority of the metals (except Mn) do not constitute a risk based on the different geochemical indices.

Fractionation of selected metals in the sediments of Cochin estuary and Periyar River (southwest coast of India) was studied by Mohan et. al. [121]. The results reveal that remobilization potential of metals bound is in the range of low to medium risk to various sedimentary phases is different and is based on bond strength. Therefore, the strength values can give a clear indication of sediment reactivity that can be used to assess the risk related with metals to the aquatic organisms.

7. RISK ASSESSMENT CODE (RAC)

The risk assessment code (RAC) mainly applies the sum of the exchangeable and carbonate bound fractions for assessing the availability of metals in sediments. These fractions are considered to be weakly bonded metals which may equilibrate with the aqueous phase and thus become more rapidly bioavailable [11, 33]. This is important because the fractions introduced by anthropogenic activities, such as agricultural runoff and tourism, are typified by the adsorptive, exchangeable, and bound to carbonate fractions, which are weakly bonded metals that could equilibrate with the aqueous phase and thus become more rapidly bioavailable [122]. According to RAC guideline (Table 8), for any metal, soil/sediment which can release in exchangeable and carbonate fractions, less than 1% of the total metal will be considered safe for the environment and soil/sediment with 11-30% carbonate and exchangeable fractions will be at medium risk to the environment. On the contrary, soil/sediment releasing in the above fractions more than 50% of the total metal has considered being highly dangerous, which can be easily enter the food chain [123].

Table 8. Criteria of Risk Assessment Code [123].

Grade	Exchangeable and bounded to carbonate metal (%)	Risk
I	<1	No risk
II	1 – 10	Low risk
III	11 – 30	Medium risk
IV	31 – 50	High risk
V	>50	Very high risk

Heavy-Metal Fractionation in surface sediments was studied by Dhanakumar et. al. [124] in the Cauvery river estuarine region, southeastern coast of India. The results revealed that most of the samples fall under the category from low- to high-risk class and from low-risk to very high-risk class in terms of labile fractions of Pb as well as Zn and Cu, respectively.

8. CONCLUSION

From the above discussion it is revealed that geochemical fractionation approach to the chemical speciation has provided a useful tool and opens a new dimension in assessing the potential mobility/bioavailability of heavy metals and metalloids in soils/sediments and opens a new dimension in the field of ecology and environmental chemistry. More efficient, non-laborious and time saving processes techniques in this field of chemical speciation are also coming up to get valid information regarding geochemical behavior of soils/sediments. Besides geochemical fractionation, Dezileau et al. [125] opined that total Fe or Fe/Al may be used to infer millennial-scales climate changes in the south eastern pacific while performing sequential extraction of Fe in marine sediments from the Chileau continental margin. However, the chemical partitioning should be carefully used in the assessment of environmental pollution as large amount of metals may naturally occur as anthropogenic fractions (including loosely bonded ions, sulfide ions and metals associated with sediments).

9. ACKNOWLEDGEMENTS

The authors gratefully acknowledge full support and cooperation of the Springer press, UK for extending permission in publishing the research paper of the journal Environmental Monitoring & Assessment, vol. 184(12), pp:7561-77, 2012.

This study was partially supported by the European Fund for Economic and Regional Development (FEDER) through the Program Operational Factors of Competitiveness (COMPETE) and National Funds through the Portuguese Foundation for Science and Technology (PEST-C/MAR/UI 0284/2011, FCOMP 01 0124 FEDER 022689).

REFERENCES

1. Gleyzes C, Tellier S, Astruc M. Fractionation studies of trace elements in contaminated soils and sediments: a review of sequential extraction procedures. Trends in Analytical Chemistry 2002; 21 (6) 451-467.

2. Tessier A, Campbell PGC, Bisson M. Sequential extraction procedure for the speciation of particulate trace metals. Analytical Chemistry 1979; 51 844-851.

3. Sahuquillo A, Rigol A, Rauret G. Overview of the use of leaching/extraction tests for risk assessment of trace metals in contaminated soils and sediments. Trends in Analytical Chemistry 2003; 22(3) 152-159.

4. Calmano W, Hong J, Förstner U. Binding and mobilization of heavy metals in contaminated sediments affected by pH and redox potential. Water Science and Technology 1993; 28(8/9) 53-58.

5. Eggleton J, Thomas KV. A review of factors affecting the release and bioavailability of contaminants during sediment disturbance events. Environment International 2004; 30 973-980.

6. Ure AM, Quevauviller P, Muntau H, Griepink B. Speciation of heavy metals in solids and harmonization of extraction techniques undertaken under the auspices of the BCR of the Commission of the European Communities. International Journal of Environmental Analytical Chemistry 1993; 51 135-151.

7. Florence TM. The speciation of trace elements in waters. Talanta 1982; 29 345-69.

8. Lung W. Speciation analysis – why and how? Fresenius' Journal of Analytical Chemistry 1990; 337 557-564.

9. Schroeder WH. Development in the speciation of mercury in natural waters. Trends in Analytical Chemistry 1989; 8 339-342.

10. Rauret G, Rubio R, Lopez-Sanchez JF, Cassassas E. Determination and speciation of copper and lead in sediments of a Mediterranean River (River Tenes, Catalonia, Spain). Water Research 1988; 22(4) 449-51.

11. Pardo R, Barrado E, Perez L, Vega M. Determination and association of heavy metals in sediments of the Pisuerga River. Water Research 1990; 24(3) 373-379.

12. Jardo CP, Nickless G. Chemical association of Zn, Cd, Pb and Cu in soils and sediments determined by the sequential extraction technique. Environmental science & Technology Letters 1989; 10 743-52.

13. Tessier A, Campbell PGC, Bisson M. Heavy metal speciation in the Yamaska and St. Francois rivers (Quebec). Canadian Journal of Earth Sciences 1980; 17 90-105.

14. Elsokkary LH, Muller G. Assessment and speciation of chromium, nickel, lead and chromium in the sediments of the river Nile, Egypt. Science of the Total Environment 1990; 97/98 455-63.

15. Ure AM. Single extraction schemes for soil analysis and related applications. Science of the Total Environment 1996; 178(1/3) 3-10.

16. Rauret G. Extraction procedures for the determination of heavy metals in contaminated soil and sediment. Talanta 1998; 46(3) 449-455.

17. Quevauviller PH, Lachica M, Barahona E, Rauret G, Ure A, Gomez A, Muntau H. Interlaboratory comparision of EDTA and DTPA procedures prior to certification of extractable trace elements in calcareous soil. Science of the Total Environment 1996; 178(1/3) 127-132.

18. Quevauviller PH, van der Slootr HA, Ure A, Muntau H, Gomez A, Rauret G. Conclusions of the workshop: harmonization of leaching/extraction tests for environmental risk assessment. Science of the Total Environment 1996; 178(1/3) 133-139.

19. Bacon JR, Davidson CM. Is there a future for Sequential chemical extraction? Analyst 2008; 133 25-46

20. Förstner U, Lechsber R, Davis RA, Hermitte PL. (eds.), Chemical methods for assessing bioavailable Metals in Sludges. Elsevier, London, 1985.

21. Meguellati M, Robbe D, Marchandise P, Astruc M, Proceedings International Conference on Heavy Metals in the Environment, Heidelberg CEP Consultants, Edinburgh, 1983, p. 1090.

22. Filgueiras AV, Lavilla I, Bendicho C. Chemextraction for metal partitioning in environmental solid samples, Journal of Environmental Monitoring 2002; 4 832-857.

23. Lopez-sanchez JF, Rubio R, Rauret G. Comparison of two sequential extraction procedures for trace metal partitioning in sediments. International Journal of Environmental Analytical Chemistry 1993; 51 113-121.

24. Hlavay J, Polyak K. Chemical speciation of elements in sediment samples collected at Lake Balaton. Microchemical Journal 1998; 58 281-290.

25. Tokalioglu S, Kartal S, Elçi L. Determination of heavy metals and their speciation in lake sediments by flame atomic absorption spectrometry after a four-stage sequential extraction procedure, Analytical Chimica Acta 2000; 413 33-40.

26. Belazi AU, Davidson CM, Keating GE, Littlejohn D. Determination and speciation of heavy metals in sediments from the Cumbrian coast, NW England, UK. Journal of Analytical Atomic Spectrometry 1995; 10 233-240.

27. Mester Z, Cremisini C, Ghiara E, Morabito R. Comparison od two sequential extraction procedures for metal fractionation in sediment samples. Analytical Chimica Acta 1998; 259 133-142.

28. Gomez-Ariza JL, Giraldez I, Sanchez-Rodas D, Morales E. Comparison of the feasibility of three extraction procedures for trace metal partitioning in sediments from southwest Spain, Science of Total Environment 2000; 246 271-283.

29. Zhang T, Shan X, Li F. Comparison of two sequential extraction procedures for speciation analysis of metals in soils and plant availability. Communications in Soil Science and Plant Analysis 1998; 29 1023-1034.

30. Albores AF, Cid BP, Gomez P, Lopez EF. Comparison between sequential extraction procedures and single extrations for metal partitioning in sewage sludge samples. Analyst 2000; 125 1353-1357.

31. Ho MD, Evans GJ. Operational speciation of cadmium, copper, lead and zinc in the NIST standard reference materials 2710 nad 2711 (Montana soil) by the BCR sequential extraction procedure and flame atomic spectrometry. Analytical Communications 1997; 34 353-364.

32. Zemberyova M, Bartekov J, Hagarov I. The utilization of modified BCR three-step sequential extration procedure for the fractionation of Cd, Cr, Cu, Ni, Pb and Zn in soil reference materials of different origins, Talanta 2006; 70 973-978.

33. Singh KP, Mohan D, Singh VK, Malik A. Studies on distribution and fractionation of heavy metals in Gomti river sediments- a tributary of the Ganges India. Journal of Hydrology 2005; 312 14-27.

34. Davidson CM, Duncan AL, Littlejohn D, Ure AM, Garden LM. A critical evaluation of the three-stage BCR sequential extraction procedure to assess the potential mobility and toxicity of heavy metals in industrially-contaminated land. Analytica Chimica Acta 1998; 363(1) 45-55.

35. McLaren RG, Crawford D. Studies on soil copper I. The fractionation of copper in soils. Journal of Soil Science 1973; 24(2), 172-181.

36. Kersten M, Förstner U. Speciation of trace elements in of water-soluble organic components. sediments and combustion waste.In Ure AM, Davidson CM (ed.) Chemical speciation in the environment. Blackie Academic and Professional, Glasgow, UK. 1995; 234-275.

37. Ianni C, Magi E, Rivaro P, Ruggieri N. Trace metals in Adriatic coastal sediments: distribution and speciation pattern. Toxicological and Environmental Chemistry 2000; 78 73-92.

38. Ianni C, Ruggieri N, Rivaro P, Frache R. Evaluation and comparison of two selective extraction procedures for heavy metal speciation in sediments. Analytical Sciences 2001; 17 1273-1278.

39. Ramirez M, Massolo S, Frache R, Correa J. Metal speciation and environmental impact on sandy beaches due to El Salvador copper mine, Chile. Marine Pollution Bulletin 2005; 50 62-72.

40. Borovec Z, Tolar V, Mraz L. Distribution of some metals in sediments of the central part of the Labe (Elbe) River, Czech Republic. Ambio 1993; 22 200-205.

41. Campanella L, D'Orazio D, Petronio BM, Pietrantonio E. Proposal for a metal speciation study in sediments. Analytica Chimica Acta 1995; 309 387-393.

42. Zdenek B. Evaluation of the concentrations of trace elements in stream sediments by factor and cluster analysis and the sequential extraction procedure. Science of the Total Environment 1996; 177 237-250.

43. Gomez-Ariza JL, Giraldez I, Sanchez-Rodas D, Moralesm E. Metal sequential extraction procedure optimized for heavily polluted and iron Oxide rich sediments. Analytica Chimica Acta 2000; 414 151-164.

44. Lopez-Sanchez JF, Sahuquillo A, Fiedler HD, Rubio R, Rauret G, Muntau H, Quevauviller P. CRM 601, a stable material for its extractable content of heavy metals. Analyst 1998; 123 1675-1677.

45. Usero J, Gamero M, Morillo J, Gracia I. Comparative study of three sequential extraction procedures for metals in marine sediments. Environment International 1998; 24 478-496.

46. Martin R, Sanchez DM, Gutierrez AM. Sequential extraction of U, Th, Ce, La and some heavy metals in sediments from Ortigas River, Spain. Talanta 1998; 46 1115-1121.

47. Agnieszka S, Wieslaw Z. Application of sequential extraction and the ICPAES method for study of the partitioning of metals in fly ashes. Microchemical Journal 2002; 72 9-16.

48. Templeton DM, Ariese F, Cornels R. IUPAC guidelines for terms related to chemical Speciation and Fractionation of elements. Pure and Applied Chemistry 2001; 72 1453-1470.

49. Bordas F, Bourg ACM. A critical evaluation of sample for storage of Contaminated Sediments to Be investigated for the potential mobility of their heavy metal load. Water. Air. Soil. Pollution 1998; 103 137-149.

50. Kersten M, Frostner U. Chemical fractionation of heavy metals in anoxic estuarine and coastal sediments. Water Science and Technology 1986; 18 121-130.

51. Dold B. Speciation of the most soluble phases in a sequential extraction procedure adapted for geochemical studies of copper sulfide mine waste. Journal of Geochemical Exploration 2003; 80 55-68.

52. Favas PJC, Pratas J, Gomes MEP, Cala V. Selective chemical extraction of heavy metals in tailings and soils contaminated by mining activity: Environmental implications. Journal of Geochemical Exploration 2011; 111 160-171.

53. Pardo R, Vega M, Debán L, Cazurro C, Carretero C. Modelling of chemical fractionation patterns of metals in soils by two-way and three-way principal component analysis. Analytica Chimica Acta 2008; 606 26-36.

54. Cuong TD, Obbard JP. Metal speciation in coastal marine sediments from Singapore from Singapore using a modified BCR-sequential extraction procedure. Applied Geochemistry 2006; 21 1335-1346.

55. Dapaah RK, Takano N, Ayame A. Solvent extraction of Pb (II) from acid medium with zinc Hexamethylenedithiocarbamate followed by back-extraction and subsequent determination by FAAS. Analytica Chimica Acta 1999; 386 281-286.

56. Gomez-Ariza JL, Giraldez I, Sanchez-Rhodes D, Morales E. Metal readsorption and re-distribution during analytical fractionation of trace elements in toxic estuarine sediments. Analytica Chimica Acta 1999; 399 295-307.

57. Cheam V, Lechner J, Sekerka I, Desrosiers R, Nriagu J. Development of laser- excited atomic fluorescence spectrometer and a method for the direct determination of lead in Great Lake waters. Analytica Chimica Acta 1992; 269 129-136.

58. Fischer E, Van D, Berg CMG. Anodic Stripping Voltammetry of Pb and Cd using a Hg film electrode and thiocyanate. Analytica Chimica Acta 1999; 385 273-280.

59. Morales MM, Mart P, Llopis A, Compos L, Sagrado S. An environmental study by factor Analysis of surface sea waters in the Gulf of Valencia (Western Mediterranean). Analytica Chimica Acta 1999; 394 109-117.

60. Hirade M, Chen Z, Sugimoto K, Kawaguchi H. Co precipitation with tin (IV) hydroxide followed by removal of tin carrier for the Determination of trace heavy metals by graphite-furnace atomic absorption Spectrometry. Analytica Chimica Acta 1980; 302 103-107.

61. Ridout PS, Jones HR, Williams JG. Determination of trace elements in a marine reference material of lobster hepatopancreas (TORT-1) using inductively coupled plasma mass spectrometry. Analyst 1988; 113 1383-1386.

62. Sakao SY, OgawaY, Uchida H. Determination of trace elements in seaweed samples by inductively coupled plasma mass spectrometry. Analytica Chimica Acta 1999; 355 121-127.

63. Baeyens W, Monteny F, Leermakers M, Bouillon S. Evalution of sequential extractions on dry and wet sediments. Analytical and Bioanalytical Chemistry 2003; 376 890- 901.

64. Guevara-Riba A, Sahuquillo A, Rubio R, Rauret G. Assessment of metal mobility in dredged harbour sediments from Barcelona, Spain. Science of the Total Environment 2004; 321 241-255.

65. Idris AM, Eltayeb MAH, Potgieter-Vermaak SS, Grieken R, Potgieter JH. Assessment of heavy metal pollution in Sudanese harbours along the Red Sea coast. Microchemical. Journal 2007; 87 104-112.

66. Wepener V, Vermeulen LA. A note on the concentrations and bioavailability of selected metals in sediments of Richards Bay Harbour, South Africa. Water SA 2005; 31 589-595.

67. Esslemont G. Heavy metals in seawater, marine sediments and corals from the Townsville section, Great Barrier Reef Marine Park, Queensland. Marine Chemistry 2000; 71 215- 231.

68. Coetzee PP. Determination and speciation of heavy metals in sediments of the Hartebeespoort Dam By sequential extraction. Water SA 1993; 19 291-300.

69. Salmons W, Förstner U. Trace metal analysis on polluted sediments. Part II:evaluation of Environmental impact. Environmental science and Technology letters 1980; 1 14-24.

70. Li X, Thornton I. Chemical partitioning of trace and major elements in soils contaminated by mining and smelting activities. Applied geochemistry 2000; 16 1693-1706.

71. Kiratli N, Ergin M. Partitioning of heavy metals in surface Black Sea sediments. Applied Geochemistry 1996; 11 775-788.

72. Rauret G, Lopez-Sanchez JF. New sediment and soil CRMs for extractable Trace metal content. International Journal of Environmental Analytical Chemistry 2001; 79 81-95.

73. Salmons W. Adoption of common schemes for single and sequential extractions of Trace metals in soil and sediments. International Journal of Environmental Analytical Chemistry 1993; 51 3-4.

74. Fiedler HD, Lopez-Sanchez JF, Rubio R, Rauret G, Quevauviller PH. Study of the stability of extractable trace metal contents in a river sediment using Sequential extraction. Analyst 1994; 119 1109-1114.

75. Ramos L, Hernandez LM, Gonzalez MJ. Sequential fraction of copper, lead, copper, Cadmium and zinc in soils from or near Donana National Park. Journal of Environmental Quality 1994; 23 7-50.

76. Tu Q, Shan XZ, Ni Z. Evaluation of a sequential extraction procedure for the Fractionationation of amorphous iron and manganese oxides and organic matter in soils. The Science of The total Environment 1994; 151 159-165.

77. Wang S, Jia Y, Wang S, Wang X, Wang H. Fractionation of heavy metals in shallow marine sediments from Jinzhou Bay, China. Journal of Environmental Science (China) 2010 22 23-31.

78. Yuan CG, Shi JB, He B, Liu JF, Liang LN. Speciation of heavy metals in marine sediments from the East China Sea by ICP-MS with sequential extraction. Environment International 2004; 30 769-783.

79. Jones B, Turki A. Distribution and Speciation of heavy metals in surficial sediments from the Tees Estuary, North – East England. Marine Pollution Bulletin 1997; 34 768-779.

80. Ajay SO, Van Loon GW. Studies on redistribution during the analytical fractionation of metals in sediments. The Science of the Total Environment 1989; 87 171-187.

81. Pempkowiak J, Sikora A, Biernacka E. Speciation of heavy metals in marine sediments vs their bioaccumulation by mussels. Chemosphere 1999; 39 313- 321.

82. Forstner U, Wittmann GTW. Metal Pollution in the Aquatic Environment, Springer- Verlag, Berlin. Springer-Verlag, Heidelberg; 1981.

83. Fedotov PS, Zavarzina, a AG, Spivakov BYa, Wennrich, b R, Mattusch J, De K, Titzeb PC, Demin VV. Accelerated fractionation of heavy metals in contaminated soils and sediments using rotating coiled columns, Journal of Environmental Monitoring 2002; 4 318-324.

84. Nemati K, Kartini N, Bakar A, Abas MR, Sobhanzadeh E. Speciation of heavy metals by modified BCR sequential extraction procedure in different depths of sediments from Sungai Buloh, Selangor, Malaysia, Journal of Hazardous Materials 2011; 192(1) 402-410.

85. Mossop KF, Davidson CM. Comparison of original and modified BCR sequential extraction procedures for the fractionation of copper, iron, lead, manganese and zinc in soils and sediments. Analytica Chimica Acta 2003; 478 (1) 111-118.

86. Walkey A, Black TA. An examination of the Dugtijaraff method for determining soil organic matter and proposed modification of the chronic and titration method. Soil Science 1934; 37 23-38.

87. Farnham IM, Johannesson KH, Singh AK, Hodge VF, Stetzenbach KJ. Factor analytical approaches for evaluating groundwater trace element chemistry data. Analytica Chimica Acta 2003; 490 123-138.

88. Ragno G, De Luca M, Ioele G. An application of cluster analysis and multivariate cassification methods to spring water monitoring data. Microchemical Journal 2007; 87 119-127.

89. Jonathan MP, Ram Mohan V. Heavy metals in sediments of the inner shelf off the Gulf of Mannar, Southeast coast of India. Marine Pollution Bulletin 2003; 46 263-268.

90. Sunil Kumar R. Distribution of organic carbon in the sediments of Cochin mangroves, south west coast of India. Indian Journal of Marine Science 1996; 25 274-276.

91. Janaki-Raman D, Jonathan MP, Srinivasalu S, Armstrong-Altrin J S, Mohan SP, Ram-Mohan V. Trace metal enrichments in core sediments in Muthupet mangroves, SE coast of India: Application of acid leachable technique. Environmental Pollution 2007; 145 245-257.

92. Sarkar SK, Bilinski SF, Bhattacharya A, Saha M, Bilinski H. Levels of elements in the surficial estuarine sediments of the Hugli river, northeast India and their environmental implications. Environment International 2004; 30 1089-1098.

93. Tam NFY, Wrong YS. Spatial variation of heavy metalsin surfe sediments of Hong, Kong mangrove swamps. Environmental Pollution 2002; 110 195-205.

94. Subramanian V, Mohanachandran G. Heavy metals distribution and enrichment in the sediments of southern east coast of India. Marine Pollution Bulletin 1990; 21 324-330.

95. Chatterjee M, Massolo S, Sarkar SK, Bhattacharya AK, Bhattacharya BD, Satpathy KK, Saha S. An assessment of trace element contamination in intertidal sediment cores of Sunderban mangrove wetland, India for evaluating sediment quality guidelines. Environmental Monitoring and Assessment 2009; 150 307-322.

96. Datta DK, Subramanian V. Distribution and fractionation of heavy metals in the surface sediments of the Ganges–Brahmaputra–Meghna river system in the Bengal Basin. Environmental Geology 1998; 36 93-101.

97. Müller G. Schwermetalle in den sedimenten des Rheins-Veranderungen seit. Umschan Verlag 1979; 79 133-149.

98. Salomon W, Förstner U. Metals in the hydrocycle. Berlin: Springer; 1984.

99. Buccolieri A, Buccolieri G, Cardellicchio N, Dell'atti A, Leo AD, Maci A. Heavy metals in marine sediments of Taranto Gulf (Ionian Sea, Southern Italy). Marine Chemistry 2006; 99 227-235.

100. Dowling CB, Poreda RJ, Basu AR, Aggarwal PK. Geochemical study of arsenic release mechanisms in the Bengal Basin groundwater. Water Resources Research 2002; 38(9) 1173-1190.

101. Acharyya SK, Lahiri S, Raymahashay BC, Bhowmilk A. Arsenic toxicity of groundwater in parts of the Bengal basin in India and Bangladesh: the role of Quaternary stratigraphy and Holocene sea-level fluctuation. Environmental Geology 2000; 39 231-238.

102. Takarina ND, Browne DR, Risk MJ. Speciation of heavy metals in coastal sediments of Semarang, Indonesia. Marine Pollution Bulletin 2004; 49 854-874.

103. Pickering WF. Metal ion speciation—soil and sediments (a review). Ore Geology Reviews 1986; 1 83-146.

104. Ngiam LS, Lim PE. Speciation patterns of heavy metals in tropical estuarine anoxic and oxidized sediments by different sequential extraction schemes. Science of the Total Environment, 2001; 275 53-61.

105. Dassenakis M, Adrianos H, Depiazi G, Konstantas A, Karabela M, Sakellari A, Scoullos M. The use of various methods for the study of metal pollution in marine sediments, the case of Euvoikos Gulf, Greece. Applied Geochemistry 2003; 18 781-794.

106. Morillo J, Usero J, Gracia I. Heavy metal distri-bution in marine sediments from the southwest coast of Spain. Chemosphere 2004; 55 431-442.

107. Thomas RP, Ure AM, Davidson CM, Littlejoh D, Rauret G, Rubio R, López-Sánchez JF. Three-stage sequential extraction procedure for the deter-mination of metals in river sediments. Analytica Chimica Acta 1994; 286 423-429.

108. Panda D, Subramanian V, Panigrahy RC. Geochemical fractionation of heavy metals in Chilka Lake (east coast of India) – a tropical coastal lagoon. Environmental Geology 1995; 26 199-210.

109. Dawson EJ, Macklin MG. Speciation of heavy metals in floodplain and flood sediments: a reconnaissance survey of the Aire Valley, West Yorkshire, Great Britain Environmental Geochemistry and Health 1998; 20 67-76.

110. Ramos L, González M, Hernández L. Sequential extraction of copper, lead, cadmium, and zinc in sediments from Ebro River (Spain): relationship with levels detected in earthworms. Bulletin of Environmental Contamination and Toxicology 1999; 62 301-308.

111. Caille N, Tiffreau C, Leyval C, Morel JL. Solubility of metals in an anoxic sediment during prolonged aeration. Science of the Total Environment 2003; 301 239-250.

112. Canfield DE. Reactive iron in marine sediments. Geochimica et Cosmochimica Acta 1989; 53 619-632.

113. Petersen W, Wallman K, Li PL, Schroeder F, Knauth HD. Exchange of trace elements at the sediment– water interface during early diagenesis processes. Marine and Freshwater Research 1995; 46 19-26.

114. Monbet P. Mass balance of lead through a small macrotidal estuary: the Morlaix River estuary (Brittany, France). Marine Chemistry 2006; 98 59-80.

115. Adriano DC. Trace elements in terrestrial environments. New York: Springer; 1986.

116. Sarkar SK, Saha M, Takada H, Bhattacharya A, Mishra P, Bhattacharya B. Water quality management in the lower stretch of the river Ganges, east coast of India: an approach through environmental education. Journal of Cleaner Production 2007; 15 1559-1567.

117. Banerjee ADK. Heavy metal levels and solid phase speciation in street dusts of Delhi, India. Environmental Pollution 2003; 123(1) 95-105.

118. Long ER, Ingersoll CG, MacDonald DD. Calculation and uses of mean sediment quality guideline quotients: a critical review. Environmental Science and Technology 2006; 40 1726-1736.

119. Venkatramanan S, Ramkumar T, Anithamary I, Jonathan MP. Speciation of selected heavy metals geochemistry in surface sediments from Tirumalairajan river estuary, east coast of India. Environmental Monitoring and Assessment 2013; 185(8) 6563-6578.

120. Ranjan RK, Singh G, Routh J, Ramanathan AL. Trace metal fractionation in the Pichavaram mangrove–estuarine sediments in southeast India after the tsunami of 2004. Environmental Monitoring and Assessment 2013 (article in press).

121. Mohan M, Augustine T, Jayasooryan KK, Chandran MSS, Ramasamy EV. Fractionation of selected metals in the sediments of Cochin estuary and Periyar River, southwest coast of India, Environmentalist 2012; 32 383-393.

122. Hseu ZY. Extractability and bioavailability of zinc over time in three tropical soils incubated with biosolids. Chemosphere 2006; 63 762-771.

123. Perin G, Craboledda L, Lucchese M, Cirillo R, Dotta L, Zanette ML, Orio AA. Heavy metal speciation in the sediments of Northern Adriatic Sea-a new approach for environmental toxicity determination, in: T.D. Lekkas (Ed.), Heavy Metal in the Environment 1985; 2 454-456.

124. Dhanakumar S, Murthy KR, Solaraj G, Mohanraj R. Heavy-Metal Fractionation in Surface Sediments of the Cauvery River Estuarine Region, Southeastern Coast of India, Arch Environmental Contamination Toxicology 2013; 65 14-23.

125. Dezileau L, Pizarro C Rubio MA. Sequential extraction f iron in marine sediments from the Chilen continental margin. Marine Geology 2007; 241 111-116.

Chapter 3

Spatial Variation of Trace Elements in The Peri-Urban Soil of Madrid

Antonio Vázquez de la Cueva [1], Ben P. Marchant [2], José Ramón Quintana[3], Ana de Santiago [3], Antonio López Lafuente [3], Richard Webster [2]

[1.] *INIA, Forest Research CentreMadridSpain*

[2.] *Rothamsted ResearchHarpendenUK*

[3.] *Departamento de Edafología, Facultad de FarmaciaUCMMadridSpain*

ABSTRACT

Purpose

The peri-urban region to the south east of Madrid contains a mixture of housing, manufacturing industry and farming, some of which disperse metals, in particular cadmium, copper, lead, and zinc, into the soil. We have mapped the concentrations of these elements and identified the major influences on their distributions.

Material and methods

We sampled the topsoil at 125 sites across 1,050 km^2 of peri-urban land to the south east of the city on two grids, one nested inside the other. At each site, we measured the current contents of the four trace elements in the soil. We used robust geostatistical methods to model the complex spatial distributions of the data as mixtures of fixed and random effects. The empirical best linear unbiased predictor was used to map the elements. Site descriptors (lithology, land cover, cultivation, relief, erosion, and stoniness) were then included as covariates to identify significant effects on trace element concentrations.

Results and discussion

The complex spatial distributions of the elements seem to arise from several sources. The concentrations generally increase from southeast to northwest, i.e., with increasing proximity to Madrid itself, the main potential source of pollution. This pattern is clear for lead and similar for copper and zinc, though with "hot spots" at or near industrial sites. The spatial pattern of cadmium is more complex and depends on varied lithology, industry, and land use such as irrigation and cultivation. In general, the concentrations of the four elements appear to decrease with increases in stoniness and erosion, and to be largest on the valley floors.

Conclusions

Robust geostatistical methods enabled us to analyze and map the complex patterns of spatial variation of trace elements in a peri-urban region of Madrid. They show that distance to the city center, lithology, manufacturing industry, and cultivation all play their parts in loading the soil with lead, copper, zinc, and cadmium. In the event, none of the metals has yet exceeded the legislative thresholds, but some concentrations are already substantially greater than would arise from natural sources, especially closest to Madrid itself.

Keywords: Peri-urban Robust geostatistics Soil contamination Trace metals

1 INTRODUCTION

Many big cities are expanding into their peri-urban surroundings to accommodate more and more people. The soil of these peri-urban regions is, in many instances, already polluted with potentially toxic heavy metals from city waste, industrial effluent, smoke, and even manures and fertilizers used in agriculture. Urbanization risks adding to the load and exposing a larger population to the pollution. It is important therefore to know what the current load is and to map its distribution so that (a) the risk to human health can be assessed and, if necessary, remedial action can be taken to protect people and (b) the sources of pollution can be identified and perhaps eliminated.

Numerous sources of contamination by heavy metals have been identified in peri-urban regions, and these sources act over disparate spatial scales. Point sources are factories, gas works, waste deposits, and applications of compost. Cities produce large amounts of sewage, much of which contains heavy metals. The sewage is often concentrated into sludge or is composted and spread on the land nearby to minimize the cost of transport. Businelli et al. (2009), for example, found enhanced concentrations of copper (Cu), lead (Pb), and zinc (Zn) in composts added to commercial vegetable plots close to cities. Alloway (2004) reported that contamination of soil by cadmium (Cd) has been caused by the use of phosphate fertilizer containing the metal and by the disposal of ash from the combustion of coal. Copper salts are used as fungicides in vineyards, and they contribute to the large concentrations of Cu in the soil (e.g. Pietrzak and McPhail 2004; Martín et al. 2007; Saby et al. 2011). More diffuse atmospheric pollution can be attributed to traffic, industrial emissions, street dust, and the burning of fossil fuels (de Miguel et al. 1999). Nevertheless, it still tends to be local. Heavy metals from natural sources tend to vary in concentration over coarser scales according to the geological parent material. The various sources combine to form complex patterns of variation.

Cadmium, Cu, Pb, and Zn are the most widespread pollutant metals in soil. For example, Wei and Yang (2010) found that these four elements had the largest ratios of urban-to-rural concentrations in Chinese soils. El Khalil et al. 2008 found that the gradient from urban-to-suburban occupation matched the gradients in the contents of Cu and Zn in soil surrounding Marrakech. Zinc is essential for both plant and human nutrition, and Spain (BOE 1990) and some European countries (European Commission 1986) set fairly large legislative thresholds of 450 and 300 mg kg^{-1}, respectively. Although Cu is also an essential nutrient, the legislative thresholds are less (Spain 210 mg kg^{-1}; European countries 140 mg kg^{-1}) because an excess can induce deficiencies of other trace elements in livestock (Rawlins et al. 2012). Lead is biologically non-essential and toxic, but the threshold of 300 mg kg^{-1} in Spain and other European countries probably reflects the ubiquitous presence of Pb contamination, particularly in urban areas. Cadmium is highly toxic to both humans and livestock, and the two legislative limits are set at 3 mg kg^{-1}. We might note that Switzerland has a much smaller guide value of 0.8 mg kg^{-1}(FOEFL 1987).

We have been concerned by potentially toxic contamination of the peri-urban region to the south east of Madrid, Spain, an area of 1,050 km^2 into which population is spreading. A survey of the region around Madrid by de Miguel et al. (2002) found it to be the most polluted. De Miguel et al. (1999) suggested that emissions from traffic were a major source of Cu, Pb, and Zn and that construction and erosion of building materials were major sources of Cd and Zn. De Miguel et al. (1998) had also observed that the widespread use of composted sewage as fertilizer in parks and the atmospheric fallout of urban particulate material has substantially increased the concentrations of several trace elements in the urban soils of Madrid. They found that concentrations of Cu, Pb, and Zn in "undisturbed" urban soil exceeded local natural background concentrations by factors of 2.3 to 4.0, while "compost modified" soils contained 5.3 to 8.3 times more. In the south east region, the mean concentrations were found to be 0.10 mg kg^{-1} for Cd, 9.48 for Cu, 17.55 for Pb, and 38.53 for Zn (de Miguel et al. 2002).

We have re-sampled this south east peri-urban region to provide a more detailed picture of the pollution (a) to map the underlying distribution of the same four metals in the soil and (b) to identify the sources of the metals. We had, in addition, the hypothesis that metal concentrations increase with decreasing distance from the city center mainly because of denser traffic and industrial activities. We analyzed the data geostatistically using recently developed robust methods (Marchant et al. 2011) to quantify the spatial covariance structure and then to predict and map the concentrations of each metal across the region. These methods allow us to estimate the underlying structure largely free of the distortion that isolated hot spots of some metals from point sources might cause. We also attempt to identify attributes of the survey sites, namely lithology, land cover, erosion, stoniness, relief, and cultivation that might explain the observed variation. This has required the estimation of more than 100 linear mixed models (LMMs) by the unbiased residual maximum likelihood (REML) estimator. Such a procedure would have been impracticable until recent enhancements in computer processing power (Lark2000).

2 MATERIAL AND METHODS

2.1 Study region, sampling design, and analytical procedure

The region studied is a rectangle 30 × 35 km (1,050 km^2) in the south east of the metropolitan area of Madrid. Its cover ranges from residential and industrial in the northwest to predominantly cultivated and natural vegetation in the south east. Figure 1 shows its limits within the Community of Madrid and the locations of the 125 sampled sites.

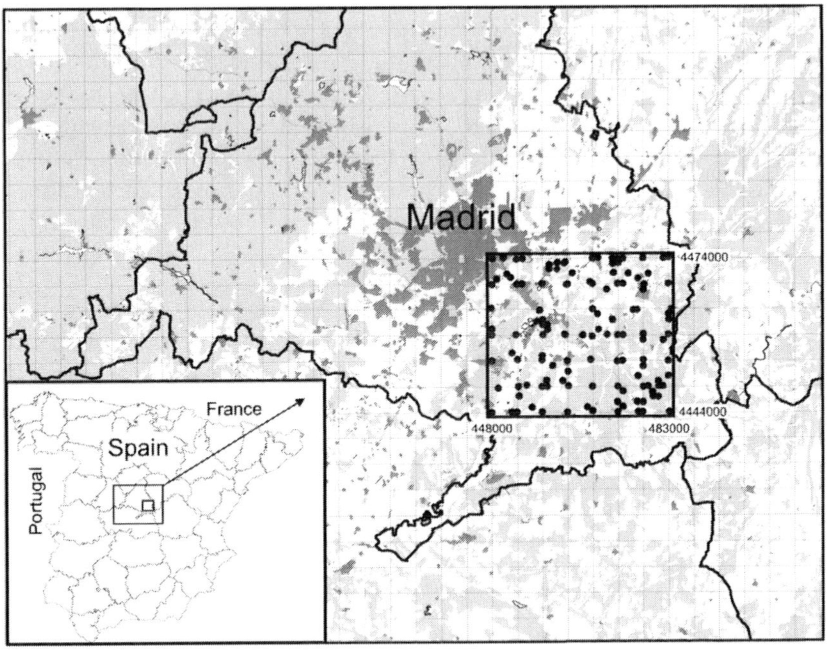

Figure 1. Limits of the 30 × 35 km study region within the community of Madrid administrative boundaries and locations of the sampled points. *Dark gray* represent urban areas, *light gray* natural vegetation, and*white* cultivated areas, according to the CORINE land cover information. The overlaid 5-km grid is based in the UTM projection (zone 30)

A non-preferential survey was made between during 2010 and 2011. The sampling design was based on two nested grids aligned with the Universal Transverse Mercator (UTM) projection system. The first was a 5 × 5 km grid. Forty eight of the 56 nodes of the grid were sampled. The second was a 1 × 1 km grid. Seventy four of the nodes of this grid were selected at random and sampled. On seven occasions, the selected node coincided with the 5 × 5 km grid, and on each occasion, the observation was made 1 km to the east. Three of the 122 observations made at grid nodes were selected at random, and for each, an additional observation was made 200 m away.

Each observation site was classified according to lithology, land cover, erosion, stoniness, relief, and cultivation. The class names and the number of elements within each class are listed in Table 1. The expected lithology and land cover were determined from a map prior to the sampling. The lithologic information was based on the Instituto Geológico Minero de España (IGME) lithologic map (IGME 2012). The land cover information was derived from the CORINE Land Cover European project (NATLAN 2000) and aggregated into three classes (background of Fig. 1).

Table 1. Names and number of elements, *n*, of classes for each site description category

Lithology	*n*	Land cover	*n*	Cultivation	*n*
Limestone	62	Forest	6	Undisturbed	46
Conglomerate	11	Shrubland	38	Cultivated	60
Urban debris	9	Grassland	8	Urban debris	13
Gravel and sand	22	Olive trees	17	Abandoned cult.	6
Gypsum	21	Cult. Irrigated	5		
		Cult. Dryland	35		
		Urban	7		
		Vineyards	9		
Relief	*n*	Erosion	*n*	Stoniness	*n*
Flat	10	None	32	None	8
Valley bottom	23	Low	65	Low	41
Alluvial terrace	4	Medium	20	Medium	35
Slope	83	Abundant	8	Abundant	25
Top	5			Very abundant	16

The surveyor sought to sample sites with the expected lithology and land cover attributes at the exact location of the grid node. If access to a site was denied or the site clearly did not possess the attributes expected, then an alternative location was chosen within 200 m of the specified location. If this was not possible, the location was discarded.

Soil samples were taken under homogeneous land cover. Locations were recorded with a GPS with typically less than 3 m of error. Field characteristics such as the actual lithological substrate, the land cover, the slope, and the aspect were recorded at each location. Other variables recorded were the physiographic characteristics of the location and evidence of changes in the soil horizons. Finally, each sampled location was assigned to a qualitative scale of erosion and of stoniness.

At each location, we took a composite sample of five subsamples of topsoil (0–15 cm): one in the center of the plot and the others 5 m away in the four cardinal directions. The mass of each composite sample was around 2.5 kg. This soil was air dried and later sieved to pass 2 mm. A subsample was sent to the Eurofins Analytico laboratories for the determination of the concentrations of the four metals, Cd, Cu, Pb, and Zn. The samples were subject to a full destruction with a pre-treatment based in aqua regia and microwave digestion before the analysis by inductively coupled plasma mass spectrometry and the protocol NEN-EN-ISO 17294–2. These tests are accredited by the “Dutch Accreditation Council (RvA)”.

2.2 Statistical analyses

2.2.1 Spatial prediction of trace element concentrations

The variation of the concentration of each element across the region was represented by a LMM. A LMM splits the variation of a property of interest into fixed effects and a spatially correlated random effect. The fixed effects are linear relationships between recorded covariates and the property. The LMM is written as

$$z = M\beta + u,$$

where z is a vector of n observations of the property of interest, M is a design matrix of size $n \times q$ which contains values of the q covariates at the observation sites, β is a vector of q-fixed effects coefficients, and u is realization of a second-order stationary spatially correlated random function U at the n observation sites. The random function U has zero mean and covariance matrix C. It is often assumed to have a Gaussian distribution, though this assumption is not essential. The elements of the covariance matrix were determined from authorized (Webster and Oliver 2007) parametric functions $C(h)$ of the separation in distance and direction between pairs of observations (the lag denoted h) where the parameter values were estimated from the available data. It is common in the geostatistical literature for the spatial covariance of a random function to be expressed in terms of the variogram

$$\gamma(h) = \frac{1}{2}E\left[\{U(x) - U(x+h)\}^2\right],$$

where $x = (x, y)$ is the location with easting x and northing y. For a second order stationary random variable

$$C(h) = C(0) - \gamma(h).$$

We had too few data to identify and estimate any anisotropy, and so we treated the lag as a separation in distance only, i.e., $h = |h|$. Once each LMM had been determined, it was used to predict the element z, at unsampled locations by universal kriging (Webster and Oliver 2007). Universal kriging yields a prediction denoted Zˆ(x0)Z^(x0) and a corresponding prediction variance σˆ2(x0)σ^2(x0).

The observations of elements included outliers close to local sources of pollution. These can distort the estimation of model parameters and hence, the predictions. Therefore, before estimating a LMM, we used robust geostatistical methods to identify and censor outliers. For each element, provisional LMMs were estimated by Matheron's method of moments (Webster and Oliver 2007) and also by the robust estimators suggested by Cressie and Hawkins (1980), Dowd (1984), and Genton (1998). The distributions of all the elements were highly skewed, and the observations were transformed to their natural logarithms before these models were estimated. The fixed effects consisted of a constant and a nested nugget-plus-exponential covariance function (Webster and Oliver 2007) was assumed for the random term. Leave-one-out cross validation was done for each fitted LMM, and the squared standardized prediction error (SSPE),

$$\theta_i = \frac{\left[\widehat{Z}(x_i) - z(x_i)\right]^2}{\widehat{\sigma}^2(x_i)},$$

was calculated at each observation site $x_{i,} i = 1, 2 \ldots, n$. If the observed data were a realization of the fitted LMM, then the ϑ_i would be realizations of a X^2 distribution with 1 ° of freedom. The mean of this distribution is equal to 1, and the median is equal to 0.455. The mean of the ϑ_i is greatly influenced by outliers, but the median is much more robust (Lark 2002). Therefore, the LMM for which the median θ~θ~ was closest to 0.455 was assumed to be the best representation of the underlying variation. Any observation with $\vartheta_i > 9$ for this LMM was assumed to be an outlier, and its magnitude was reduced such that $\vartheta_i = 9$. An observation near to an outlier might itself be falsely classified as an outlier because of the large discrepancy between its value and that of its neighbor. Therefore, where a pair of outliers were identified within 250 m of each other, the cross-validation procedure was repeated with both potential outliers removed from the dataset (Table 2). If the recalculated ϑ_i at either of the two sites was less than 9, then that observation was returned to the dataset. The final LMM was fitted to these censored observations.

Various candidates for the final LMM were considered. The fixed effects were either (1) a constant, (2) linear relationships with the eastings x and northings y, or (3) a quadratic trend (i.e., linear relationships with x, y, x^2, xy and y^2). The covariance model was either pure nugget (i.e., no spatial correlation) or a nested nugget and Matérn model (Matérn 1960). This model is written:

Table 2. Summary and cross-validation statistics

	Cd	Cu	Pb	Zn
Mean/mg kg^{-1}	0.14	12.95	27.95	42.75
Median/mg kg^{-1}	0.11	9.50	17.00	40.00
Std dev.mg kg^{-1}	0.08	21.33	98.73	22.04
Min/mg kg^{-1}	0.01	3.03	5.60	13.00
Max/mg kg^{-1}	0.48	240.00	1100.00	180.00
Skew	1.71	9.80	10.43	2.91
Skew of $ln(z)$	−1.11	1.93	3.72	0.38
Matheron $\tilde{\theta}$	0.25	0.25	0.20	0.39
Best estimator	C–H*	Dowd	Genton	Matheron
Robust $\tilde{\theta}$	0.42	0.49	0.41	0.39
Large outliers	1	1	3	1
Small outliers	2	1	1	0
λ	−0.02	−0.26	−0.08	−0.07

*Cressie-Hawkins

$$c(h) = \begin{cases} c_0 + c_1 & \text{if } h = 0 \\ c_1 \left\{ \frac{1}{2^{v-1}\Gamma(v)} \left(\frac{h}{a}\right)^v K_v\left(\frac{h}{a}\right) \right\} & \text{otherwise.} \end{cases}$$

where h is the distance separating two observations, c_0 is the nugget variance, c_1 is the sill variance of the spatially correlated component, a is a distance parameter, v is a smoothness parameter which yields flexibility in how the covariance function behaves for small h, $\Gamma(v)$ is the Gamma function and K_v is a modified Bessel function of the second kind of order v(Abramowitz and Stegun 1972). If $v = 0.5$ then the Matérn model is equivalent to the exponential model (Webster and Oliver 2007). We transformed the observed values by the Box–Cox method (Diggle and Ribeiro 2007) so that they were consistent with the assumed Gaussian random effects of the LMM. The Box–Cox transform is

$$z* = \begin{cases} \frac{(z^\lambda - 1)}{\lambda} & \text{if } \lambda \neq 0 \\ ln\ (z) & \text{if } \lambda = 0 \end{cases}$$

where λ is a parameter.

The parameters of the LMM, $\alpha = (c_0, c_1, a, v, \beta, \lambda)$, were estimated by the maximum likelihood (ML) estimator described by Diggle and Ribeiro (2007).

This selects the model parameters that maximize the likelihood that the data arose from the model. The different fitted candidate models were compared by calculation of the Akaike Information Criterion (AIC; Akaike 1973):

$$ALC = -2\mathrm{L}\left(\widehat{\alpha}|z\right) + 2p,$$

where p is the number of parameters in the LMM and L(α^|z)L(α^|z) is the log likelihood that the observed data arose from the LMM with estimated parameters α^α^. The model with the smallest AIC strikes the best balance between complexity (number of parameters) and closeness of fit to the data (likelihood). There is a small bias in the ML parameter estimates of a LMM when the fixed effects are not constant. Therefore, once the model with the smallest AIC had been found, it was re-fitted by the REML estimator (Diggle and Ribeiro 2007) which minimizes this bias. Then, this model was used to predict the concentration across the region by universal trans-Gaussian kriging (Diggle and Ribeiro 2007).

2.2.2 Identification of major influences on trace element concentrations

A stepwise regression algorithm incorporating likelihood ratio tests was used to determine site descriptor classes that had significant effects on concentrations. We note that these descriptors could not be used to predict the concentrations across the region because they were known only at the sampling sites. The 33 classes of the six location classifications (see Table 1) were treated as indicator variables which described the presence or absence of the class at each site. The eastings and northings, x and y formed two further potential covariates. Some of these potential covariates are strongly correlated. This means that the order in which covariates are added to the LMM can influence whether or not a particular covariate is included because a correlated fixed effect might already be included. Therefore, an automated iterative algorithm was implemented.

Initially, a LMM with constant mean and nested nugget and Matérn covariance structure was fitted to the element being investigated. Further models were then fitted where the fixed effects contained an additional covariate. Each of the potential covariates was considered in turn. The covariate that increased the likelihood by the largest amount was added to the fixed effects if the likelihood ratio test showed that the increase was significant at the $p = 0.15$ level. This process was the forward selection step. The addition of a covariate to a model might have meant that one of the covariates already included in the fixed effects was no longer required. Therefore a backward elimination step was then applied. In this step, the model was re-fitted with one covariate removed from the fixed effects. This was repeated for all of the covariates in the fixed effects, and the covariate removal that led to the smallest reduction in the likelihood was found. If this reduction in the likelihood was not significant at the $p = 0.15$ level, then the covariate was removed from the model and a further backward elimination step was made. Otherwise, a further forward selection step was made. The algorithm continued until neither a forward selection nor a backward elimination step led to a change in the fixed effects.

Such a stepwise procedure does not necessarily determine the best covariates to include in the fixed effects of a model. Other fixed effects might have performed similarly, and when the aim of a study is to determine the best model for predictive purposes, it is important that one takes into account one's knowledge of the covariates. However, the procedure did indicate some covariates that influence concentrations in an objective manner.

3 RESULTS

Table 2 summarizes the statistics for the observed concentrations. The frequency distributions of all elements are positively skewed: they have skewness coefficients greater than 1. This skew is reduced by a log transform (Fig. 2). In the case of cadmium, the log transform produced a strong negative skew. However, the histogram of log Cd suggests that this negative skew might be the result of three unusually small measurements. When the variograms fitted to the concentrations of Cd, Cu, and Pb by Matheron's estimator were cross-validated, the median SSPEs (θ~)(θ~) were less than 0.3. This indicates that outliers are present amongst the observations of these properties. When robust variogram estimators were used, the median SSPEs were between 0.4 and 0.5. Different robust variogram estimators performed best on each of these three properties. The median SSPE for the Matheron variogram of Zn was 0.39, and this model fitted the data better than those estimated by robust methods.

Six large outliers (one of Cd, one of Cu, three of Pb, and one of Zn) were identified and censored. They are all in the urban northwest corner of the region (Fig. 3). A further four unusually small outliers were identified. We considered these to be less relevant than the large outliers because they required only small censoring adjustment in the untransformed units (mg kg^{-1}).

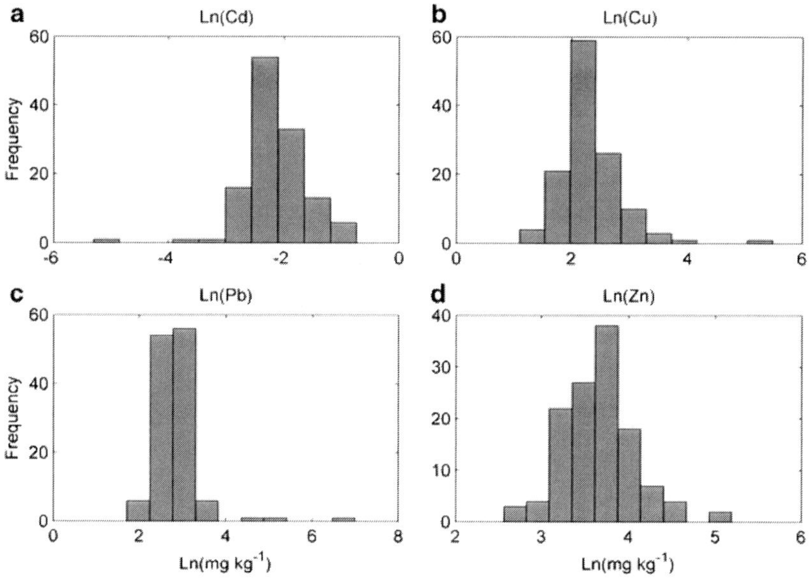

Figure 2. istograms of log transformed observations of trace elements within study region

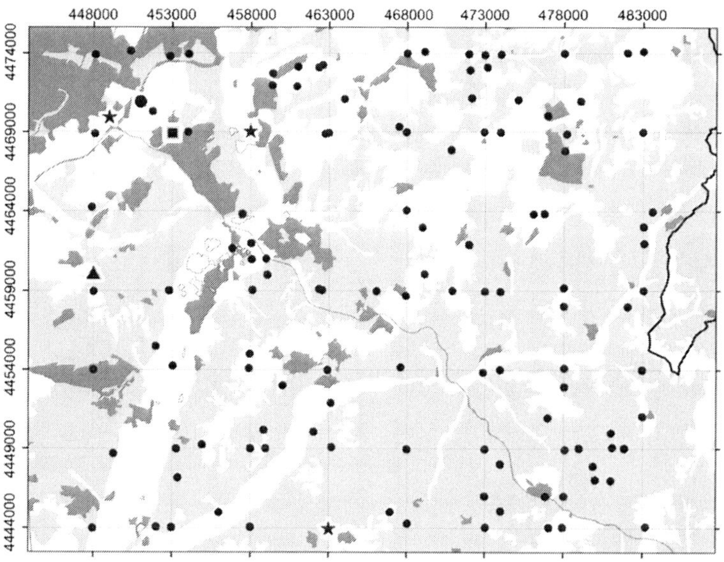

Figure 3. Sample design and location of large outlying observations of Cd (*black square*), Cu (*big disc*), Pb (*black star*), and Zn (*black triangle*). *Dark gray* represent urban areas, *light gray* natural vegetation, and *white cultivated areas*, according to the CORINE land cover information. The overlaid 5-km grid is based in the UTM projection (zone 30)

The calculated AIC values (Table 3) showed that the best fitting LMMs for each of the four censored elements included a linear trend in the eastings and northings. The models for Cd, Cu, and Zn included a Matérn term in the random effect, whereas there was no evidence of spatial correlation in the model for Pb. The REML-estimated variograms for the random effects of these models are shown in Fig. 4. The variograms for Cu and Zn are very similar and have a smaller nugget-to-sill ratio and smaller range than the model for Cd.

The predictions (see the maps in Fig. 5) of Cu, Pb, and Zn, decrease in concentrations from the urban northwest corner to the rural south east corner. For Cd, the trend is more from the west to east. The patterns of variation for Cu and Zn are fairly similar with several fairly small areas of contamination. The variation of Cd is smoother. The absence of spatial correlation in the LMM of Pb means that the spatial trend is the only cause of variation.

Table 3. AIC for models fitted by maximum likelihood. The smallest AIC value for each metal is shown in bold

Trend	Variogram	Cd	Cu	Pb	Zn
Constant	Matérn	−160.57	−180.91	−148.63	−178.76
Linear	Matérn	**−166.54**	**−194.03**	−154.57	**−194.20**
Quadratic	Matérn	−164.55	−191.58	−149.71	−190.99
Constant	Nugget	−157.67	−169.79	−149.64	−159.16
Linear	Nugget	−164.55	−191.31	**−155.86**	−192.63
Quadratic	Nugget	−164.76	−190.18	−150.87	−189.50

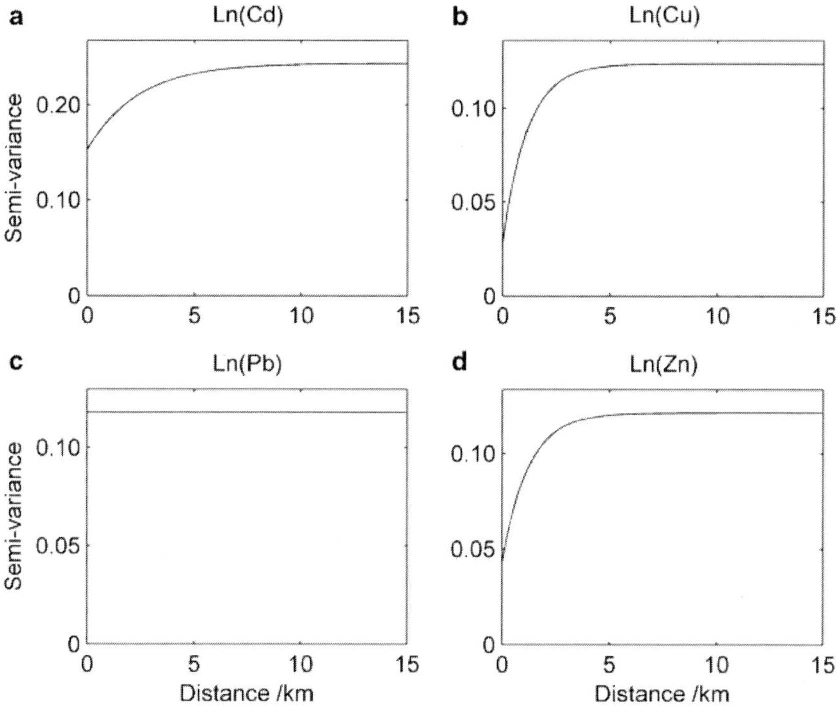

Figure 4. Estimated variograms for best fitting models

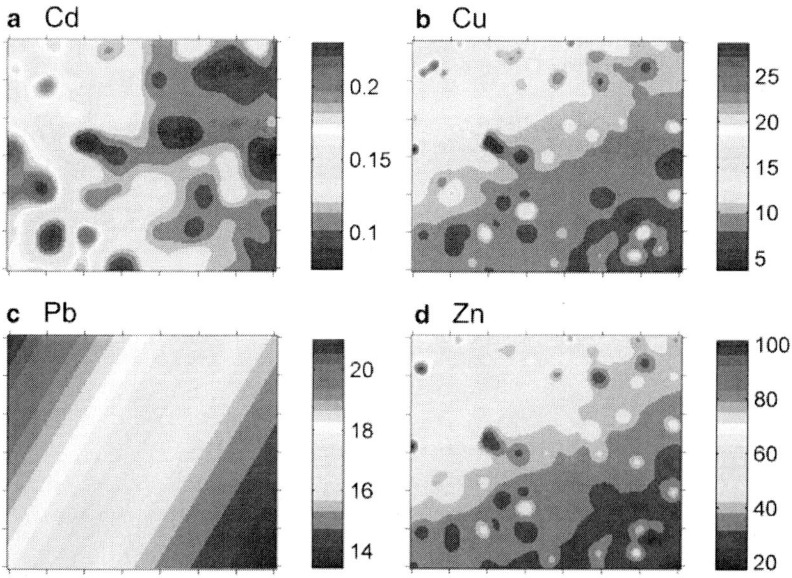

Figure 5. Spatial prediction of metal concentrations (in milligrams per kilogram) across the study region

Table 4. Site descriptor classes included in the best-fitting LMM for each metal and effect of membership of class upon metal concentration

Cd			Cu		
Category	Class	Effect	Category	Class	Effect
Lithology	Gypsum	+	Cultivation	Undisturbed	–
Relief	Bottom	+	Stoniness	None	+
Erosion	Medium	–	Spatial	Northing	+
Spatial	Easting	–	Relief	Top	–
Lithology	Conglomerate	–	Lithology	Gravel and sand	–
Relief	Alluvial terrace	–	Land cover	Irrigated cult	+
Lithology	Limestone	–	Spatial	Easting	–
Spatial	Northing	+	Relief	Slope	–
Land cover	Shrub land	+	Cultivation	Cultivated	–
Cultivation	Cultivated	–	Land cover	Grassland	–
Erosion	None	+	Relief	Alluvial terrace	–
Pb			**Zn**		
Category	Class	Effect	Category	Class	Effect
Spatial	Easting	–	Stoniness	None	+
Erosion	None	+	Stoniness	Low	+
Land cover	Unirrigated cult	–	Spatial	Northing	+
Stoniness	Abundant	–	Relief	Flat	+
Relief	Alluvial terrace	–	Spatial	Easting	–
Cultivation	Abandoned	+	Lithology	Gravel and sand	–
Land cover	Irrigated cult	+	Relief	Valley bottom	+
Lithology	Gravel/sand	–	Relief	Slope	+
Stoniness	Medium	–	Erosion	Medium	–
Spatial	Northing	+	Cultivation	Abandoned	+
			Erosion	Abundant	–

The most effective indicators to include in the fixed effects are listed in Table 4. Membership of some of the classes has a similar effect on the concentration of all four trace elements. For example, concentrations appear to decrease with increases in stoniness and erosion, and valley floors have large concentrations. Copper, Pb, and Zn all have small concentrations on gravel and sand. Land cover classes feature fewer times. This might be an artifact caused by the large number of these classes. There are some differences between the four elements. More lithology classes are included in the model for Cd than for the others, and gypsum is the first class to be included for it. Stoniness appears to be important for Zn.

4 DISCUSSION AND CONCLUSIONS

There are some similarities in the spatial patterns of the four trace elements we have studied. The concentrations of all are generally largest in the northwest of the region. This result is likely to be caused by the fallout of airborne particles originating in urban Madrid further to the northwest and from road traffic on the highways across this part of the region (de Miguel et al.1999; Manta et al. 2002; Li et al. 2004). The concentrations of all four elements appear to decrease with increases in stoniness and erosion, and to be largest in the fine sediments on the valley floors. This accords with general knowledge: trace metals tend to be adsorbed predominantly on clays (Sillanpää 1972), and the fine fraction of soil mobilized by erosion accumulates with adsorbed metals on valley floors.

The spatial patterns of Cu and Zn, in particular, are similar. Other authors (Adachi and Tainosho 2004; Li et al. 2004; Yang et al. 2011) have described how these metals in car components and lubricants can be emitted into the atmosphere as small particles and then be deposited on the ground. Such emissions are probably the major source of contamination in the northwest of the region. Hot spots of Cu and Zn are evident further from Madrid. These coincide with industrial sites such as a large urban waste incineration plant, a sepiolite mining installation, and other construction treatment plants. Morton-Bermea et al. (2009) observed a similar pattern of Cu and Zn contamination in Mexico City. There, the underlying pattern was attributed to emissions from motor traffic with local hot spots close to industrial sites.

The first factor in the best-fitting LMM of Cu (see Table 4) was undisturbed (non-cultivated) soils, which suggest that the concentrations on cultivated soils are larger than elsewhere. This might be because Cu salts have been used as fungicides, as other investigators have found (Brun et al. 1998; Micó et al. 2007).

The map of lead shows only a decreasing trend from the northwest. Lead used to be added to petrol (the practice ceased in 1999; Xia et al. 2011), and emissions in exhaust are likely to have been the main source of lead in the region. No other features can be seen in the map because the random effects were not spatially correlated.

The variation of Cd across the region seems to depend on several factors. Table 4 indicates the importance of lithology. Previous studies (e.g. Atteia et al. 1995; Marchant et al. 2010) have also found that Cd in the parent material of the soil contributes substantially to that in the soil itself. In the region of our study, Cd is most concentrated in the two main areas of gypsum. The Cd concentrations are also largely close to the urban center which includes river valleys where intensive and irrigated agriculture is practiced. Other authors have found a close relation between Cd concentrations and intensive agriculture caused by the use of phosphate fertilizers and irrigation water contaminated by urban or industrial effluents (Hu et al. 2006; Chen et al.2009; Acosta et al. 2011). Cadmium hotspots are also evidently close to the plant for incinerating urban waste.

The four elements vary over disparate spatial scales, and we presume they are affected by several factors. Local sources of contamination such as industrial emissions, construction materials, and uncontrolled waste deposits produced outliers. The natural variation seems to be related largely to lithology (Adriano 1986; Kabata-Pendias and Pendias 2001). There is great lithological diversity in the sedimentary rocks and evaporites. It is for this reason that standard linear geostatistical models were not suitable for analyzing the data and that robust methods were required to identify outliers and reduce their influence. Similarly, the determination of the factors controlling the variation was far from trivial. We iteratively built up a LMM for each element using an automated stepwise algorithm. The results of such an approach should be interpreted with caution. The site descriptors considered were highly correlated, and therefore the inclusion of one factor within the model might mask the effect of another. In contrast to previous studies, there was little evidence that composts were a substantial source of the metals. Indeed, when cultivation does appear in the LMMs in Table 4, it has a negative effect. This might be only relative to soil containing urban debris. Despite these concerns, some general causes of variation can be discerned from the fitted LMMs such as the effects of erosion, relief, and stoniness discussed above.

The concentrations of the four elements were generally less than the legal limits established for agriculture by the European Commission (1986) and Spain (BOE 1990). Only one observation of Pb and one of Cu exceeded these limits. The mean values of all four metals were less than those reported in other peri-urban areas in towns with longer industrial histories (Kelly et al.1996; Frangi and Richard 1997; Manta et al. 2002; Imperato et al. 2003). The contamination in Madrid has occurred relatively recently and therefore, the enrichment in metals of the soil is more similar to that found in towns of developing countries with more recent industrialization (Hu et al. 2006; Yang et al. 2009; Xia et al. 2011). Note that de Miguel et al. (1998) found larger concentrations in the centre of Madrid. Nevertheless, the concentrations in this peri-urban region exceed the natural concentrations, and we should beware of further contamination which might lead to toxic quantities entering the food chain and eventually causing ill health.

ACKNOWLEDGMENTS

We thank the CARESOIL_CM network (Project founded by the Comunidad Autónoma de Madrid, P2009/AMB-1648). B.P. Marchant's contribution was funded by the Biotechnology and Biological Sciences Research Council (BBSRC).

REFERENCES

1. Abramowitz M, Stegun IA (eds) (1972) Handbook of mathematical functions with formulas, graphs, and mathematical tables. Dover Publications, New York

2. Acosta JA, Faz A, Martínez-Martínez S, Arocena JM (2011) Enrichment of metals in soils subjected to different land uses in a typical Mediterranean environment (Murcia City, southeast Spain). Appl Geochem 26:405–414

3. Adachi K, Tainosho Y (2004) Characterization of heavy metal particles embedded in tire dust. Environ Int 30:1009–1017

4. Adriano DC (1986) Trace elements in the terrestrial environment. Springer, New York

5. Akaike H (1973) Information theory and an extension of the maximum likelihood principle. In: Petrov BN, Csáki F (eds) Second International Symposium on Information Theory. Akadémiai Kiadó, Budapest, pp 267–281

6. Alloway BJ (2004) Contamination of soils in domestic gardens and allotments: a brief overview. Land Contam Reclam 12:179–187

7. Atteia O, Thélin P, Pfeifer HR, Dubois JP, Hunziker JC (1995) A search for the origin of cadmium in the soil of the Swiss Jura. Geoderma 68:149–172

8. BOE (1990) Real Decreto 1310/1990 de 29 de octubre por el que se regula la utilización de los lodos de depuración en el sector agrario.

Boletín Oficial del Estado, núm. 262 de 1 de noviembre de 1990, pp 32339–32340, BOE-A-1990-26490

9. Brun LA, Maillet J, Richarte J, Herrmann P, Remy JC (1998) Relationships between extractable copper, soil properties and copper uptake by wild plants in vineyard soils. Environ Pollut 102:151–161

10. Businelli D, Massaccesi L, Said-Pullicino D, Gigliotti G (2009) Long-term distribution, mobility and plant availability of compost-derived heavy metals in a landfill covering soil. Sci Total Environ 407:1426–1435

11. Chen T, Liu X, Li X, Zhao K, Zhang J, Xu J, Shi J, Dahlgren RA (2009) Heavy metal sources identification and sampling uncertainty analysis in a field-scale vegetable soil of Hangzhou, China. Environ Pollut 157:1003–1010

12. Cressie N, Hawkins D (1980) Robust estimation of the variogram. J Int Ass Math Geol 12:115–125

13. De Miguel E, de Grado UMJ, Llamas JF, Martín-Dorado A, Mazadiego LF (1998) The overlooked contribution of compost application to the trace element load in the urban soil of Madrid (Spain). Sci Total Environ 215:113–122

14. De Miguel E, Llamas JF, Chacón E, Mazadiego LF (1999) Sources and pathways of trace elements in urban environments: a multi-elemental qualitative approach. Sci Total Environ 235:355–357

15. De Miguel E et al (2002) In: Barettino D, Callaba A (eds) Determinación de niveles de fondo y niveles de referencia de metales pesados y otros elementos traza en suelos de la Comunidad de Madrid. Instituto Geológico y Minero de España, Madrid

16. Diggle PJ, Ribeiro PJ (2007) Model-based geostatistics. Springer, New York

17. Dowd PA (1984) The variogram and kriging: robust and resistant estimators. In: Verly G, David M, Journal AG, Marechal A (eds) Geostatistics for Natural Resources Characterization. Part 1. Reidel, Dordrecht, pp 91–106

18. El Khalil H, Schwartz C, Elhamiani O, Kubiniok J, Morel JL, Boularbah A (2008) Contribution of technic materials to the mobile fraction of metals in urban soils in Marrakech (Morocco). J Soils Sediments 8:17–22

19. European Commission (1986) European Commission Council Directive 86/278/EEC of 12 June 1986 on the protection of the environment, and in particular of the soil, when sewage sludge is used in agriculture. Off J Eur Comm L 181:6–12

20. FOEFL (1987) Commentary on the ordinance relating to pollutants in soil (VSBo of June 1986). FOEFL (Federal Office of Environment, Forest and Landscape), Bern

21. Frangi JP, Richard D (1997) Heavy metal soil pollution cartography in northern France. Sci Total Environ 205:71–79

22. Genton MG (1998) Highly robust variogram estimation. Math Geol 30:213–221

23. Hu KL, Zhang FR, Li H, Huang F, Li BG (2006) spatial patterns of soil heavy metals in urban–rural transition zone of beijing. Pedosphere 16:690–698

24. IGME (2012) Instituto Geológico y Minero de España. Cartografía Geológica MAGNA (in digital format). http://www.igme.es. Accessed May 2012

25. Imperato M, Adamo P, Naimo D, Arienzo M, Stanzione D, Violante P (2003) Spatial distribution of heavy metals in urban soils of Naples city (Italy). Environ Pollut 124:247–256

26. Kabata-Pendias A, Pendias H (2001) Trace elements in soils and plants, 2nd edn. CRC Press, Boca Ratón, Florida

27. Kelly J, Thornton I, Simpson PR (1996) Urban geochemistry: a study of the influence of anthropogenic activity on the heavy metal content of soils in traditionally industrial and non-industrial areas of Britain. Appl Geochem 11:363–370

28. Lark RM (2000) Estimating variograms of soil properties by the method-of-moments and maximum likelihood. Eur J Soil Sci 51:717–728

29. Lark RM (2002) Modelling complex soil properties as contaminated regionalized variables. Geoderma 106:1731–1790

30. Li XD, Lee SL, Wong SC, Shi WZ, Thornton L (2004) The study of metal contamination in urban soil of Hong Kong using a GIS-based approach. Environ Pollut 129:113–124

31. Manta DS, Angelone M, Bellanca A, Neri R, Sprovieri M (2002) Heavy metals in urban soils: a case study from the city of Palermo (Sicily), Italy. Sci Total Environ 300:229–243

32. Marchant BP, Saby NPA, Lark RM, Bellamy PH, Jolivet CC, Arrouays D (2010) Robust analysis of soil properties at the national scale: cadmium content of French soils. Eur J Soil Sci 61:144–152

33. Marchant BP, Tye AM, Rawlins BG (2011) The assessment of point-source and diffuse soil metal pollution using robust geostatistical methods: a case study in Swansea (Wales, UK). Eur J Soil Sci 62:359–370

34. Martín JAR, de la Cueva AV, Corbí JMG, López-Arias M (2007) Factors controlling the spatial variability of copper in topsoils of the northeastern region of the Iberian Peninsula, Spain. Water Air Soil Pollut 186:311–321

35. Matérn B (1960) Spatial variation. Meddelanden från Statens Skogsforskningsinstitut, 49, No 5, 2nd edn. Springer, New York, Lecture Notes in Statistics No. 36

36. Micó C, Peris M, Recatala L, Sánchez J (2007) Baselines values for heavy metals in agricultural soils in an European Mediterranean region. Sci Total Environ 378:13–17

37. Morton-Bermea O, Hernández-Álvarez E, González-Hernández G, Romero F, Lozano R, Beramendi-Orosco LE (2009) Assessment of heavy metal pollution in urban topsoils from the metropolitan area of Mexico City. J Geochem Explor 101:218–224

38. NATLAN (2000) CLC 1990, CORINE Land Cover 250m. European Environment Agency, Copenhagen. http://www.eea.europa.eu/data-and-maps. Accessed May 2011

39. Pietrzak U, McPhail DC (2004) Copper accumulation, distribution and fractionation in vineyard soils of Victoria, Australia. Geoderma 122:151–166

40. Rawlins BG, McGrath SP, Scheib AJ, Breward, N Cave, M, Lister TR, Ingham, M, Gowing, C, Carter, S (2012) The advanced soil geochemistry atlas of England and Wales. British Geological Survey, Keyworth. www.bgs.ac.uk/gbase/advsoilatlasEW.html. Accessed Jan 2012

41. Saby NPA, Marchant BP, Lark RM, Jolivet CC, Arrouays D (2011) Robust geostatistical prediction of trace elements across France. Geoderma 162:303–311

42. Sillanpää M (1972) Trace elements in soils and agriculture. Soils Bulletin 17. FAO, Rome, 67 pp

43. Webster R, Oliver MA (2007) Geostatistics for environmental scientists, 2nd edn. John Wiley and Sons, Chichester

44. Wei BG, Yang LS (2010) A review of heavy metal contaminations in urban soils, urban road dusts and agricultural soils from China. Microchem J 94:99–107

45. Xia X, Chen X, Liu R, Liu H (2011) Heavy metals in urban soils with various types of land use in Beijing, China. J Hazard Mater 186:2043–2050

46. Yang P, Mao R, Shao H, Gao Y (2009) An investigation on the distribution of eight hazardous heavy metals in the suburban farmland of China. J Hazard Mater 167:1246–1251

47. Yang Z, Lu W, Long Y, Bao X, Yang Q (2011) Assessment of heavy metals contamination in urban topsoil from Changchun City, China. J Geochem Explor 108:27–38

Chapter 4

Contamination of Paddy Soil and Rice with Arsenic

Khageshwar Singh Patel[1], Bharat Lal Sahu[1], Shobhana Ramteke[1], Elza Bontempi[2]

[1]*School of Studies in Chemistry/Environmental Science, Pt. Ravishankar Shukla University, Raipur, India*

[2]*INSTM and Chemistry for Technologies Laboratory, University of Brescia, Brescia, Italy*

ABSTRACT

The drinking water and food are main pathway entry of the As in humans and animals. Their intakes cause diseases i.e. skin cancer, vascular disorder, etc. A wide variety of the rice is cultivated in the central India. The field soil and rice cultivated in the summer season at Koudikasa village, central India were selected for the As contamination studies. The concentration (n = 20) of total- As (AsT) in the field soil, rice grain, husk, straw and root was ranged from 44 - 270, 0.17 - 0.72, 0.40 - 1.58, 2.5 - 5.9 and 204 - 354 mg/kg with mean value of 126 ± 28, 0.47 ± 0.07, 0.83 ± 0.15, 4.2 ± 0.5 and 276 ± 21 mg/kg, respectively. The total arsenic, monomethylarsenonate, dimethylarsinite and inorganic As in the rice grain are quantified.

Keywords: Arsenic, Soil, Rice, Accumulation, Speciation

1. INTRODUCTION

Arsenic is a toxic metalloid, and its long exposure causes diseases i.e. hypopigmentation, melanosis, keratosis, skin, bladder, lung, and prostate cancer, etc. [1] . Rice is a predominant staple food in many countries i.e. Asia, Pacific, America and Africa. The rice husk and straw are used as fodder for feeding cattle and fuel. The exposure of arsenic from rice has been reported a global health issue [2] - [4] . The elevated levels of As in the field soils in various regions of the World were reported [5] - [12] . Arsenic is a bioactive toxic element, accumulated in rice of several regions of the World [13] - [19] . In the present work, the contamination of As in the paddy soil, rice grain, husk, straw and root grown in the contaminated environment, Kaudikasa village, Ambagarh Chowki, central India is described. The speciation and translocation of As in various parts of the rice plants are discussed.

2. METHODS AND MATERIALS

2.1. Choice of Study Area

The Ambagarh Chowki block, Rajnandgaon district, Chhattisgarh state, India is a hot spot for the As contamination research due to the huge mineralization of As in the environment [20] [21] . The studied area falls in a tribal belt with population of ≈0.1 million over 155 villages. Among them, Koudikasa village (area » 5 km^2) was selected for the proposed studies due to As contamination of the environment at the hazardous levels.

2.2. Sampling of Soil and Rice Samples

The sampling network for collection of field water, soil and rice is presented in Figure 1. The water samples were collected as prescribed in the literatures [22] . Twenty water samples (once in a month) in duplicate from August-November, 2012 were taken from 20 different rice fields. A total of 4 samples, from each field from the period: August-November, 2012 was taken for the investigation. They were collected directly from rice field in duplicate, and placed in 250-mL polypropylene plastic bottle. Bottles were first rinsed thrice with the water and then, completely filled with the same water. The first sample was acidified with concentrated nitric acid (0.1%, v/v) for analysis of the As. The second sample was left free for analysis of the ions. The physical parameters i.e. pH and electrical conductivity (EC) of the water were measured at the spot.

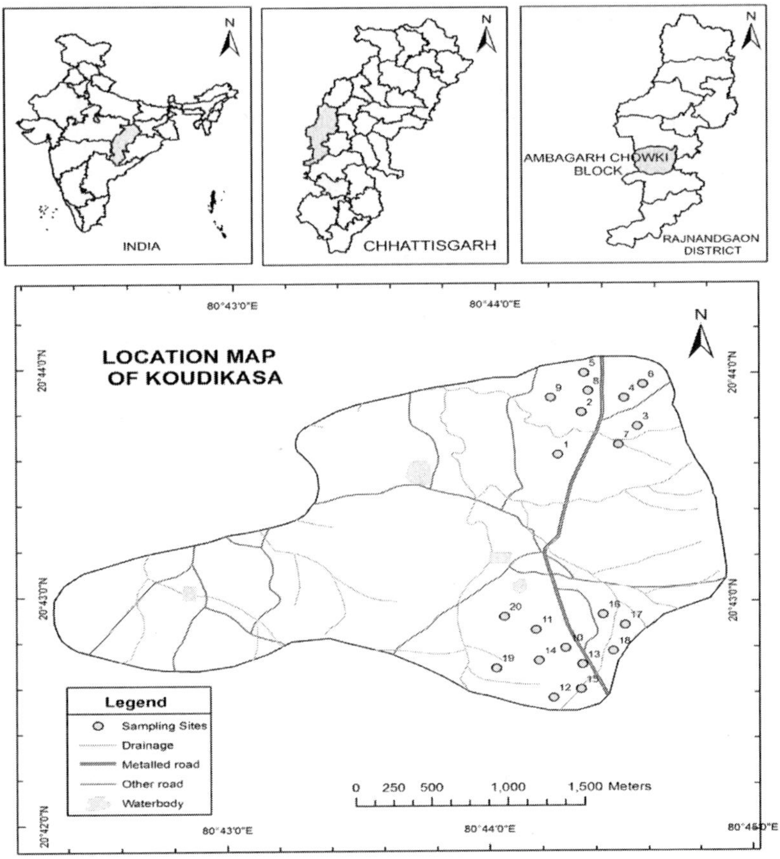

Figure 1. Sampling net-work for collection of the field soil and rice samples in Koudikasa village.

Twenty composite soil samples (0 - 10-cm depth) were collected after harvesting of rice paddy (December, 2009) from 20 fields of Koudikasa village as prescribed in the literature [23] . The samples were stored in polyethylene bottles and dried in open air under diffused sunlight followed by drying in oven at 50°C for 24 hr. Rice grains were separated from the plants by hand picking in December, 2012 from the field in the polyethylene bag. They were dried in the oven at 50°C for 24 hr and their husk was separated manually. Similarly, the straw and root of the rice plant were collected. They were washed with the deionized water several times to remove the soil particles. The dried soil, rice grain, husk, straw and root samples were ground to a fine powder with mortar and passed through a mesh sieve of <1 mm.

2.3. Analysis

The Bruker S2 Picofox TXRF portable spectrometer was used for the analysis of the elements in soil. A suspended solution was prepared by mixing 10 mg of soil sample with 10 ml of a water solution containing 1% (w/v) triton and 10 μg/mL Ga in ultrasonic bath for 15 min. For each measurement, 10 μL of sample solution was sprayed on the quartz filter with subsequent drying. The X-ray source was focused on the filter for quantification of the elements. The peak area of the signal was computed. The three replicate measurements for each sample were carried out. The content of 14 elements (i.e. P, K, Ca, As, Ti, V, Cr, Mn, Fe, Ni, Cu, Zn and Pb) in each soil was analyzed. The standard soil sample (NCS DC 73382CRM) was used for the quality control.

The As content in the soil and rice samples were analysed by the ICP-MS and AFS techniques. The samples were digested with nitric acid and H_2O_2 in Perkin-Elmer microwave system. The total, inorganic and organic As species in the rice grain samples were quantified by using method proposed by Williams et al. [24] . The Rice flour SRM 1568 was used for the quality control.

3. RESULTS AND DISCUSSION

3.1. Rice Morphology

The details of the rice morphology (i.e. cultivation period, height of plant, grain and husk content) are summarized in Table 1. The cultivation period, plant height and yield of rice varies by variety and environmental conditions, ranging from 80 - 145 day, 90 - 140 cm and 33 - 57 Q/ha with mean value of 124 day, 110 cm and 44 Q/ha, respectively. Among them, the plant height was found to be partially correlated (r = 0.42) with the cultivation period. The single grain weight, rice and husk content were varied from 14 - 30 mg, 65% - 87% and 13% - 35% with mean value of 21 mg, 74% and 26%, respectively.

3.2. Characteristics of Field Water

The mean value (n = 4) of the water parameters in 20 fields is summarized in Table 2. The pH value of the water was found to be neutral, ranging from 6.9 - 7.6 with mean value of 7.3 ± 0.1. The moderate EC values of the water were observed, ranging from 180 - 472 μS/cm with mean value of 316 ± 34 μS/cm.

The concentration of F^-, Cl^-, SO_4^{2-}, NO_3^-, Na^+, K^+ and As was ranged from 1.7 - 3.4, 10 - 43, 32 - 150, 1.9 - 7.9, 14 - 50, 2.6 - 9.8 and 0.029 - 0.090 mg/L with mean value of 2.4 ± 0.2, 20 ± 4, 78 ± 12, 4.2 ± 0.7, 32 ± 7, 5.2 ± 0.8 and 0.058 ±

0.007 mg/L, respectively. Among them, SO_4^{2-} showed the highest content in the water. The concentration of F^- and As was found to higher than the recommended value of 1.5 and 0.001 mg/L, respectively [25] .

3.3. Concentration of As and Other Elements in Soils

The distribution of As and other elements i.e. P, K, Ca, Cr, Mn, Fe, Ni, Cu, Zn and Pb in the field soil is shown in Table 3. The concentration of elements i.e. As, P, K, Ca, Cr, Mn, Fe, Ni, Cu, Zn and Pb in the field soil (n = 20) was ranged from 44 - 270, 823 - 2653, 29,182 - 76,242, 12,173 - 116,031, 574 - 1306, 1623 - 14,419, 57,111 - 157,542, 245 - 548, 237 - 441, 444 - 817 and 9 - 79 mg/kg with mean value of 126 ± 28, 1516 ± 195, 45192 ± 5305, 35292 ± 12135, 883 ± 73, 8395 ± 1587, 94804 ± 10650, 392 ± 39, 343 ± 26, 582 ± 50 and 32 ± 10 mg/kg, respectively. Among them, the highest concentration of Fe was observed in all soil samples. They were found to occur in the following increasing order: Pb < As < Cu < Ni < Zn < Cr < P < Mn << Ca < K < Fe. Among them, a good correlation (r = 0.73) of the As with the K was observed, indicating their existence as K_3AsO_4 in the soil, Table 4. The Fe showed good correlation with the heavy metals i.e. Cr, Mn, Ni, Cu and Pb in the soil, indicating origin from similar sources, Table 4. The As concentration in the soil of studied area was found to be higher than reported in other regions of the country and World [5] - [12] .

Table 1. Rice morphology in Koudikasa village.

S. No.	Type	Cultivation Period, day	Height of rice plant, cm	Yield Q/Ha	Grain type	Weight of grain, mg	Rice %	Husk %
1	IR-64	100	90	43	LS	22.6	78	22
2	Culture	80	110	43	MS	14.5	72	28
3	Shyamla	130	90	33	LS	25.0	76	24
4	G. Gurmatia	140	140	37	LB	29.6	81	19
5	Masuri	145	120	47	MS	14.3	87	13
6	Purnima	100	90	37	LS	21.6	79	21
7	Mahamaya	125	110	57	LB	28.8	83	17
8	Kalinga	100	120	37	LS	17.1	82	18
9	Luchai	145	120	42	MS	13.6	75	25
10	Safari	140	140	47	MS	16.3	72	28
11	Ek Hazar Das	115	98	50	LS	26.0	70	30
12	Sarna	140	110	50	MS	22.0	65	35
13	HMT	130	110	43	LS	20.0	65	35
14	M2	140	105	45	LS	23.0	68	32
15	Zero JR	125	100	45	MS	22.0	65	35
16	Sonam	130	100	43	LS	24.0	70	30

LS = Long slender, LB = Long Bold, MS = Medium size.

Table 2. Characteristics of field water in Koudikasa village.

Field no.	pH	EC	F^-	Cl^-	SO_4^{2-}	NO_3^-	Na	K	As
		μS/cm	mg/L						
Fi1	7.3	382	2.1	19	150	5.2	40	5.2	60
Fi2	7.4	180	2.7	25	85	3.4	17	3.9	53
Fi3	7.1	310	1.9	10	56	3.5	23	3.2	48
Fi4	7.3	221	2.0	12	32	4.3	40	6.5	65
Fi5	7.4	312	1.8	13	47	3.8	50	8.2	49
Fi6	7.1	229	2.7	31	72	7.2	14	2.6	29
Fi7	7.5	325	3.2	24	76	2.8	31	4.7	56
Fi8	7.2	310	3.4	43	92	1.9	21	4.3	90
Fi9	7.1	305	1.8	11	103	2.8	23	3.4	49
Fi10	7.2	439	2.3	18	105	5.7	34	6.2	88
Fi11	6.9	207	2.9	25	101	3.7	29	4.7	68
Fi12	7.2	472	2.1	10	69	3.9	24	3.8	62
Fi13	7.4	254	1.8	12	37	4.7	45	7.8	64
Fi14	7.5	374	1.7	13	59	4.2	46	9.8	63
Fi15	7.3	248	2.6	30	103	7.9	26	3.1	76
Fi16	7.1	389	2.5	24	83	3.1	35	5.6	49
Fi17	7.5	372	2.9	42	44	2.1	24	5.2	54
Fi18	7.6	351	2.0	10	71	3.1	25	4.1	34
Fi19	7.4	311	2.8	14	89	4.6	31	4.7	62
Fi20	7.2	319	2.2	17	92	5.2	38	7.7	41

Fi = Field.

Table 3. Chemical characteristics of field soil in Koudikasa village, mg/kg.

S. No.	As	P	K	Ca	Cr	Mn	Fe	Ni	Cu	Zn	Pb
1	148	1528	42,773	40,453	878	12,065	100,476	413	437	648	64
2	110	1425	36,673	38,177	811	14,292	104,938	436	418	606	79
3	90	1761	46,996	54,885	838	10,286	98,347	410	357	584	62
4	122	1560	35,543	37,082	922	9280	98,261	398	379	488	21
5	132	2113	49,238	97,179	961	11,212	157,542	475	441	817	64
6	44	1368	29,182	15,357	882	8077	90,405	320	279	455	46
7	186	914	60,933	15,381	673	3560	71,388	245	314	533	10
8	270	1133	76,242	13,477	963	1673	69,027	401	310	601	21
9	112	2653	56,127	116,031	1099	10,472	140,329	495	425	768	66
10	250	1235	65,635	12,173	879	2509	57,111	278	255	536	9
11	140	1219	51,707	12,236	717	3846	68,913	262	335	561	16
12	104	1318	48,651	14,209	640	5226	68,909	301	283	470	16
13	110	1268	41,638	53,294	782	14,419	95,301	466	377	526	46
14	104	1939	42,435	35,031	947	7287	87,998	355	300	491	10
15	234	1173	35,658	27,387	574	7925	72,626	256	237	444	20
16	70	1389	32,930	20,603	986	9676	107,638	425	365	547	28
17	74	1877	33,457	22,435	1306	11,632	111,356	497	359	768	25
18	56	1534	34,941	23,668	948	7722	96,383	548	283	526	12
19	91	823	42,334	31,345	836	8201	95,473	411	363	514	11
20	78	2093	40,740	25,427	1014	8534	103,649	439	341	761	14

3.4. Distribution of As in Rice Grain, Husk, Straw and Root

The distribution of As in rice grain, husk, straw and root is summarized in Table 5. The concentration of As in the rice grain, husk, straw and root (n = 20) was ranged from 0.17 - 0.72, 0.40 - 1.58, 2.5 - 5.9 and 204 - 354 mg/kg with mean value of 0.47 ± 0.07, 0.83 ± 0.15, 4.2 ± 0.5 and 276 ± 21 mg/kg, respectively. The As content in husk was found to be higher than the rice grain, may be due to external contamination from the environment. The high yield rice varieties i.e. Kalinga, IR-64, G. Gurmatia, Shyamla, Ek Hazar Das, M2, etc. were found to be more sensitive to the As-accumulation. The concentration of As in the rice of the studied area was found to be higher than reported in the other region of the country and World [13] - [19] . The As content in the straw and root was found to be comparable to the values reported in the Taiwanese rice plants [14] .

3.5. Biological Absorption Coefficient and Concentration Factor

The biological absorption coefficient, BAC (i.e. plant to soil metal ratio) and concentration factor, CF (i.e. plant to water soluble metal ratio) of As are presented in Table 6. The CF and BAC values depend on the physical

Table 4. Correlation matrix for elements in field soil.

	P	K	Ca	Cr	Mn	Fe	Ni	Cu	Zn	As	Pb
P	1.00										
K	−0.12	1.00									
Ca	**0.71**	0.04	1.00								
Mn	0.41	−0.64	0.54	0.28	1.00						
Fe	**0.73**	−0.30	0.82	0.57	**0.69**	1.00					
Ni	0.55	−0.28	0.51	**0.70**	**0.64**	**0.74**	1.00				
Cu	0.44	−0.09	**0.68**	0.37	**0.69**	**0.77**	0.58	1.00			
Zn	**0.70**	0.15	0.57	**0.66**	0.34	**0.72**	0.58	**0.64**	1.00		
As	−0.36	**0.73**	−0.14	−0.35	−0.54	−0.48	−0.50	−0.29	−0.12	1.00	
Pb	0.42	−0.14	**0.67**	0.13	**0.71**	**0.63**	0.40	**0.71**	0.43	−0.19	1.00

Table 5. Distribution of As in various parts of rice in Koudikasa village, mg/kg.

S. No.	Rice type	Grain	Husk	Straw	Root
1	IR-64	0.46	1.13	5.4	354
2	Culture	0.25	0.40	2.8	204
3	Shymla	0.56	0.84	4.1	282
4	G.Gurmatia	0.62	1.11	5.2	306
5	Kalinga	0.68	1.58	5.9	348
6	Masuri	0.35	0.55	3.1	222
7	Purnima	0.27	0.47	2.8	216
8	IR-64	0.62	1.29	5.6	318
9	Ek Hazar Das	0.72	0.94	4.6	282
10	Culture	0.48	0.76	3.8	258
11	Luchai	0.50	1.36	5.7	330
12	Safari	0.17	0.46	2.5	210
13	Mahamaya	0.38	0.51	3.2	246
14	Sarna	0.51	0.72	4.7	276
15	HMT	0.48	0.68	3.8	258
16	M2	0.64	1.02	5.1	294
17	HMT	0.32	0.51	2.8	240
18	Masuri	0.22	0.44	2.7	234
19	Zero JR	0.62	0.98	4.9	330
20	Sonan	0.57	0.83	4.4	312

and chemical composition of the field water, type of rice species and their morphology. The CF values in the rice grain, husk, straw and root was ranged from 3 - 15, 7 - 32, 40 - 120 and 2934 - 7656 with mean value 9 ± 2, 15 ± 3, 75 ± 11 and 5065 ± 1646, respectively. Relatively very low BAC values for the rice grain, husk, straw and root were observed, ranging from 0.001 - 0.009, 0.003 - 0.015, 0.015 - 0.073 and 1.0 - 5.0 with mean value of 0.004 ± 0.001, 0.008 ± 0.002, 0.039 ± 0.007 and 2.7 ± 0.5, respectively. The remarkably higher FC and BAC values with new rice variety i.e. Kalinga, Masuri, Ek Hazar Das and M2 were seen, may be due to smaller grain size and higher cultivation period, Figure 2 and Figure 3. In addition, the FC and BAC values for different plant parts were found to be fairly correlated (r = 0.76 - 0.96). The As was found to be poorly translocated from the root to grain and its concentration was found to increase in order: grain < husk < straw << root. The As content was found 1.9 ± 0.3, 9 ± 1 and 108 ± 15 folds higher in the rice husk, straw and root with respect to the respective grain. The higher As content in the husk, straw and root of the rice species i.e. Masuri, IR-64, Luchai, Safari and HMT was observed.

Table 6. BAC and CF value of rice grain, husk, straw and root.

S. No.	Rice	CF				BAC			
		Grain	Husk	Straw	Root	Grain	Husk	Straw	Root
1	IR-64	8	19	90	5898	0.003	0.008	0.036	2.4
2	Culture	5	8	53	3852	0.002	0.004	0.025	1.9
3	Shymla	12	18	85	5874	0.006	0.009	0.046	3.1
4	G.Gurmatia	10	17	80	4710	0.005	0.009	0.043	2.5
5	Kalinga	14	32	120	7104	0.005	0.012	0.045	2.6
6	Masuri	12	19	107	7656	0.008	0.013	0.070	5.0
7	Purnima	5	8	50	3858	0.001	0.003	0.015	1.2
8	IR-64	7	14	62	3534	0.002	0.005	0.021	1.2
9	Ek Hazar Das	15	19	94	5754	0.006	0.008	0.041	2.5
10	Culture	5	9	43	2934	0.002	0.003	0.015	1.0
11	Luchai	7	20	84	4854	0.004	0.010	0.041	2.4
12	Safari	3	7	40	3390	0.002	0.004	0.024	2.0
13	Mahamaya	6	8	50	3846	0.003	0.005	0.029	2.2
14	Sarna	8	11	75	4380	0.005	0.007	0.045	2.7
15	HMT	6	9	50	3396	0.002	0.003	0.016	1.1
16	M2	13	21	104	6000	0.009	0.015	0.073	4.2
17	HMT	6	9	52	4446	0.004	0.007	0.038	3.2
18	Masuri	6	13	79	6882	0.004	0.008	0.048	4.2
19	Zero JR	10	16	79	5322	0.007	0.011	0.054	3.6
20	Sonan	14	20	107	7608	0.007	0.011	0.056	4.0

3.6. Speciation of As in Rice

The As species concentration in four new fast growing rice grains is summarized in Table 7. The main As species detected in the rice extract were monomethylarsenonate (MMA) and dimethylarsinite (DMA) and As(III). Total arsenic concentration of the four rice samples was varied from 207 - 548 mg/kg. The inorganic As(III) concentration was ranged from 46% - 66%. The higher DMA fraction was observed in the rice grain containing higher As content. Amongst them, the lowest concentration of the MMA was marked. The inorganic and organic Certified value of As in the CMR = 290 ± 30 mg/kg.

As content was found to be fairly correlated with the AsT content in the grain. The distribution pattern of As- species in the rice grain of this region is observed similar to Bangladeshi rice [24] .

3. Results and Discussion

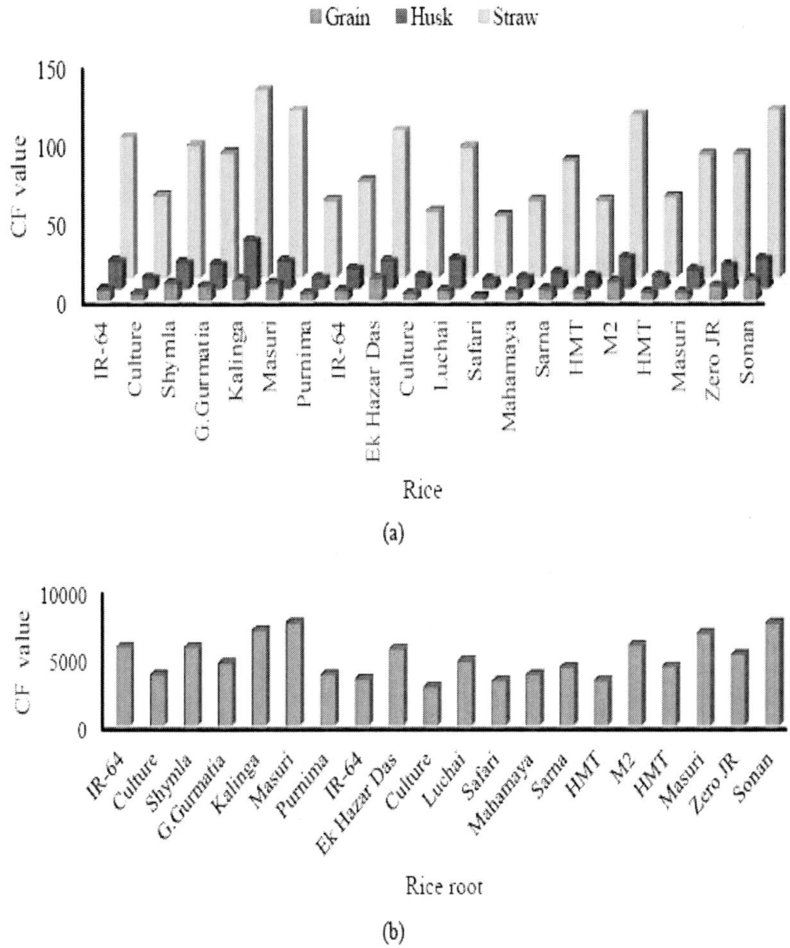

Figure 2. The FC value of rice plants in Koudikasa village.

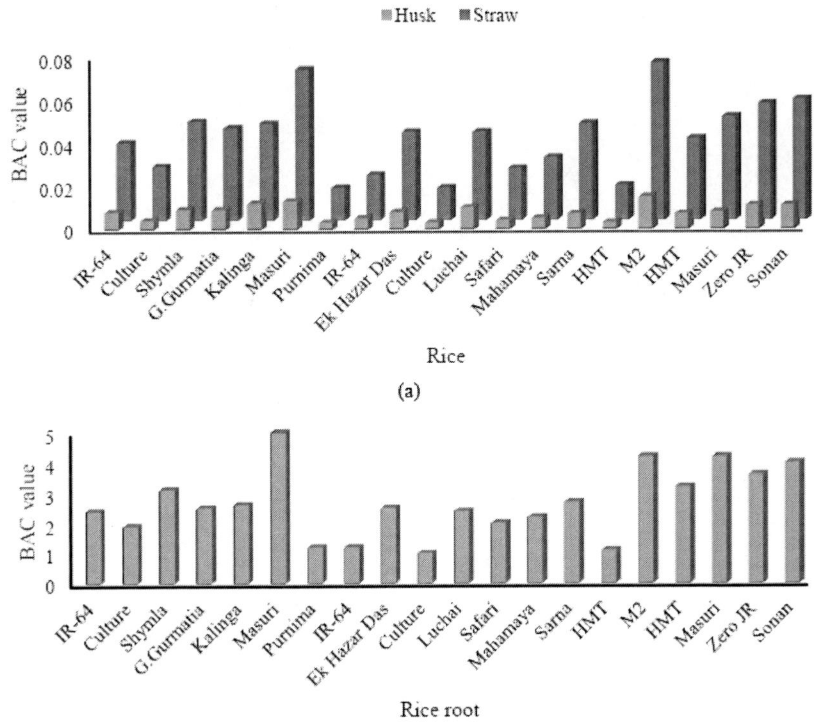

Figure 3. The BAC value of rice plants in Koudikasa village.

Table 7. Arsenic speciation in rice grain, mg/kg.

S. No.	Sample	AsT	DMA	MMA	As(III)
1	CMR rice	296 ± 21	165 ± 3	10 ± 1	97 ± 5
2	Masuri	207 ± 39	65	<3	115
3	Sonan	548 ± 14	207	16	253
4	HMT	309 ± 34	92	5	204
5	Zero JR	509 ± 29	95 ± 3	15 ± 5	270 ± 5

Certified value of As in the CMR = 290 ± 30 μg/kg.

4. CONCLUSION

The field water and soil were found to be contaminated with As at dangerous levels, may be due to geogenic origins. The high yield rice variety i.e. IR-64, Kalinga, Ek Hazar Das, M2, Zero JR, etc. was found to be more sensitive to the

As accumulation. Several folds higher As contamination of the straw and root of the rice plants than the grain was marked. The feeding of straw to the domestic animals seems to be a potential pathway entry of As. The rice root was marked as hyper phytoextractants for accumulating As from the surface soil.

ACKNOWLEDGEMENTS

We are thankful to the Alexander von Humboldt Foundation for award of fellowship to KSP. The sincere thanks is extended to Prof. J. Feldmann, College of Physical Sciences—Chemistry, Trace Element Speciation Laboratory Aberdeen, Scotland for As speciation studies of the rice grain.

REFERENCES

1. Naujokas, M.F., Anderson, B., Ahsan, H., Aposhian, H.V., Graziano, J.H., Thompson, C. and Suk, W.A. (2013) The Broad Scope of Health Effects from Chronic Arsenic Exposure: Update on a Worldwide Public Health Problem. Environmental Health Perspectives, 121, 295-302.

2. Sohn, E. (2014) Contamination: The Toxic Side of Rice. Nature, 514, S62-S63.

3. Meharg, A.A. (2004) Arsenic in Rice—Understanding a New Disaster for South-East Asia. Trends in Plant Science, 9, 415-417.

4. Zhu, Y.G., Williams, P.N. and Meharg, A.A. (2008) Exposure to Inorganic Arsenic from Rice: A Global Health Issue? Environmental Pollution, 154, 169-171.

5. Meharg, A.A. and Rahman, M.M. (2003) Arsenic Contamination of Bangladesh Paddy Field Soils: Implications for Rice Contribution to Arsenic Consumption. Environmental Science and Technology, 37, 229-234.

6. Brammer, H. (2009) Mitigation of Arsenic Contamination in Irrigated Paddy Soils in South and South-East Asia. Environment International, 35, 856-863.

7. Dittmar, J., Voegelin, A., Roberts, L.C., Hug, S.J., Saha, G.C., Ali, M.A., Badruzzaman, A.B. and Kretzschmar, R. (2007) Spatial Distribution

and Temporal Variability of Arsenic in Irrigated Rice Fields in Bangladesh. 2. Paddy Soil. Environmental Science and Technology, 41, 5967-5972.

8. Stroud, J.L., Norton, G.J., Islam, M.R., Dasgupta, T., White, R.P., Price, A.H., Meharg, A.A., McGrath, S.P. and Zhao, F.J. (2011) The Dynamics of Arsenic in Four Paddy Fields in the Bengal Delta. Environmental Pollution, 159, 947-953.

9. Seyfferth, A.L., McCurdy, S., Schaefer, M.V. and Fendorf, S. (2014) Arsenic Concentrations in Paddy Soil and Rice and Health Implications for Major Rice-Growing Regions of Cambodia. Environmental Science and Technology, 48, 4699-4706.

10. Dahal, B.M., Fuerhacker, M., Mentler, A., Karki, K.B., Shrestha, R.R. and Blum, W.E. (2008) Arsenic Contamination of Soils and Agricultural Plants through Irrigation Water in Nepal. Environmental Pollution, 155, 157-163.

11. Smith, E., Smith, J., Smith, L., Biswas, T., Correll, R. and Naidu, R. (2003) Arsenic in Australian Environment: An Overview. Journal of Environmental Science and Health: Part A—Toxic/Hazardous Substances & Environmental Engineering, 38, 223-239.

12. Liao, X.Y., Chen, T.B., Xie, H. and Liu, Y.R. (2005) Soil As Contamination and Its Risk Assessment in Areas near the Industrial Districts of Chenzhou City, Southern China. Environment International, 31, 791-798.

13. Bhattacharya, P., Samal, A.C., Majumdar, J. and Santra, S.C. (2010) Accumulation of Arsenic and Its Distribution in Rice Plant (Oryza sativa L.) in Gangetic West Bengal, India. Paddy and Water Environment, 8, 63-70.

14. Lin, S.C., Chang, T.K., Huang, W.D., Lur, H.S. and Shyu, G.S. (2015) Accumulation of Arsenic in Rice Plant: A Study of An Arsenic-Contaminated site in Taiwan. Paddy and Water Environment, 13, 11-18.

15. Rahman, M.A., Hasegawa, H., Mahfuzur Rahman, M., Nazrul Islam, M., Miah, M.A.M. and Tasmin, M.A. (2007) Arsenic Accumulation in Rice (Oryza sativa L.) Varieties of Bangladesh: A Glass House Study. Water, Air, and Soil Pollution, 185, 53-61.

16. Jayasumana, C., Paranagama, P., Fonseka, S., Amarasinghe, M., Gunatilake, S. and Siribaddana, S. (2015) Presence of Arsenic in Sri Lankan Rice. International Journal of Food Contamination, 2, 1.

17. 17. Azizur Rahman, M. and Hasegawa, H. (2011) High Levels of Inorganic Arsenic in Rice in Areas Where Arsenic- Contaminated Water Is Used for Irrigation and Cooking. Science of the Total Environment, 409, 4645-4655.

18. Sommella, A., Deacon, C., Norton, G., Pigna, M., Violante, A. and Meharg, A.A. (2013) Total Arsenic, Inorganic Arsenic, and Other Elements Concentrations in Italian Rice Grain Varies with Origin and type. Environmental Pollution, 181, 38-43.

19. Adomako, E.E., Williams, P.N., Deacon, C. and Meharg, A.A. (2011) Inorganic Arsenic and Trace Elements in Ghanaian Grain Staples. Environmental Pollution, 159, 2435-2442.

20. Patel, K.S., Shrivas, K., Brandt, R., Jakubowski, N., Corns, W. and Hoffmann, P. (2005) Arsenic Contamination in Water, Soil, Sediment and Rice of Central India. Environmental Geochemistry and Health, 27, 131-145.

21. Sahu, B.L., Ramteke, S., Rajhans, K.P., Patel, K.S., Wysocka, I. and Jaron, I. (2016) Contamination of Pond with Fluoride and Heavy Metals in Central India. Water Resources, Submitted.

22. APHA (2005) Standard Methods for the Examination of Water and Wastewater. 21st Edition, APHA, AWWA and WEF, Washington DC.

23. Tan, K.H. (2005) Soil Sampling, Preparation and Analysis. 2nd Edition, CRC Press, Boca Raton.

24. Williams, P.N., Price, A.H., Raab, A., Hossain, S.A., Feldmann, J. and Meharg, A.A. (2005) Variation in Arsenic Speciation and Concentration in Paddy Rice Related to Dietary Exposure. Environmental Science and Technology, 39, 5531-5540.

25. WHO (2011) Guidelines for Drinking-Water Quality. 4th Edition, World Health Organization, Geneva.http://www.haceclick.com.uy/documentos/GuIa_OMS%202011_4aEd.pdf

Chapter 5

Risk Assessment of Heavy Metals Contamination in Paddy Soil, Plants, and Grains (Oryza sativa L.) at the East Coast of India

Deepmala Satpathy,[1] M. Vikram Reddy,[1] and Soumya Prakash Dhal[2]

[1]Department of Ecology and Environmental Sciences, Pondicherry University, Pondicherry 605 014, India

[2]Department of Physics, Pondicherry University, Pondicherry 605 014, India

ABSTRACT

Heavy metals known to be accumulated in plants adversely affect human health. This study aims to assess the effects of agrochemicals especially chemical fertilizers applied in paddy fields, which release potential toxic heavy metals into soil. Those heavy metals get accumulated in different parts of paddy plant (Oryza sativa L.) including the grains. Concentrations of nonessential toxic heavy metals (Cd, Cr, and Pb) and the micronutrients (Cu, Mn, and Zn) were measured in the paddy field soil and plant parts. Mn and Cd are found to be accumulated more in shoot than in root. The metal transfer factors from soil to rice plant were significant for Pb, Cd, Cu, Cr, Mn, and Zn. The ranking order of bioaccumulation factor (BAF) for heavy metals was Zn > Mn > Cd > Cu > Cr > Pb indicating that the accumulation of micronutrients was more than that of nonessential toxic heavy metals. The concentrations of heavy metals were found to be higher in paddy field soils than that of the nearby control soil but below permissible limits. The higher Health Index (HI) values of rice consuming adults (1.561) and children (1.360) suggest their adverse health effects in the near future.

1. INTRODUCTION

Heavy metals from natural and anthropogenic sources accumulate in soil and plants and as a consequence represent important environmental contamination problems. Nevertheless, food safety issues and adverse health risks make this one of the most serious environmental issues [1]. Soils are considered to be an excellent media to monitor and assess heavy metal pollution because anthropogenic heavy metals are usually deposited in the top soils [2]. Heavy metal contaminated soil adversely affects the whole ecosystem when these toxic heavy metals migrate into groundwater or are taken up by flora and fauna, which may result in great threat to ecosystems due to translocation and bioaccumulation [3]. Heavy metals are potentially toxic to crop plants, animals, and human beings when the contaminated soils are used for crop production [4]. Environmental contamination of the biosphere with heavy metals due to intensive agricultural and other anthropogenic activities poses serious problems for safe use of agricultural land [5]. Contemporary agriculture with indiscriminate use of agrochemicals such as fertilizers and pesticides, along with mechanical cultivation, for higher crop productivity contaminates agriculture soils with potentially nonessential and essential heavy metals [6, 7]. Human health is directly affected through intake of crops grown in polluted soils. There is clear evidence that human renal dysfunction is related with contamination of rice with Cd in subsistence farms in Asia [8]. Indeed, in Asia, rice has been identified as one of the major sources of Cd and Pb for human beings [9–11]. In Japan, rice was found to be the main source of Cd contamination in human beings [12].

It has also been reported that crop plants have different abilities to absorb and accumulate heavy metals in their body parts and that there is a broad difference in metal uptake and translocation between plant species and even between cultivars of the same plant species [13–20]. Plants absorb heavy metals from the soil, and the surface 25 cm zone of soil is mostly affected by such pollutants resulting from anthropogenic activities. Heavy metals are adsorbed and accumulated in this soil layer probably due to relatively high organic matter. The plant parts of interest for direct transfer of heavy metals to human body are the edible parts such as the rice grain, which may consequently become a threat to human health. Nevertheless, heavy metals in the environment, consequently, are of immense concern, because of their persistence nature, bioaccumulation, and biomagnification characters causing ecotoxicity to plants, animals, and human beings [21].

The micronutrients such as Zn, Mn, and Cu are required in small but critical concentrations for both plants and animals, and these have vital role in physical growth and development of crop plants such as paddy. The deficiency of Zn in soil casts a conspicuous adverse effect, with stunted growth of crop plants like paddy and groundnut [22, 23] reducing the overall productivity. Generally, the monitoring and assessment of total heavy metal concentrations in agricultural soils are required to evaluate the potential risk of paddy soils contaminated due to toxic heavy metals—Cd, As, and Pb [7, 24]. Heavy metals are known to accumulate in living organisms [25]. There is a tendency of plants to take up heavy metals that may subsequently transfer into the food chain. Use of polluted soil or water for crop cultivation mainly results in decrease of overall productivity and contaminates food grains and vegetables, which adversely affect human health too [26]. A number of reports on concentrations of toxic metal such as Cd, and Pb in rice and paddy soils in Japan, China, and Indonesia are available [27–31]. However, such studies are very few in India, with little information on toxic heavy metal contamination of paddy fields and risk assessment [32], though rice is the most important staple food for Indian people. The objectives of the present study, therefore, are primarily risk assessment of potential toxic and nonessential heavy metals—Cd, As, Pb, and Cr, in the surface soil of paddy fields at the predominantly paddy-cultivated area nearby Kalpakkam (Tamil Nadu). Concentrations of the toxic heavy metals were assessed in soil, root, shoot, and grains of paddy crop to assess the bioaccumulation factor and transfer factor. Risk assessment was made assessing the potential risk factor for the local residents consuming rice, the staple food.

2. Materials and Methods

2.1. Study Site

The study site is located at about 30 km away from Pondicherry towards east on the East Coast Road (ECR) at Swarnabhoomi, near Kalpakkam in Tamil Nadu, India (12°22′21.5″ and 12°21′32.4″ N to 80°04′36.6″ and 80°04′ 04.2″ E0 (Figure 1)). The agricultural watershed is of about 50 ha, and it is transected at one end by the ECR. About 30 ha of the watershed comprising of paddy fields applied with chemical fertilizers and pesticides, shed their water into adjacent rivulet which finally joins a small lagoon located towards south-west. About 20 ha of the paddy fields is on the same plane as that of the rivulet, from where the water is taken for irrigation of the fields. Sampling of the soil and rice plants with grains was carried out during crop season (winter 2012) in order to investigate the concentrations and spatial distribution of potentially toxic heavy metals such as Pb, Cd, Cu, Cr, Mn, and Zn originated from the agricultural activities of the agriculture watershed. The study site and sampling locations of the study site are shown in Figure 1.

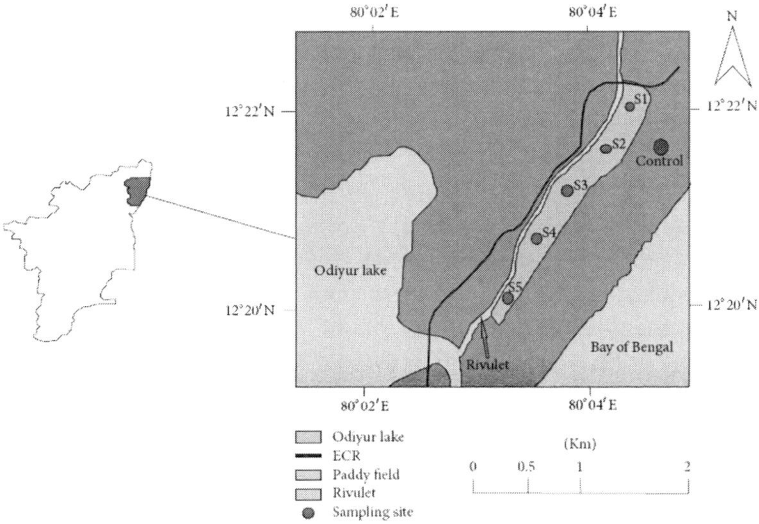

Figure 1. Location of sampling area.

2.1.1. Sampling of Soil, Rice Plants, and Grain and Sample Preparation

Soil samples were collected from six different sites of the paddy fields. Out of these one was the control site where no crop cultivation was done and other five sites (S-1, S-2, S-3, S-4, and S-5) were selected from paddy fields. At each sampling site, a composite of five soil samples was collected separately by a random selection, from each field, from surface (0–15 cm soil layer) with a small core sampler and mixed to make one composite sample. Samples were collected from centre of the fields in order to avoid the edge effect. Each soil sample of about 500 mg was collected from the 0–15 cm layer, which represented the plough layer. Rice plant samples were collected from the corresponding soil sampling site of the paddy field for computing correlations between heavy metal concentrations of soil and plant. All soil and rice plants along with grain samples were kept in clean polyethylene bags and brought to the laboratory for analyses. During plant sampling, it was ensured that plant samples were of the same physiological age and identical size. Paddy crop plants were collected and washed thoroughly with deionized water. Paddy plant was cut and separated into root, shoot, and grain subsamples. All subsamples were oven-dried at 60°C for 24 h, and the dried samples were weighed, then pulverized, and stored in Petri dishes. The soil samples were air-dried at room temperature for several days, then pulverized, and sieved through a 01 mm stainless steel mesh. Rice grain samples were washed with deionized water and hulls were removed. The rice grain samples without hull were oven-dried at 70°C for 72 h and then ground with an agate mortar to fine powder.

2.2. Sample Analysis

Soil pH and conductivity were determined by using a digital pH meter. For heavy metal analysis, one gm of soil and 1 g of rice grain samples were digested after adding 15 mL of triacid mixture (HNO_3, H_2SO_4, and $HClO_4$in a 5 : 1 : 1 ratio) with three replicates at 80°C until a transparent solution was obtained [33]. After cooling, the digested sample was diluted up to 30 mL with 2% HNO_3 and concentrations of Pb, Cd, Cu, Cr, Mn, and Zn were determined by AAS (GBC make—Model Avanta PM).

For plant samples, 1 g of dried sample was digested with HNO_3 and $HClO_4$ in a 5 : 1 ratio until a transparent solution was obtained, and the plant digests were filtered and diluted to 30 mL, with distilled water (Reddy et al.) [32]. The filtrates of plant were then assessed by using atomic absorption spectroscopy (AAS; GBC make—Model Avanta PM) for analysis of Pb, Cd, Cu, Cr, Mn, and Zn. The AAS value of blank (without sample) of each metal was deducted from the sample value for final calculations [26]. All the analyses were done with three replications.

2.3. Bioaccumulation Factor (BAF)

The BAF (bioaccumulation factor, the ratio of the concentration of the element in the grain to that in the corresponding soil) was calculated for each rice sample to quantify the bioaccumulation effect of rice on the uptake of heavy metals from the soils [34]. The BAF was computed as

$$\mathrm{BAF} = \frac{\mathrm{Cr}}{\mathrm{Cs}}, \tag{1}$$

where Cr and Cs represent the heavy metal concentrations in rice grain and soils, respectively.

2.4. Translocation Factor (TF) and Enrichment Factor (EF)

Translocation factor (TF) or mobilization ratio [35, 36] was calculated to determine relative translocation of metals from soil to other parts (root, shoot, or grain) of the plant species as follows:

$$\mathrm{TF} = \frac{\text{Concentration of metal in plant tissue}}{\text{Concentration of metal in corresponding soil or root or shoot}}. \quad (2)$$

The enrichment factor (EF) has been calculated to derive the degree of soil contamination and heavy metal accumulation in soil and in plants growing on contaminated site with respect to soil and plants growing on uncontaminated soil [37] as follows:

$$\mathrm{EF} = \frac{\text{Concentration of metals in soil at contaminated site}}{\text{Concentration of metals in soil at uncontaminated site}}. \quad (3)$$

2.5. Risk Assessment

The Health Risk Index (HRI) was calculated as the ratio of estimated exposure of rice and oral reference dose (ORD) [7]. ORDs were 2×10^{-2}, 1×10^{-3}, and 4×10^{-2} and 1.5 mg/kg/day for Pb, Cd, Cu, Cr, and Zn, respectively [38, 39]. Estimated exposure is obtained by dividing the daily intake (DI) of heavy metals by their safe limits. An index value >1 is considered unsafe for human health [39]. DI was calculated by the following equation:

$$\mathrm{DI} = \frac{\mathrm{C} \times \mathrm{Con} \times \mathrm{EF} \times \mathrm{ED}}{\mathrm{Bw} \times \mathrm{AT}}, \quad (4)$$

where C (in milligrams per kilogram) is the concentration of heavy metals in the rice, Con (in grams per person per day) is the daily average consumption of rice in the region, Bw (in kilograms per person) represents body weight, EF is exposure frequency (365 days/year), ED is exposure duration (70 years, equivalent to the average lifespan), and AT is average time (365 days/year number of exposure years, assuming 70 years in this study). The average daily rice intake of adults and children was considered to be 389.2 and 198.4 g/person/day, respectively [40]. Average adult and child body weights were taken to be 55.9 and 32.7 kg, respectively, as used in many previous studies [7, 40–42]. The health risk for adult and children is considered separately since the contact pathway with each exposure way changes with age. There may be a certain amount of discrepancy in health risk between age groups and locality of inhabitants (Wang et al. [41]). Harrison and Chirgawi [43] reported that exposure of two or more pollutants may result in additive or interactive effects. Wang et al. [41], Chien et al. [44], Zheng et al. [40], and Hang et al. [7] have studied hazard index for different diets. Assuming the additive effect here HRI can, therefore, be summed across the constituents to calculate the HI for a specific receptor (e.g., diet) combination. The HI is calculated to evaluate the potential risk of adverse health effects from a mixture of chemical constituents in rice. The HI was calculated through daily average consumption of rice for a human being (adults and children) and is as follows:

$$HI = \sum_{i=1}^{n} HRI. \tag{5}$$

2.6. Data Analysis

Arithmetical means ± standard deviation (SD;) were used to assess the contamination levels of heavy metals in soils, root, shoot, and grains. Coefficient correlation analysis was done to find out the heavy metals characteristics in agricultural field soil and grain samples.

3. RESULTS AND DISCUSSION

3.1. Heavy Metal Concentration in Soil

3. Results and Discussion

It was found that, in the essential heavy metals in the paddy soil, Mn concentration ranged from 12.5 to 53.9 μg g^{-1}, Zn concentration ranged from 3.8 to 33.8 μg g^{-1}, Cu concentration ranged from 0.03 to 2.9 μg g^{-1} in the paddy field soils, and in the concentrations of nonessential toxic metals, Pb ranged from 5.3 to 19.8 μg g^{-1}, Cr ranged from 1.3 to 7.8 μg g^{-1}, and Cd from 0.02 to 0.6 μg g^{-1} (Table 1). Among these metals, Cd and Cr are highly toxic, while Pb is moderately toxic and Zn, Mn, and Cu are essential elements and micronutrients [45]. The ranking order of occurrence of the heavy metals in the paddy field soils was Mn > Zn > Pb > Cr > Cu > Cd indicating that Mn followed by Zn was in the maximum concentrations and Cd was in minimum concentration. Concentrations of the heavy metals are higher in the paddy field soils compared to that of nearby control field soil. However, the ranking order of concentration of the metals is different from that of the paddy soil (Cd > Mn > Zn > Cu > Pb) of a predominantly paddy cultivated area at Bahour near Puducherry, with Cd in maximum concentration and Pb in minimum concentration [32]. The concentrations of Pb and Cd were higher in S-5 (.3 μg g^{-1} and μg g^{-1}), respectively, followed by S-4 (μg $g^{-1}$and μg g^{-1}), respectively, the concentrations of Pb were μg g^{-1} at S-2 and μg g^{-1} at S-1, and Cd concentration was (μg g^{-1}) at S-1, showing the ranking order of Pb concentration in paddy soil, S-5 > S-4 > S-3 > S-2 > S-1. Concentration of Cd was higher in S-5 followed by S-4, S-3, and S-2 (μg g^{-1}) showing the ranking order of S-5 > S-4 > S-3 > S-1 > S-2, which is attributable to spatial difference in fertilizer broadcasting and consequential input on the soil surface.

Table 1. Mean concentrations of heavy metals along with standard deviation in soil and different plant parts across the sampling sites.

Heavy metals	Soil range ($\mu g\, g^{-1}$)	Root range ($\mu g\, g^{-1}$)	Shoot range ($\mu g\, g^{-1}$)	Grain range ($\mu g\, g^{-1}$)
Pb	5.3 ± 0.4–19.8 ± 1.3	3.6 ± 0.2–5.3 ± 0.04	0.3 ± 0.01–1.2 ± 0.01	0.01 ± 0.001–1 ± 0.02
Cd	0.02 ± 0.005–0.6 ± 0.04	0.11 ± 0.008–0.2 ± 0.01	0.2 ± 0.01–0.3 ± 0.01	0.02 ± 0.001–0.05 ± 0.002
Cu	0.03 ± 0.004–5.4 ± 1.5	0.2 ± 0.02–0.5 ± 0.04	0.04 ± 0.008–0.3 ± 0.03	0.1 ± 0.008–0.3 ± 0.01
Cr	1.3 ± 0.01–7.8 ± 0.3	0.6 ± 0.02–1.7 ± 0.04	0.4 ± 0.01–0.9 ± 0.04	0.1 ± 0.02–0.6 ± 0.01
Mn	12.5 ± 0.2–53.9 ± 1.5	14.4 ± 0.9–21.9 ± 0.3	25 ± 2.8–32.9 ± 1.9	5.6 ± 0.04–7.5 ± 0.03
Zn	3.8 ± 1.7–33.8 ± 1.3	4.7 ± 0.1–16.9 ± 0.9	2.3 ± 0.01–6 ± 0.2	3.2 ± 0.05–7.2 ± 0.008

Concentration of Cu was higher in S-2 (μg g^{-1}) followed by S-3 (μg g^{-1}), S-4 (μg g^{-1}), S-5 (μg g^{-1}), and S-1 (μg g^{-1}) showing the ranking order of S-2 > S-3 > S-4 > S-5 > S-1. Concentration of Cr was higher in S-5 (μg g^{-1}) followed by S-4 (μg g^{-1}), S-3 (μg g^{-1}), S-2 (μg g^{-1}), and S-1 (μg g^{-1}) showing the ranking order of S-2 > S-3 > S-4 > S-5> S-1. Concentration of Mn was higher in S-3 (μg g^{-1}) followed by S-2 (μg g^{-1}), S-4 (μg g^{-1}), S-5 (μg g^{-1}), and S-1 (μg g^{-1}) showing the ranking order of S-3 > S-2 > S-4 > S-5 > S-1. Concentration of Zn was higher in S-

5 (ppm) followed by S-4 (μg g^{-1}), S-2 (μg g^{-1}), S-3 (μg g^{-1}), and S-1 (μg g^{-1}) showing the ranking order of S-5 > S-4 > S-2 > S-3 > S-1.

The toxic heavy metals Cd and Pb and the micronutrients Zn, Cu, and Mn accumulated in the soil of paddy fields which was higher than that of the control soil. The heavy metals were in a ranking order of Mn > Zn > Pb > Cr > Cu > Cd. The concentrations of Pb, Cd, Cu, Cr, and Zn in the paddy soils are comparable to those for worldwide normal soils (i.e., within the range of published values) [46, 47]. Only concentration of Mn was higher than the value of uncontaminated soil, critical soil concentration [48], and worldwide normal soils (Table 2).

Table 2. Mean values of heavy metals (Pb, Cd, Cu, Cr, Mn, and Zn) for uncontaminated paddy soils, mean values for worldwide normal surface soils, critical concentrations for contaminated soils, Indian standards, and European Union standards, compared with the values of present study.

Elements (μg g^{-1})	Mean values for paddy soils[a]	Mean values for worldwide normal surface soils[b]	Critical soil concentration[c]	Indian standards[d]	European Union standards[e] (EU 2002) [69]	Present study
Pb	23.3	22–44	100–400	250–500	300	5.3–19.8
Cd	0.34	0.37–0.78	3–8	3–6	3	0.2–0.6
Cu	20.7	13–24	60–125	135–270	140	1.10–2.9
Cr	64	12–83	75–100	—	—	1.3–7.8
Mn	0.39	0.27–0.53	1.5–3	—	—	12.5–33.9
Zn	61	45–100	70–400	300–600	300	3.8–33.8

[a] Mean of total concentrations of elements in uncontaminated paddy soils. Data are from Wong et al. (2002) [4]; Wang et al. (2003) [66]; Chandrajith et al. (2005) [67].
[b] Mean of total concentrations of elements in the surface horizon of worldwide normal soil, and normal means that the data do not include contaminated or mineralized soils. Data are from Kabata-Pendias (2001) [46]; Essington (2004) [68].
[c] Higher concentrations may be toxic to plants depending on speciation (Alloway, 1995) [48].
[d] Indian standards (Awashthi, 2000) [47] for agricultural soils.
[e] European standards (EU 2002) [69] for agricultural soils.

3.2. Heavy Metal Concentration in Different Plant Parts

The mean concentrations of heavy metals in the paddy plant parts (Table 1) showed that most of the metals accumulated more in the roots than in other plant parts, shoots, and grains and ranged from 14.4–21.9 μg $g^{-1}$for Mn, 4.7–16.9 μg g^{-1} for Zn, 3.6–5.3 μg g^{-1} for Pb, 0.6–1.7 μg g^{-1} for Cr, 0.2–0.5 for Cu, and 0.1-0.2 μg $g^{-1}$for Cd among the five sites (Table 1). It indicated that the Cd concentrations were minimum in the paddy soil, in contrast to the Cd concentrations of paddy soil at Bahour in Puducherry [32]. The mean concentrations of heavy metals in the paddy plant parts showed that most of the metals accumulated more in the roots than in other parts. In general metal uptake was higher for the micronutrients; Mn and Zn in the roots were followed by Pb, Cr, Cu, and Cd. In the present study concentration of Pb was found to be higher in roots that in shoots and grains. Calluna vulgaris L. Hull (common heather) and Agrostis vinealis, harvested from an abandoned Pb mine in UK, contained 320 and 2930 mg/kg dry wt., respectively, in shoot tissue, while Pb values for root were 9610 and 9740 mg/kg, indicating high plant availability of the Pb in the soil as well as its limited mobility inside the plant [49]. Cu was also found to be more in roots than that in shoots and grains, which is in corroboration with findings of earlier workers [50, 51]. Yang et al. [52] reported that accumulation of Cu was more in roots, while a small fraction (10%) of absorbed Cu was translocated to stem. The Cu and Zn accumulated at their highest concentration in roots of the rice plants followed by shoots and grains. Most metals, Fe, Mn, Zn, and Cu, that were found profusely in the paddy plants were the micronutrients that are required for various enzyme activities and play important roles in photosynthesis and growth of the plant [53, 54].

It was seen that Mn and Cd were accumulated more in shoot than in root and found in the ranges of 25–32.9 $\mu g\ g^{-1}$ for Mn, 2.3–6 $\mu g\ g^{-1}$ for Zn, 0.4–0.9 $\mu g\ g^{-1}$ for Cr, 0.3–1.2 $\mu g\ g^{-1}$ for Pb, 0.2-0.3 $\mu g\ g^{-1}$ for Cd, and 0.05–0.3 $\mu g\ g^{-1}$ for Cu among the five sites (Table 1). In the shoots, concentrations of Mn and Cd were higher than their concentrations in roots and grains. Jarvis, Jones, and Hopper [50] reported that Cd was easily taken up by plants and transported to different parts, although it is nonessential and is of no beneficial effects on plants and animals. Moreover, Cd is toxic to animals and plants, and plants when exposed to this metal show reduction in photosynthesis and uptake of water and nutrient [55]. Higher concentration of Mn in leaves of both the plants indicated its high mobility [56], as leaf chlorophyll content requires Mn for photosynthesis. In contrast, Gupta and Sinha [57] reported higher accumulation of Mn in roots followed by leaves in Chenopodium. The mean concentrations of heavy metals in the grains were found in the ranges of 5.6–7.5 $\mu g\ g^{-1}$ for Mn, 3.2–7.2 $\mu g\ g^{-1}$ for Zn, 0.1–0.6 $\mu g\ g^{-1}$ for Cr, 0.1–0.3 $\mu g\ g^{-1}$ for Cu, 0.02–0.05 $\mu g\ g^{-1}$ for Cd, and 0.01–1 $\mu g\ g^{-1}$ for Pb among the five sites (Table 1). In grains, among all metals, Mn and Zn were in more elevated concentrations than Cr, Cu, Cd, and Pb, but their concentrations were less compared to that of roots and shoots. Concentration of Zn ranged between 3.2 $\mu g\ g^{-1}$ and 7.2 $\mu g\ g^{-1}$, which did not exceed the maximum permissible limit of 50 (Pilc et al.) [58]. Concentration of Pb ranged between 0.01 $\mu g\ g^{-1}$ and 1 $\mu g\ g^{-1}$. The highest Pb content was found in S-4 (1 $\mu g\ g^{-1}$) and S-5 (0.9 $\mu g\ g^{-1}$), which exceeds the values given by Pilc et al. [58] or the corresponding limit defined by the Commission Regulation Directive EC [59]. However, the concentrations of Cr, Cu, and Cd ranged between 0.1 $\mu g\ g^{-1}$ and 0.6 $\mu g\ g^{-1}$, 0.1 $\mu g\ g^{-1}$ and 0.2 $\mu g\ g^{-1}$, and 0.02 $\mu g\ g^{-1}$ and 0.05 $\mu g\ g^{-1}$, respectively, which did not exceed the values defined by the Commission Regulation Directive EC [59] or the corresponding limits given by Pilc et al. [58] and FAO/WHO [60] (Table 3). Concentration of Cd was found within the limit defined by CODEX

[61] and European Commission [62] (Table 3). Table 3 lists the maximum allowable concentrations (MAC) of Pb, Cd, Cu, Cr, and Zn in foods recommended by the Chinese National Standard Agency. The mean concentrations of all the elements in the rice grain were below their maximum allowable levels except for Pb. The results indicate that the concentration of Pb in rice grain may have been affected by various anthropogenic activities such as use of tractor for farming and use of chemical fertilizers and pesticides. Apart from this, the agricultural watershed is transected on one end by ECR (East Coast Road), a state highway (State Highway number 49) which is used by more than 10,000 vehicles daily. On the other end, Maduvankarai Road, a normal street road connecting the ECR to the boat house via a bridge, also serves as mode of transportation. Both the roads may be contributing to the increase of Pb concentration that may have come from the vehicular emission. Boating, fishing by motor boats, and other recreational activities taking place nearby lagoon could be also a reason for elevated concentration of Pb in rice grain. Concentration of Cr is slightly higher but below MAC. It may be due to some anthropogenic activities such as use of chemical fertilizers and pesticides and other industrial activities near Kalpakkam which comes as runoff and would therefore reflect contamination by the element.

Table 3. The total concentrations of Pb, Cd, Cu, Cr, Mn, and Zn in rice (Oryza sativa L.) samples from the experimental sites compared with the standard values.

Elements ($\mu g\, g^{-1}$)	1	2	3	4	5	6	Present study
Pb	0.2	—	1.0	—	—	0.2	0.01–1.0
Cd	0.1	—	0.15	0.4	0.2	0.2	0.02–0.05
Cu	—	10	—	—	—	10	0.1–0.23
Cr	—	—	—	—	—	1.0	0.13–0.56
Mn	—	—	—	—	—	—	5.58–7.47
Zn	—	—	50	—	—	50	3.23–7.24

Maximum permitted levels for heavy metals in food crops: (1) Commission Regulation Directive EC (2001) [59]; (2) FAO/WHO (1992) [60]; (3) Pilc et al. (1994) [58]; (4) CODEX Alimentarius Commission (2006) [61] and (5) European Food Safety Authority. EFSA [62].
(6) maximum levels of contaminants in foods (GB 13106-1991; GB 2762-2005; GB 15199-1994; GB 4810-1994) [7].

3.3. Bioaccumulation Factor

Bioaccumulation factors (BAFs) for the heavy metal transfer from soils to rice are shown in Figure 2. The BAF values of the heavy metals such as Zn, Mn, Cd, Cu, Cr, and Pb were found to be in the ranges of 0.2 to 0.5, 0.1 to 0.2, 0.05 to 0.2, 0.04 to 0.1, 0.04 to 0.07, and 0.001 to 0.06, respectively. The trend in the BAF for heavy metals in the study sites was in the ranking order of Zn > Mn > Cd > Cu > Cr > Pb. Among the heavy metals, BAF values were found to be higher for Zn, Mn, and Cd, whereas relatively lower BAF values were found in Cu, Cr, and Pb. The food chain (soil-plant-human) is mainly known as one of the major pathways for exposure of human to soil contaminants. Soil-to-plant transfer is one of the key processes of human exposure to toxic heavy metals through the food chain [63]. When BCF < 1 or BAF = 1, it denotes that the plant only absorbs the heavy metal but does not accumulate when BCF > 1, and this indicates that plant accumulates the heavy metals. BAF values of Pb, Cd, Cu, Cr, Mn, and Zn were less than one in the rice grain which indicates that plants only absorb the heavy metals.

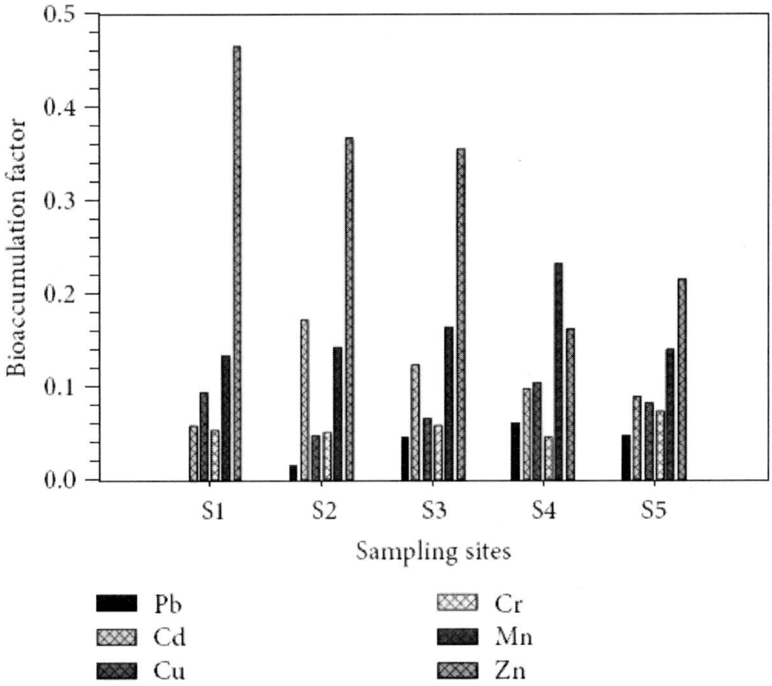

Figure 2. Bioaccumulation factor of the heavy metals across the sites.

3.4. Translocation Factor

Transfer factor is one of the main components of human exposure to toxic heavy metals through the food chain. The transfer factors (TFs) of metals from soil to root (TF_{Soil}), root to shoot (TF_{Root}), and shoot to grain (TF_{Shoot}) were calculated and given in Table 4. The average translocation values of metals in paddy soils from soil to root (TF_{Soil}) were found to be in the order of Zn (0.4 to 0.9) > Mn (0.3 to 0.7) > Cd (0.3 to 0.6) > Pb (0.2 to 0.4) > Cr (0.2 to 0.3) > Cu

(0.09 to 0.2). In the case of root (root to shoot), TF_{Root} values were found in the order of Cd (1.3 to 2.4) > Mn (1.3 to 2.3) > Cr (0.5 to 0.8) > Cu (0.2 to 0.6) > Zn (0.2 to 0.5) > Pb (0.07 to 0.3). The translocation values for shoot to grain (TF_{Shoot}) were found in the following order: Cu (1.1 to 2.5) > Zn (1 to 1.5) > Pb (0.04 to 0.8) > Cr (0.3 to 0.7) > Mn (0.2 to 0.3) > Cd (0.09 to 0.2). There was a significant difference in TFs values among the heavy metals (). Soil-to-plant transfer factor is one of the major components of human exposure to metals through the food chain, and it could reveal bioavailability of heavy metals in investigated soils. The higher the TF values are, the more mobile/available the metals are [1, 42, 64]. The TFs vary noticeably within the plant species even for an individual heavy metal [1]. In the present study transfer of Cd and Zn from roots to shoots was more than other heavy metals as the concentrations of these two were found to be more in shoots than that of roots and grains. Roots often contain more Zn than the shoot parts, but the Zn may be translocated from the roots and accumulated in the plant shoot parts [46]. Cd was also translocated more from root to shoot which is known to be relatively mobile in plants [46]. The metal translocation process in plant species is a crucial factor in determining the metal distribution in different plant tissues [65].

Table 4. Ranges of the translocation factors of the heavy metals from soil to root, root to shoot, and shoot to grain across the sampling sites.

Heavy metals	Tf (soil to root)	Tf (root to shoot)	Tf (shoot to grain)
Pb	0.2–0.4	0.1–0.3	0.04–0.8
Cd	0.2–0.6	1.4–2.4	0.1-0.2
Cu	0.1-0.2	0.2–0.6	1.1–2.5
Cr	0.2-0.3	0.5–0.8	0.3–0.7
Mn	0.3–0.7	1.3–2.3	0.2-0.3
Zn	0.4–0.9	0.3–0.5	1–1.5

3.5. Enrichment Factor

The EFs of the paddy field soils for the heavy metals were found to be in the ranges of Pb (2.2 to 3.7), Cd (8.4 to 27.1), Cu (48.6 to 204.3), Cr (2 to 6.2), Mn (3.3 to 4.3), and Zn (1.9 to 9) and in the ranking order of Cu > Cd > Zn > Cr > Mn > Pb. Moreover, there was a significant difference in EFs values among the heavy metals (). The EF values greater than 1 indicate higher availability and distribution of metals in the contaminated soil, subsequently increasing the metal accumulation in plants species grown on the soil [36, 37]. Among the metals estimated, the maximal enrichment was found in case of Cu and Cd for the paddy soils (Figure 3).

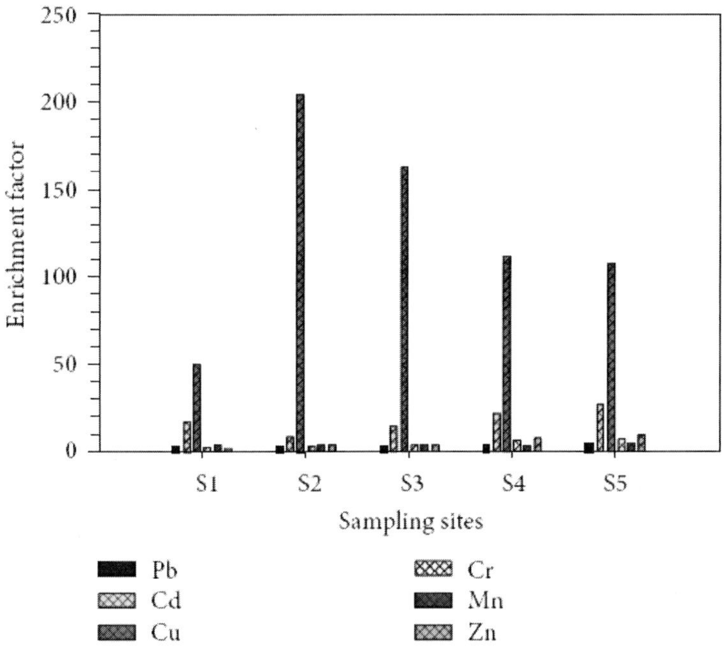

Figure 3. Enrichment factor of the heavy metals in soil across the sites.

3.6. Correlation Matrix

The correlation coefficient matrix is normally used to measure the degree of correlation between logarithms of the elemental concentrations [70], and this matrix for the heavy metals of paddy showed highly significant positive correlations between the pairs of elements of soil samples—Cd-Pb (0.65), Cr-Pb (0.95), Zn-Pb (0.96), Cd-Cr (0.83), Cd-Zn (0.69), and Cr-Zn (0.98) (Table 5(a)). Besides, highly significant positive correlations were found between the pairs of elements present in grains which are Pb-Cd (0.92), Pb-Cu (0.75), Cr-Pb (0.73), Mn-Pb (0.96), Cd-Cr (0.91), Mn-Cd (0.98), Zn-Cd (0.71), Cr-Mn (0.85), and Cr-Zn (0.72) (Table 5(b)). The significant positive correlation between the elements of surface soil grains suggests that their common source of origin is probably the agrochemicals, such as phosphate and nitrate fertilizers, broadcasted in the paddy fields [32].

Table 5. (a) Correlation of the heavy metals in paddy field soil. (b) Correlation of the heavy metals in grains.

	Pb	Cd	Cu	Cr	Mn	Zn
Pb	1					
Cd	0.654123	1				
Cu	0.193626	−0.61182	1			
Cr	0.952659	0.802794	−0.06512	1		
Mn	0.316071	0.273936	0.032893	0.175871	1	
Zn	0.967693	0.699247	0.079194	0.984368	0.192237	1

(b)

	Pb	Cd	Cu	Cr	Mn	Zn
Pb	1					
Cd	0.924811	1				
Cu	0.759248	0.629072	1			
Cr	0.733659	0.915057	0.267458	1		
Mn	0.968289	0.988837	0.716005	0.850878	1	
Zn	0.447148	0.719954	0.42386	0.722836	0.64716	1

3.7. Potential Health Risk of Heavy Metals through Rice Intake

Rice consumption has been identified as one of the major pathways of human exposure to the toxic heavy metals accumulated in rice grain. Table 6 showed the dietary intake (DI) of heavy metals via rice for adults and children in the study region as the local people consume generally rice, the staple food for the people available in the region. The DIs of Pb, Cd, Cu, Cr, and Zn through rice were estimated to be 4.02, 0.27, 1.66, 1.98, and 37.04 mg/kg/day for adults and 3.50, 0.23, 1.45, 1.73, and 32.28 mg/kg/day for children, respectively. The DIs of heavy metals for adults were found to be higher than those for children. This is most probably due to relatively higher quantity of rice consumption of adults compared to the children, which increased the DIs of heavy metals. The result is in conformity with previous studies in the neighboring country, China [40, 71].

Table 6. Health risk assessment of heavy metals via intake of rice.

Individuals	Element	ORD	DI	HRI	HI
	Pb	3.50	4.02	0.269	
	Cd	1.00	0.27	0.042	
Adults	Cu	40.00	1.66	0.001	1.561
	Cr	1500.00	1.98	0.123	
	Zn	300.00	37.04	1.126	
	Pb	3.50	3.50	0.234	
	Cd	1.00	0.23	0.036	
Children	Cu	40.00	1.45	0.001	1.360
	Cr	1500.00	1.73	0.108	
	Zn	300.00	32.28	0.981	

ORD oral reference dose (in micrograms per kilogram per day). DI daily intake (in micrograms per kilogram per day).

The HRIs of heavy metals through rice consumption are given in Table 4. The HRI of heavy metals for adults from rice consumption was in decreasing order: Zn > Pb > Cr > Cd > Cu. The HRI of heavy metals for children also has the similar trend as the adults. The Zn has the highest HRI value as it is an essential micronutrient. Hence, it may not pose a potential risk up to a certain concentration but may cause adverse effects at certain elevated level. So among the toxic heavy metals, Pb ingestion has the highest potential health risk of adverse effects for adults and children and Cu ingestion has minimum risk. Only HRI of Zn for adults was found beyond the value of 1. Therefore, the health risk of single metal exposure through rice consumption was generally assumed to be safe for the people of the region. However, local inhabitants may be at risk due to the combination of several toxic heavy metals [71]. The HI values for rice consumption of adults and children were 1.561 and 1.360, respectively. This indicates that adults and children may experience poor health effects in the near future as the heavy metal accumulation over a period of time leads to biomagnification. Our assessment was only to measure the intake of toxic heavy metals through rice consumption. In fact, humans are also exposed to heavy metals through other foods/pathways such as consumption of contaminated vegetables, fruits, fish, meat, water, and milk [40, 41, 72, 73]. Moreover, there may be the other sources such as dust inhalation and dermal contact [74, 75].

4. CONCLUSION

The present study carried out on paddy fields near Kalpakkam in Tamil Nadu, South India, determined the accumulation of essential and nonessential heavy metals in paddy soils as well as in rice plants including paddy grains. Theoccurrence of heavy metals in paddy field soils was in a ranking order of Mn > Zn > Pb > Cr > Cu > Cd. Concentrations of the heavy metals were higher in paddy field soils compared with the control soil. However, the concentrations of Pb, Cd, Cu, Cr, and Zn except for Mn in the paddy soils were comparable to those of worldwide normal soils, which were higher than the value of uncontaminated soil. The uptake of Mn and Zn was higher in the roots of paddy plants, which were followed by Pb, Cr, Cu, and Cd. Mn and Cd accumulated more in the shoots than in roots and grains. Pb content in grains exceeded the maximum permissible value in S-4 and all other metals were below the safe limits. Estimations showed that DIs of heavy metals for adults were found to be higher than those for children, which was most probably due to relatively higher quantity intake by adults. In general, no HRI values were >1 through rice consumption except for Zn for adults. The HRI values for both adults and children were 1.561 and 1.360, respectively, indicating that both adults and children may experience some adverse health effects in the future, since chemical fertilizers and pesticides are indiscriminately used by Indian farmers, which are probably the main sources of the toxic heavy metals accumulated in the paddy fields. Organic agriculture with little use of agrochemicals could be the alternative solution for reducing the contamination of toxic heavy metals particularly the toxic Cd, Cr, and Pb in the paddy fields producing rice, the staple food in India and other Southeast Asian countries.

CONFLICT OF INTERESTS

The authors declare that there is no conflict of interests regarding the publication of this paper.

ACKNOWLEDGMENTS

The present research was funded through a major Research Project (MRP/F., no. 40-137/2011 (SR) dated 04-07-2011), funded by University Grants Commission (UGC) (New Delhi). Ms. Deepmala Satpathy received a project research fellowship from the MRP.

REFERENCES

1. Y.-J. Cui, Y.-G. Zhu, R.-H. Zhai et al., "Transfer of metals from soil to vegetables in an area near a smelter in Nanning, China," Environment International, vol. 30, no. 6, pp. 785–791, 2004.

2. P. K. Govil, G. L. N. Reddy, and A. K. Krishna, "Contamination of soil due to heavy metals in the Patancheru industrial development area, Andhra Pradesh, India," Environmental Geology, vol. 41, no. 3-4, pp. 461–469, 2002.

3. G. R. Bhagure and S. R. Mirgane, "Heavy metal concentrations in groundwaters and soils of Thane Region of Maharashtra, India," Environmental Monitoring and Assessment, vol. 173, no. 1–4, pp. 643–652, 2011.

4. S. C. Wong, X. D. Li, G. Zhang, S. H. Qi, and Y. S. Min, "Heavy metals in agricultural soils of the Pearl River Delta, South China," Environmental Pollution, vol. 119, no. 1, pp. 33–44, 2002.

5. K. Fytianos, G. Katsianis, P. Triantafyllou, and G. Zachariadis, "Accumulation of heavy metals in vegetables grown in an industrial

area in relation to soil," Bulletin of Environmental Contamination and Toxicology, vol. 67, no. 3, pp. 423–430, 2001.

6. J. Liu, C.-Q. Duan, Y.-N. Zhu, X.-H. Zhang, and C.-X. Wang, "Effect of chemical fertilizers on the fractionation of Cu, Cr and Ni in contaminated soil," Environmental Geology, vol. 52, no. 8, pp. 1601–1606, 2007.

7. X. Hang, H. Wang, J. Zhou, C. Ma, C. Du, and X. Chen, "Risk assessment of potentially toxic element pollution in soils and rice (Oryza sativa) in a typical area of the Yangtze River Delta," Environmental Pollution, vol. 157, no. 8-9, pp. 2542–2549, 2009.

8. R. L. Chaney, J. S. Angle, M. S. McIntosh et al., "Using hyperaccumulator plants to phytoextract soil Ni and Cd," Zeitschrift fur Naturforschung C, vol. 60, no. 3-4, pp. 190–198, 2005.

9. S. Shimbo, Z.-W. Zhang, T. Watanabe et al., "Cadmium and lead contents in rice and other cereal products in Japan in 1998-2000," Science of the Total Environment, vol. 281, no. 1–3, pp. 165–175, 2001.

10. M.-K. Zhang and Z.-X. Ke, "Heavy metals, phosphorus and some other elements in urban soils of Hangzhou City, China," Pedosphere, vol. 14, no. 2, pp. 177–185, 2004.

11. C. Fangmin, Z. Ningchun, X. Haiming et al., "Cadmium and lead contamination in japonica rice grains and its variation among the different locations in southeast China," Science of the Total Environment, vol. 359, no. 1–3, pp. 156–166, 2006.

12. T. Tsukahara, T. Ezaki, J. Moriguchi et al., "Rice as the most influential source of cadmium intake among general Japanese population," Science of the Total Environment, vol. 305, no. 1–3, pp. 41–51, 2003.

13. H. Kurz, "Selection of cultivars to reduce the concentration of cadmium and thallium in food and fodder plants," Journal of Plant Nutrition and Soil Science, vol. 162, no. 3, pp. 323–328, 1999.

14. T. Arao and N. Ae, "Genotypic variations in cadmium levels of rice grain," Soil Science and Plant Nutrition, vol. 49, no. 4, pp. 473–479, 2003.

15. K. Q. Li, J. G. Liu, and X. L. Lu, "Uptake and distribution of cadmium in different rice cultivars," Agro-Environmental Science, vol. 22, pp. 529–532, 2003 (Chinese).

16. Z. W. Li, Y. L. Zhang, G. X. Pan, J. M. Li, X. M. Huang, and J. F. Wang, "Grain contents of Cd, Cu and Se by 57 rice cultivars and the risk significance for human dietary uptake," Environmental Science, vol. 24, pp. 112–115, 2003 (Chinese).

17. J. G. Liu, K. Q. Li, J. K. Xu, J. S. Liang, X. L. Lu, and J. L. Yang, "Interaction of Cd and five mineral nutrients for uptake and accumulation in different rice cultivars and genotypes," Field Crops Research, vol. 83, no. 3, pp. 271–281, 2003.

18. J. G. Liu, J. S. Liang, K. Q. Li et al., "Correlations between cadmium and mineral nutrients in absorption and accumulation in various genotypes of rice under cadmium stress," Chemosphere, vol. 52, no. 9, pp. 1467–1473, 2003.

19. J. G. Liu, Q. S. Zhu, Z. J. Zhang, J. K. Xu, J. C. Yang, and M. H. Wrong, "Variations in cadmium accumulation among rice cultivars and types and the selection of cultivars for reducing cadmium in the diet," Journal of the Science of Food and Agriculture, vol. 85, no. 1, pp. 147–153, 2005.

20. H. Yu, J. Wang, W. Fang, J. Yuan, and Z. Yang, "Cadmium accumulation in different rice cultivars and screening for pollution-safe cultivars of rice," Science of the Total Environment, vol. 370, no. 2-3, pp. 302–309, 2006.

21. B. J. Alloway, "Soil factors associated with zinc deficiency in crops and humans," Environmental Geochemistry and Health, vol. 31, no. 5, pp. 537–548, 2009.

22. T. Arao, S. Ishikawa, M. Murakami, K. Abe, Y. Maejima, and T. Makino, "Heavy metal contamination of agricultural soil and countermeasures in Japan," Paddy and Water Environment, vol. 8, no. 3, pp. 247–257, 2010.

23. "Deficiency in soil mars farming, Green Revolution-II. Orissa Post, (Bhubaneswar)," November 2011.

24. M. Karatas, S. Dursun, E. Guler, C. Ozdemir, and M. Emin Argun, "Heavy metal accumulation in wheat plants irrigated by waste water," Cellulose Chemistry and Technology, vol. 40, no. 7, pp. 575–579, 2007.

25. R. Singh, D. P. Singh, N. Kumar, S. K. Bhargava, and S. C. Barman, "Accumulation and translocation of heavy metals in soil and plants from fly ash contaminated area," Journal of Environmental Biology, vol. 31, no. 4, pp. 421–430, 2010.

26. R. Masironi, S. R. Koirtyohann, and J. O. Pierce, "Zinc, copper, cadmium and chromium in polished and unpolished rice," Science of the Total Environment, vol. 7, no. 1, pp. 27–43, 1977.

27. S. Suzuki, N. Djuangshi, K. Hyodo, and O. Soemarwoto, "Cadmium, copper, and zinc in rice produced in Java," Archives of Environmental Contamination and Toxicology, vol. 9, no. 4, pp. 437–449, 1980.

28. S. Suzuki and S. Iwao, "Cadmium, copper, and zinc levels in the rice and rice field soil of Houston, Texas,"Biological Trace Element Research, vol. 4, no. 1, pp. 21–28, 1982.

29. I. F. Rivai, H. Koyama, and S. Suzuki, "Cadmium content in rice and its daily intake in various countries,"Bulletin of Environmental Contamination and Toxicology, vol. 44, no. 6, pp. 910–916, 1990.

30. M. S. Fazeli, F. Khosravan, M. Hossini, S. Sathyanarayan, and P. N. Satish, "Enrichment of heavy metals in paddy crops irrigated by paper mill effluents near Nanjangud, Mysore district, Karnatake, India,"Environmental Geology, vol. 34, no. 4, pp. 297–302, 1998.

31. N. Herawati, S. Suzuki, K. Hayashi, I. F. Rivai, and H. Koyama, "Cadmium, copper, and zinc levels in rice and soil of Japan, Indonesia, and China by soil type," Bulletin of Environmental Contamination and Toxicology, vol. 64, no. 1, pp. 33–39, 2000.

32. M. V. Reddy, D. Satpathy, and K. S. Dhiviya, "Assessment of heavy metals (CD and Pb) and micronutrirnts (Cu, Mn, and Zn) of paddy (Oryza sativa L.) field surface soil and water in a predominantly paddy-cultivated area at Puducherry (Pondicherry, India), and effects of the agricultural runoff on the elemental concentrations of a receiving rivulet," Environmental Monitoring and Assessment, vol. 185, pp. 6693–6704, 2013.

33. S. E. Allen, H. M. Grimshaw, and A. P. Rowland, "Chemical analysis," in Methods in Plant Ecology, P. D. Moore and S. B. Chapman, Eds., pp. 285–344, Blackwell, London, UK, 1986.

34. H. Liu, A. Probst, and B. Liao, "Metal contamination of soils and crops affected by the Chenzhou lead/zinc mine spill (Hunan, China)," Science of the Total Environment, vol. 339, no. 1–3, pp. 153–166, 2005.

35. S. C. Barman, R. K. Sahu, S. K. Bhargava, and C. Chaterjee, "Distribution of heavy metals in wheat, mustard, and weed grown in field irrigated with industrial effluents," Bulletin of Environmental Contamination and Toxicology, vol. 64, no. 4, pp. 489–496, 2000.

36. S. Gupta, S. Nayek, R. N. Saha, and S. Satpati, "Assessment of heavy metal accumulation in macrophyte, agricultural soil, and crop plants adjacent to discharge zone of sponge iron factory," Environmental Geology, vol. 55, no. 4, pp. 731–739, 2008.

37. G. C. Kisku, S. C. Barman, and S. K. Bhargava, "Contamination of soil and plants with potentially toxic elements irrigated with mixed industrial effluent and its impact on the environment," Water, Air, and Soil Pollution, vol. 120, no. 1-2, pp. 121–137, 2000.

38. US Environmental Protection Agency (USEPA), Exposure Factors Handbook-General Factors, EPA/600/P-95/002Fa, vol. I, Office of Research and Development, National Center for Environmental Assessment, Washington, DC, USA, 1997.

39. US Environmental Protection Agency (USEPA), Region 9, Preliminary Remediation Goals, USEPA, Washington, DC, USA, 2002.

40. N. Zheng, Q. Wang, X. Zhang, D. Zheng, Z. Zhang, and S. Zhang, "Population health risk due to dietary intake of heavy metals in the industrial area of Huludao city, China," Science of the Total Environment, vol. 387, no. 1–3, pp. 96–104, 2007.

41. X. Wang, T. Sato, B. Xing, and S. Tao, "Health risks of heavy metals to the general public in Tianjin, China via consumption of vegetables and fish," Science of the Total Environment, vol. 350, no. 1–3, pp. 28–37, 2005.

42. S. Khan, Q. Cao, Y. M. Zheng, Y. Z. Huang, and Y. G. Zhu, "Health risks of heavy metals in contaminated soils and food crops irrigated with wastewater in Beijing, China," Environmental Pollution, vol. 152, no. 3, pp. 686–692, 2008.

43. R. M. Harrison and M. B. Chirgawi, "The assessment of air and soil as contributors of some trace metals to vegetable plants. III. Experiments with field-grown plants," Science of the Total Environment, vol. 83, no. 1-2, pp. 47–62, 1989.

44. L.-C. Chien, T.-C. Hung, K.-Y. Choang et al., "Daily intake of TBT, Cu, Zn, Cd and As for fishermen in Taiwan," Science of the Total Environment, vol. 285, no. 1–3, pp. 177–185, 2002.

45. N. C. Brady and R. R. Weil, The Nature and Properties of Soil, Prentice-Hall, Upper Saddle River, NJ, USA, 13th edition, 2002.

46. A. Kabata-Pendias, Trace Elements in Soils and Plants, CRC Press, Boca Raton, Fla, USA, 2001.

47. S. K. Awashthi, Prevention of Food Adulteration Act No. 37 of 1954. Central and State Rules as Amended for 1999, Ashoka Law House, New Delhi, India, 2000.

48. B. J. Alloway, "The origins of heavy metals in soils," in Heavy Metals in Soils, B. J. Alloway, Ed., pp. 29–39, Chapman & Hall, London, UK, 1995.

49. J. Pichtel and C. A. Salt, "Vegetative growth and trace metal accumulation on metalliferous wastes,"Journal of Environmental Quality, vol. 27, no. 3, pp. 618–624, 1998.

50. S. C. Jarvis, L. H. P. Jones, and M. J. Hopper, "Cadmium uptake from solution by plants and its transport from roots to shoots," Plant and Soil, vol. 44, no. 1, pp. 179–191, 1976.

51. L. Leita, M. De Nobili, C. Mondini, and M. T. Baca Garcia, "Response of Leguminosae to cadmium exposure," Journal of Plant Nutrition, vol. 16, no. 10, pp. 2001–2012, 1993.

52. X.-E. Yang, X.-X. Long, W.-Z. Ni et al., "Assessing copper thresholds for phytotoxicity and potential dietary toxicity in selected vegetable crops," Journal of Environmental Science and Health B, vol. 37, no. 6, pp. 625–635, 2002.

53. R. M. Tripathi, R. Raghunath, and T. M. Krishnamoorthy, "Dietary intake of heavy metals in Bombay city, India," Science of the Total Environment, vol. 208, pp. 149–159, 1997.

54. W. G. Hopkins, Introduction to Plant Physiology, John Wiley & Sons, New York, NY, USA, 2nd edition, 1999.

55. L. Sanità Di Toppi and R. Gabbrielli, "Response to cadmium in higher plants," Environmental and Experimental Botany, vol. 41, no. 2, pp. 105–130, 1999.

56. M. S. Li, Y. P. Luo, and Z. Y. Su, "Heavy metal concentrations in soils and plant accumulation in a restored manganese mineland in Guangxi, South China," Environmental Pollution, vol. 147, no. 1, pp. 168–175, 2007.

57. A. K. Gupta and S. Sinha, "Phytoextraction capacity of the Chenopodium album L. grown on soil amended with tannery sludge," Bioresource Technology, vol. 98, no. 2, pp. 442–446, 2007.

58. L. Pilc, J. Rosada, and J. Siepak, "Heavy metals concentration in soil, water and plants in the emission region of copper foundry 'Głogów',"

Materiaty Sesji Naukowej Instytutu Ochrony Roslin, vol. 2, pp. 228–232, 1994.

59. Commission Regulation Directive EC Commission Regulation (EC), no. 466, Official Journal of the European Communities, 2001.

60. FAO/WHO Cadmium (EHC no. 134), Geneva, Switzerland, 1992.

61. CODEX Alimentarius commission, joint FAO/WHO food standards programme; Report of the thirty ninth session international conference centre, CODEX, Geneva, Switzerland, 2006.

62. European Food Safety Authority, EFSA, "Scientific opinion-Cadmium in food,scientific opinion of the panel on contaminants in the food chain," EFSA Journal, vol. 980, pp. 1–139.

63. P. Zhuang, M. B. McBride, H. Xia, N. Li, and Z. Li, "Health risk from heavy metals via consumption of food crops in the vicinity of Dabaoshan mine, South China," Science of the Total Environment, vol. 407, no. 5, pp. 1551–1561, 2009.

64. J. R. Dean, Bioavailability, Bioaccessibility and Mobility of Environmental Contaminants, John Wiley & Sons, London, UK, 1st edition, 2007.

65. Z.-T. Xiong, "Lead uptake and effects on seed germination and plant growth in a Pb hyperaccumulator Brassica pekinensis Rupr," Bulletin of Environmental Contamination and Toxicology, vol. 60, no. 2, pp. 285–291, 1998.

66. X. C. Wang, W. D. Yan, Z. An et al., "Status of trace elements in paddy soil and sediment in Taihu Lake region," Chemosphere, vol. 50, no. 6, pp. 707–710, 2003.

67. R. Chandrajith, C. B. Dissanayake, and H. J. Tobschall, "The abundances of rarer trace elements in paddy (rice) soils of Sri Lanka," Chemosphere, vol. 58, no. 10, pp. 1415–1420, 2005.

68. M. E. Essington, Soil and Water Chemistry: An Integrative Approach, CRC Press, Boca Raton, Fla, USA, 2004.

69. European Union, Heavy metals in wastes. European Commission on Environment, 2000, http://ec.europa.eu/environment/waste/studies/pdf/heavy_metalsreport.pdf.

70. R. Garcia and E. Millán, "Assessment of Cd, Pb and Zn contamination in roadside soils and grasses from Gipuzkoa (Spain)," Chemosphere, vol. 37, no. 8, pp. 1615–1625, 1998.

71. J. Liu, X.-H. Zhang, H. Tran, D.-Q. Wang, and Y.-N. Zhu, "Heavy metal contamination and risk assessment in water, paddy soil, and rice around an electroplating plant," Environmental Science and Pollution Research, vol. 18, no. 9, pp. 1623–1632, 2011.

72. N. Sridhara Chary, C. T. Kamala, and D. Samuel Suman Raj, "Assessing risk of heavy metals from consuming food grown on sewage irrigated soils and food chain transfer," Ecotoxicology and Environmental Safety, vol. 69, no. 3, pp. 513–524, 2008.

73. E. Sipter, E. Rózsa, K. Gruiz, E. Tátrai, and V. Morvai, "Site-specific risk assessment in contaminated vegetable gardens," Chemosphere, vol. 71, no. 7, pp. 1301–1307, 2008.

74. D. Grasmück and R. W. Scholz, "Risk perception of heavy metal soil contamination by high-exposed and low-exposed inhabitants: the role of knowledge and emotional concerns," Risk Analysis, vol. 25, no. 3, pp. 611–622, 2005.

75. L. Hellström, B. Persson, L. Brudin, K. P. Grawé, I. Öborn, and L. Järup, "Cadmium exposure pathways in a population living near a battery plant," Science of the Total Environment, vol. 373, no. 2-3, pp. 447–455, 2007.

Chapter 6

A Simple Evaluation of Soil Quality of Waterlogged Purple Paddy Soils with Different Productivities

Zhanjun Liu1,[2], Wei Zhou[1] *, Jialong Lv[2] , Ping He[1] , Guoqing Liang[1] *, Hui Jin[3]

[1] *Ministry of Agriculture Key Laboratory of Crop Nutrition and Fertilization, Institute of Agricultural Resources and Regional Planning, Chinese Academy of Agricultural Sciences, Beijing 100081, China, 2 College of Resource and Environmental Science, Northwestern University of A & F, Yangling 712100, China, 3 Institute of Agricultural Resources and Economy, Shanxi Academy of Agricultural Sciences, Taiyuan, 030006, China*

ABSTRACT

Evaluation of soil quality can be crucial for designing efficient farming systems and ensuring sustainable agriculture. The present study aimed at evaluating the quality of waterlogged purple paddy soils with different productivities in Sichuan Basin. The approach involved comprehensive analyses of soil physical and chemical properties, as well as enzyme activities and microbial community structure measured by phospholipid fatty acid analysis (PLFA). A total of 36 soil samples were collected from four typical locations, with 12 samples representing high productivity purple paddy soil (HPPS), medium productivity purple paddy soil (MPPS) and low productivity purple paddy soil (LPPS), respectively. Most measured soil properties showed significant differences ($P \le 0.05$) among HPPS, MPPS and LPPS. Pearson correlation analysis and principal component analysis were used to identify appropriate soil quality indicators. A minimum data set

(MDS) including total nitrogen (TN), available phosphorus (AP), acid phosphatase (ACP), total bacteria (TB) and arbuscular mycorrhizal fungi was established and accounted for 82.1% of the quality variation among soils. A soil quality index (SQI) was developed based on the MDS method, whilst HPPS, MPPS and LPPS received mean SQI scores of 0.725, 0.536 and 0.425, respectively, with a ranking of HPPS > MPPS > LPPS. HPPS showed relatively good soil quality characterized by optimal nutrient availability, enzymatic and microbial activities, but the opposite was true of LPPS. Low levels of TN, AP and soil microbial activities were considered to be the major constraints limiting the productivity in LPPS. All soil samples collected were rich in available N, K, Si and Zn, but deficient in available P, which may be the major constraint for the studied regions. Managers in our study area should employ more appropriate management in the LPPS to improve its rice productivity, and particularly to any potential limiting factor.

INTRODUCTION

Waterlogged paddy soil is a common soil type associated with low productivity, and occupies a large area putting constraints on crop production in southwest China. Purple soil (FAO, Regosol) is widely distributed in this area, where it is susceptible to soil erosion and nutrient loss due to terrain disturbance by humans [1]. Therefore, numerous previous studies mainly focused on nutrient loss and its associated factors [2, 3]. An improved knowledge of soil quality assessment is important for developing appropriate soil anti-degradation measures and design management plans [4]. This is particularly important as little is known about soil quality status of waterlogged paddy soils, particularly in the purple soil region.

Soil quality has received more attention recently in response to an increasing public interest in sustainability and soil resources [5]. However, there is no established methodology so far to characterize soil quality based on a universal set of indicators [6]. Selecting representative variables is critical for soil quality evaluation [7], which should combine soil physical, chemical and biological properties [8]. At present, a universal set of the main indicators to evaluate soil quality is still the subject of debate [9]. While numerous studies on soil quality evaluation have concentrated on soil physical and chemical properties [10,11], biological variables can also be important and have been increasingly used as indicators of soil quality owing to their rapid response and high sensitivity to changes in soil management [12,13]. These studies illustrate the effects of specific agricultural management practices on soil quality. However, little information is available in the literature that evaluates soil quality aiming at any special soil type in terms of productivity on a regional scale [4,14,15], particularly to the purple paddy soil. In this way, a comprehensive assessment of soil quality integrating soil physical, chemical and biological properties would be greatly desired.

An accurate evaluation of soil quality requires analyzing a large number of soil parameters [16]. Usually, a minimum data set (MDS) was established based on a careful selection [12,17], and then a soil quality index (SQI) was also increasingly developed to quantify and compare the quality status of soils related to different management practices [11,15]. However, many previous studies typically did not provide crop-yield data [2,18,19] or looked further into the relationship between SQI and yield [10,17], thus making their results likely have little biological significance.

In our study, we examined soil physical, chemical and biological properties with a goal to provide a comprehensive assessment of soil quality for different productive waterlogged paddy soils in the acidic purple soil region. Our objectives were to: (*i*) establish an MDS for soil quality evaluation; (*ii*) develop a SQI to quantify soil quality status; and (*iii*) identify the limiting factors associated with the crop productivity of waterlogged acidic purple paddy soils.

MATERIALS AND METHODS

Study area

The measurements were conducted only on the most representative areas of waterlogged purple paddy soils. Four typical locations, Tongliang (106°04' E, 29°46' N), Beibei (106°23' E, 29°54' N), Luxian (105°23' E, 29°11' N) and Tianxing (105°19' E, 29°06' N) were selected accounting for the optimal analytical costs. These four areas are situated in Sichuan Basin, southwest China. The selected paddy fields are private lands, and the study was carried out following the permission of each land ower. A subtropical monsoon climate prevails in those regions, with a mean annual temperature and mean annual precipitation of 17.0°C and 1000 mm, respectively. Precipitation varies in each month, and 70% occurs from May to September. The soil type of the study area is the acidic purple paddy soil, and land use is dominantly arable with a single-harvest rice per year. The parent material is purple sandstone, and the mineral compositions include vast hydrous mica and slight vermiculite.

According to the mean annual rice yield over the past five years, the selected paddy fields were divided into three classes with high (> 7500 kg ha^{-1}, HPPS), medium (6000–7500 kg ha^{-1}, MPPS) and low (<6000 kg ha^{-1}, LPPS) productivity. Based on farmers' surveys, conventional fertilization focused on mineral fertilizers, and fertilizer types were $CO(NH_2)_2$, $Ca(H_2PO_4)_2$ and KCl for N, P, and

K, respectively. Similar fertilization was employed in HPPS and MPPS, with the mean rates of 155 kg N ha^{-1} (2:1 basal: topdressing), 60 kg P_2O_5 ha^{-1} (100% basal) and 80 kg K_2O ha^{-1} (1:1 basal: opdressing), respectively, and mean rates of 125 kg N ha^{-1} (2:1 basal: topdressing), 60 kg P_2O_5 ha^{-1}(100% basal) and 80 kg K_2O ha^{-1} (1:1 basal: topdressing) were applied to the LPPS. One tillage operation was conducted before the cropping season, and the sampling areas were considered typical waterlogged purple paddy soil regions because of their concentrated distributions, similar weather conditions, cropping systems, agricultural management (i.e. fertilization, tillage regime) and productivity levels.

Soil sampling

Based on the distribution of waterlogged purple paddy soils and their productivities, 36 waterlogged paddy fields were selected as sampling sites, and their geographical locations are shown in Fig 1. In each selected waterlogged paddy field, ten cores (5.0 cm diameter) were collected randomly from a soil depth of 0–15 cm (plow layer) and well-mixed to form a composite sample during jointing stage in May 2012. In each typical location, three composite soil samples were collected to represent each productivity class of waterlogged paddy soil. Those samples were immediately transported to the laboratory, and one sub-sample was air-dried at room temperature for physical and chemical analysis, one sub-sample was stored at 4°C for biochemical analysis and the last sub-sample was freeze-dried prior to being stored at −18°C for phospholipid fatty acids (PLFA) analysis.

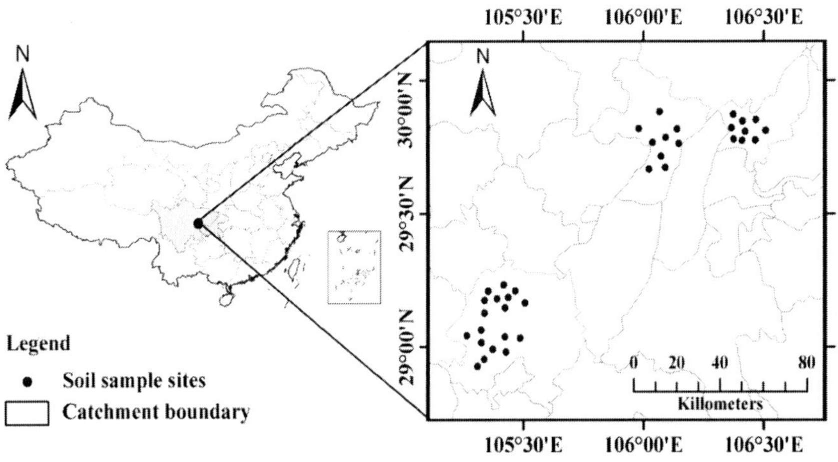

Figure 1. Geographic locations of soil sampling sites.

Soil analysis

Physico-chemical properties

Soil aggregate stability (SAS) was measured using the wet sieving method [20]. Soil texture, bulk density (BD), total nitrogen (TN), available nitrogen (AN), available phosphorus (AP), soil organic matter (SOM), pH (soil/water 1:2.5), available potassium (AK), available zinc (AZn, 0.01 M hydrochloric acid extraction), available silicon (ASi, 0.025 M citric acid extraction) and cation exchange capacity (CEC) were determined using the methods described by Page et al. (1982) [21].

Enzyme activity

The activities of β-glucosidase, acid phosphatase and peroxidase were measured according to DeForest (2009) [22], and arylsulfatase activity was essayed as described by Giacometti et al. (2014) [23], whilst urease and dehydrogenase activities were determined based on Alef and Nannipieri (1995) [24].

Microbial community structure

Microbial community structure was determined using PLFA analysis as described by Wu et al. (2009) [25]. Concentrations of PLFAs were calculated as nmol g^{-1} and mol%. PLFA bio-indicators were selected according to the procedure of Moeskops et al. (2010) [26], using only the PLFAs clearly identified by GC-FID: fatty acids iC15:0, aC15:0, iC16:0, iC17:0 and aC17:0 were used as biomarkers for Gram-positive bacteria (G+) and C16:1ω7c, C18:1ω7c and cyC17:0 for Gram-negative bacteria (G−). The sum of G+, G−, C:15, C17:0 and cyC19:0ω11,12c was assumed to represent the total bacterial community, and the sum of 10MeC16:0 and 10MeC18:0 was regarded as an indicator of actinomycetes. The fatty acids C18:2ω6,9c and C16:1ω5c were used as biomarkers for fungi and arbuscular mycorrhizal fungi (AMF), respectively.

Soil quality evaluation method

Optimizing selection of a minimum data set (MDS)

According to the indicator selection method described by Li et al. (2013) [17], those soil properties that did not show a significant Pearson correlation ($P < 0.05$) with rice yields were eliminated from the data list. A principal component analysis (PCA) was conducted with the remaining soil variables to select the most appropriate indicators. Only principal components (PCs) with eigenvalues of ≥1 and which explained at least 5% of the variation in the data were considered [11,27]. Within each principal component (PC) only highly weighted factors (i.e. with absolute values within 10% of the largest factor loading) were retained for the MDS [28]. When more than one variable was retained in a PC, each was considered important and was considered in the MDS if the two considered values were not correlated ($r < 0.60$) [18]. Among well-correlated variables within a PC, the variable with the highest correlation sum was selected for the MDS [28].

Weight assignment and indicator scoring

MDS variables were included to another PCA, and the weight for each indicator was calculated by its communality and was equal to the ratio of its communality with the sum of communalities of all MDS indicators [29].

Because of different indicator units, a standard scoring function (SSF) was used to score all MDS variables using each indicator method, and the detailed scoring functions were the same as those used by Qi et al. (2009) [14]. For the MDS variables without a certain threshold value (the optimum), every observation was normalized using a linear scoring function [30]. For "higher is better" indicators, each observation was divided by the highest observed value and thus the highest observed value received a score of 1. For "lower is better" indicators, the lowest observed value (in the numerator) was divided by each value (denominator) and thus the lowest observed value received a score of 1. For those indicators which were neither higher is better nor lower is better, the values were scored as "higher is better" up to a threshold value and then scored as "lower is better" above the threshold value [10,30].

Developing an SQI

After the MDS indicators were weighted and scored, the SQI was calculated using the following equation described by Doran and Parkin [31]:

$$SQI = \sum_{i=1}^{n} Wi \times Si \quad (1)$$

where Wi is the assigned weight of each indicator, Si is the indicator score, and n is the number of variables in the final MDS.

Data analysis

Data were subjected to statistical analysis using SPSS 18.0 (IBM Corporation). One-way analysis of variance (ANOVA) was used to test all the parameters, and the least significant difference (LSD) method at the probability level of 0.05 was used to separate mean difference of the soil attributes.

RESULTS AND DISCUSSION

Soil physico-chemical attributes

All measured soil physical and chemical properties significantly differed between HPPS, MPPS and LPPS with an exception of available N (Table 1). Relatively lower bulk density (BD) was observed in HPPS, while LPPS had significantly higher BD that may adversely affect soil nutrient circulation and rice root growth [4]. Rice husk biochar may be particularly necessary in LPPS because of its beneficial amendment for improving soil poor physical characteristics such as lowering soil bulk density [11,32].

Table 1. Summary statistics of measured soil physical and chemical properties of high- (HPPS), medium- (MPPS) and low- (LPPS) productivity purple paddy soils (mean ± standard deviation and range of variation).

Soil parameters	HPPS (n = 12)		MPPS (n = 12)		LPPS (n = 12)	
	Mean	Range	Mean	Range	Mean	Range
Bulk density (g cm^{-3})	1.15±0.06 c	1.05–1.25	1.23±0.05 b	1.15–1.30	1.30±0.03 a	1.27–1.37
pH	4.73±0.41 b	4.28–5.50	4.94±0.25 b	4.59–5.44	5.46±0.48 a	4.64–5.95
Soil organic C (g kg^{-1})	27.3±3.06 a	23.4–33.3	20.7±4.52 b	16.6–29.3	20.6±3.09 b	16.9–26.2
Total N (g kg^{-1})	1.03±0.09 a	0.90–1.18	0.90±0.09 b	0.74–1.09	0.84±0.07 b	0.75–0.98
Available N (mg kg^{-1})	218±33.3 a	159–256	227±32.8 a	172–265	229±34.9 a	160–275
Available P (mg kg^{-1})	8.10±2.02 a	5.13–11.4	7.10±1.32 a	5.34–8.99	4.70±1.70 b	1.67–6.71
Available K (mg kg^{-1})	157±34.5 a	114–217	107±56.8 b	76.8–201	95.4±31.9 b	67.8–152
CEC (cmol kg^{-1})	12.9±3.39 c	8.28–18.8	16.0±3.72 b	11.3–21.7	20.4±1.30 a	18.0–21.9
Available Si (mg kg^{-1})	144±38.3 a	101–187	116±14.4 b	94.7–138	107±6.26 b	95.6–116
Available Zn (mg kg^{-1})	5.84±0.46 a	5.02–6.46	4.65±1.25 b	3.04–6.48	3.98±1.01 b	3.00–5.58

Means for the same property with different letters indicate significant differences at $p \leq 0.05$.

Soil pH values varied from 4.40 to 5.87, as typical for acidic purple paddy soils. Soil pH showed lower values in HPPS and MPPS compared to LPPS, which were lower by 0.7 and 0.5, respectively. This is probably due to the relatively greater application of inorganic N fertilizer in HPPS and MPPS. This is consistent with the conclusion that recent soil acidification in China has resulted mainly from high N fertilizer inputs [33]. Soil acidification can result in nutrient deficiency such as K, Na, Ca and Mg and thus decrease crop production [34]. Rice straw addition beneficial for pH improvement should be applied to mitigate soil acidification [35], and the widely used N fertilizer type of urea could be replaced by $(NH_4)_2SO_4$ because of its capacity of inhibiting the increase of ammonia-oxidizing bacteria and thus decreasing nitrification and soil acidification [36]. Soil organic carbon (SOC) plays a pervasive role in ensuring soil fertility and promoting soil ecological functions [37]. MPPS and LPPS had

SOC contents without any significant differences, in contrast to HPPS characterized by larger SOC. The mean SOC showed a remarkable increase compared to the value of 12 g kg^{-1} obtained by SSSC (1996) [38]. In the last twenty years, an increasing number of farmers have restored rice straw to paddy fields, thereby contributing to the rapid increase of SOC [39].

Purple paddy soil is characterized by low TN content, and only 0.75 g kg^{-1} was observed in SSSC (1996) [38]. Table 1 points to significant differences among HPPS, MPPS and LPPS, with their mean contents of TN typically higher than 0.75 g kg^{-1}, and which increased by 37.3%, 20.0% and 12.0%, respectively. This is inconsistent with the findings of Wang et al. (2009) [2], who concluded that the remarkable increase in TN may be attributable to the continued application of N fertilizer. In addition, soil available N (AN) varied from 159–275 mg kg^{-1}, which is substantially higher than the threshold value of 100 mg kg^{-1} [40], while no statistically significant differences were observed among HPPS, MPPS and LPPS. Our results indicate that the rate of basic N fertilizer may be overused and, therefore, the basal: topdressing ratio (2:1) needs adjustment to make N fertilizer application more timed to coincide with the period of rapid N uptake by plants.

No significant differences were observed for available P between HPPS and MPPS, but their mean contents were significantly higher than LPPS (i.e. by 72.3% and 51.1%, respectively). Although the total input of P fertilizer was applied as basal, mean contents of AP were all lower than the critical value of 10 mg P kg^{-1} [41], indicating a serious deficiency of AP. This may be related to the low soil temperature during the seedling stage that adversely affects P availability [42]. Therefore it seems that the deficiency in soil P may be one of the most important constraints limiting rice growth. It is recommended that straw mulch is carried out during winter fallow season to improve soil optimum temperature and thus increase P bioavailability [43]. He (2003) reported that the effect of K application on increasing the crop productivity in purple soil was not significant when soil AK was over 80 mg kg^{-1} [40]. Our whole study area was sufficient in AK because their mean contents were all higher than the critical threshold value. As suggested by Li (1991) the sufficient AK may be attributable to inheriting high K from the parent materials of purple soils [44].

CEC can be a vital predictor of soil quality [45] and has been assessed as an indicator of soil erosion and degradation [46]. Significant differences were found among the different productivity classes of waterlogged purple paddy soils, giving ranking of HPPS < MPPS < LPPS (Table 1). The result is inconsistent with the observation of Tesfahunegn et al. (2011) [4], who concluded that soils with high productivity are usually characterized by high CEC. Soil CEC is related to its parent material [46], which is why serious soil erosion has occurred in HPPS and MPPS probably due to their significant reduction compared to LPPS. On the other hand, our studied paddy fields are located in hillslopes, and the concentrated precipitation (from May to October) could lead to more severe erosion and nutrient loss for HPPS and MPPS beacuse of their relative high-elevations [2]. This supports the previous conclusion and accounts for the notable low levels of CEC in HPPS. Black carbon, which has been demonstrated to be beneficial in increasing CEC [47], seemed particularly necessary in HPPS.

As shown in Table 1, the studied paddy fields were sufficient in soil available silicon (ASi), because their concentrations were all higher than the critical value of 60 mg Si kg^{-1}[48]. Similarly, the whole study area presented abundance of available zinc (AZn) whose concentrations were always higher than the critical value of 1.0 mg Zn kg^{-1} [49], although the contents of AZn in MPPS and LPPS were typically lower than those in HPPS. This is consistent with the finding of Liu (1994) [50], who reported that AZn is not a limited nutrient in acidic purple soils. And soil AZn abundance might be attributable to their parent materials and frequent inputs coupled to the application of mineral fertilizers [51].

Enzyme activities

Enzyme activities play an essential role in nutrient cycling, changes in soil quality and the degree of soil degradation due to their sensitivities to small changes in soils [52]. In our study, six enzymes were measured to estimate the microbial activity (dehydrogenase, peroxidase), C cycle (β-Glucosidase), N cycle (urease), P cycle (acid phosphatase) and S cycle (arylsulphatase). In Table 2 summarized are the statistics of soil enzymatic activities used as soil quality indicators in all of the waterlogged purple paddy fields. Significant differences were observed among HPPS, MPPS and LPPS for all measured soil enzymes. Urease, dehydrogenase, β-glucosidase, arylsulphatase, acid phosphatase and peroxidase were significantly lower in LPPS than in HPPS (i.e. by 42.0%, 25.9%, 42.6%, 49.1%, 26.6% and 30.3%, respectively). Urease and β-glucosidase were lower by 27.1% and 28.8% in LPPS compared to the MPPS, respectively; but no significant difference was found between LPPS and MPPS with respect to dehydrogenase, acid phosphatase, arylsulphatase and peroxidase. The amount of dehydrogenase and acid phosphatase in MPPS was not significantly different than in HPPS, while the activities of urease, β-glucosidase, arylsulphatase and peroxidase were typically lower in MPPS than HPPS (i.e. by 20.2%, 19.5%, 28.1% and 27.1%, respectively). Overall, HPPS presented high enzymatic activities, indicating a substantial capacity of nutrient supply and a better soil quality. Opposite was true for LPPS where application of fly ash [53] or organic amendment [54] could be beneficial to maintaining nutrient balance and increasing soil enzyme activities.

Table 2. Summary statistics for measured enzymes of high- (HPPS), medium- (MPPS) and low- (LPPS) productivity purple paddy soils (mean ± standard deviation and range of variation).

Soil parameters	HPPS (n = 12)		MPPS (n = 12)		LPPS (n = 12)	
	Mean	Range	Mean	Range	Mean	Range
Urease (mg NH_4^+-N kg^{-1} h^{-1})	20.8±2.95 a	16.8–27.2	16.6±4.49 b	11.7–27.1	12.1±2.41 c	7.49–15.5
Dehydrogenase (μg TPF g^{-1})	148±52.6 a	73.9–241	126±38.3 ab	75.2–180	110±37.6 b	45.4–159
β-glucosidase (nmol g^{-1} h^{-1})	128±15.3 a	102–149	103±32.7 b	51.4–148	73.3±24.8 c	32.2–106
Arylsulfatase (nmol g^{-1} h^{-1})	7.48±2.88 a	4.02–11.2	5.38±2.24 b	2.25–9.08	3.81±1.01 b	2.58–5.92
Acid phosphatase (nmol g^{-1} h^{-1})	521±103 a	385–672	433±156 ab	172–585	382±136 b	207–534
Peroxidase (nmol g^{-1} h^{-1})	8.82±1.28 a	6.34–10.6	6.43±2.18 b	2.67–10.0	6.15±2.43 b	2.80–12.1

Means for the same property with different letters indicate significant differences at $p < 0.05$.

Microbial community structure

PLFA analysis showed its total concentrations ranging from 19.7 to 52.6 nmol g^{-1}, and the selected microbial groups all presented statistically significant differences among HPPS, MPPS, and LPPS (Table 3). Compared to the HPPS which was chosen as a benchmark for soil quality, the concentrations of total PLFAs, Gram-positive bacteria, Gram-negative bacteria, total bacteria, actinomycetes, fungi and AMF were significantly lower in LPPS (i.e. by 36.0%, 30.3%, 45.2%, 37.1%, 32.9%, 37.8% and 46.6%, respectively); and were also significantly lower in MPPS (i.e. by 35.6%, 28.1%, 43.6%, 34.7%, 38.3%, 21.9% and 32.9%, respectively). With the exception of fungi, MPPS contained levels of total PLFAs and the selected marker PLFAs similar to those of LPPS.

Table 3. Concentrations of total and marker PLFAs (mean±standard deviation and range of variation, nmol g^{-1} dry soil).

Soil parameters	HPPS (n = 12)		MPPS (n = 12)		LPPS (n = 12)	
	Mean	Range	Mean	Range	Mean	Range
Total PLFAs	38.1±8.13 a	26.7–52.6	24.6±3.96 b	19.7–31.9	24.4±3.23 b	20.9–31.3
Gram-positive	7.52±1.63 a	4.81–9.92	5.41±0.82 b	4.00–6.75	5.25±0.71 b	4.24–6.59
Gram-negative	7.66±1.61 a	5.87–10.8	4.32±0.68 b	3.15–5.30	4.20±0.57 b	3.24–5.09
Total bacteria	16.2±1.41 a	14.3–18.3	10.5±1.69 b	7.60–12.7	10.2±1.13 b	8.41–12.4
Actinomycetes	3.79±1.40 a	1.86–5.93	2.34±0.31 b	1.92–2.82	2.54±0.72 b	1.57–3.70
Fungi	1.12±0.10 a	0.94–1.29	0.87±0.22 b	0.52–1.21	0.70±0.15 c	0.48–0.89
AMF	0.96±0.30 a	0.73–1.54	0.64±0.14 b	0.47–0.90	0.51±0.11 b	0.37–0.72

Means for the same property with different letters indicate significant differences at $p < 0.05$.

Canonical analysis of the PLFA data indicated that the microbial community structures of each productivity class of waterlogged purple paddy soil showed only small differences among the four typical locations, and HPPS were well separated from MPPS and LPPS (Fig 2A). The PC1 and PC2 accounted for 65.0% and 11.6% of the total variation, respectively, whereas the PC loadings of each PLFA showed that almost all PLFA bioindicators were characterized by high concentrations in HPPS (Fig 2B).

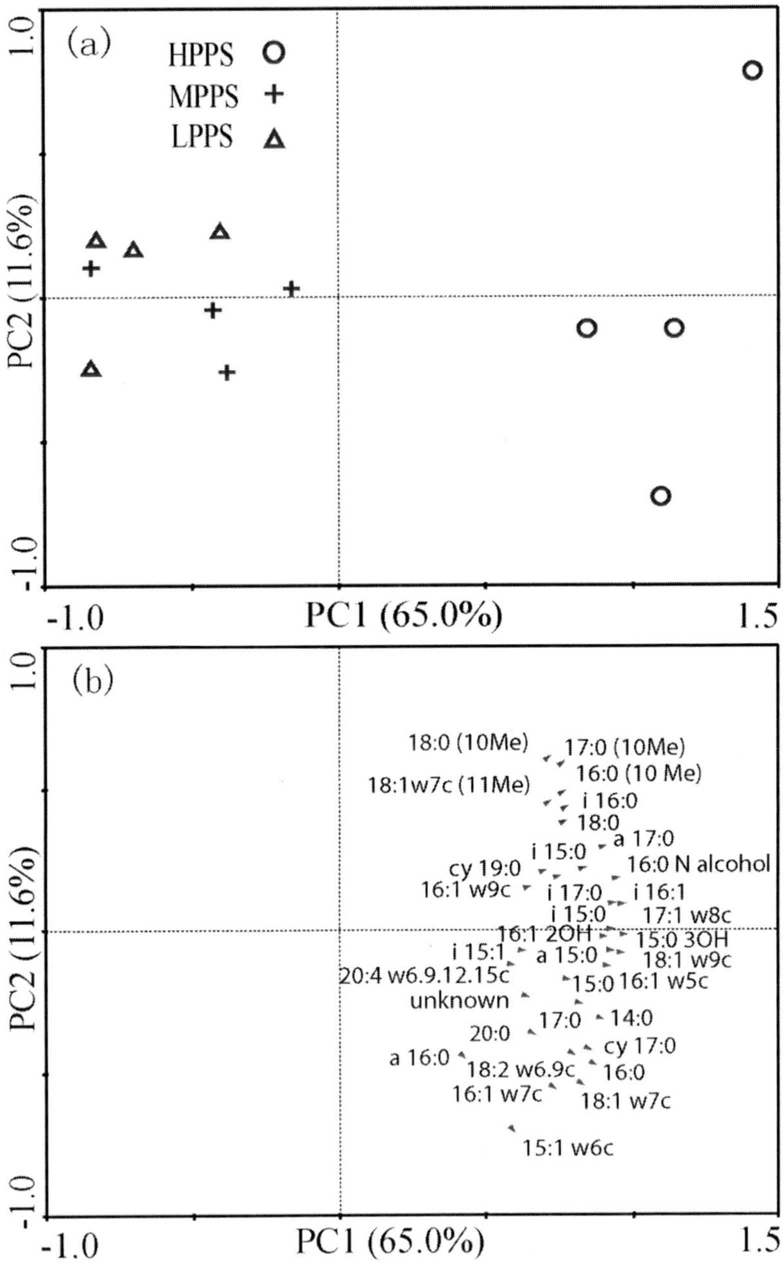

Figure 2. Plot of the first two principal components (PC1 and PC2) grouped in HPPS, MPPS and LPPS (a) and plot of the PC1 and PC2 for 35 PLFAs (b).

Microbial indicators can serve as early and sensitive markers of soil degradation or improvement [55], and also can indicate the degree of soil fertility and the quality of agricultural management [56]. MPPS and LPPS characterized by relatively low PLFA concentrations were associated with poor soil fertility and quality. In this way, manure amendment could be carried out because of its beneficial to improve soil biofertility and quality in purple soil [57].

Correlation analysis

Correlation analysis was conducted with 24 parameters including soil properties and rice yield (Table 4). Rice yield was significantly correlated with all measured soil parameters except for AN, AK and dehydrogenase. The high levels of AN and its non-significant correlation with rice yield suggested that AN is not the factor limiting crop production. The significant correlations between selected soil microbial communities and rice yield as well as their relatively high correlation coefficients support the conclusion established by Lima et al. (2013) [12], who reported that biological indicators are the most sensitive in indicating differences in soil quality under rice production systems. SOC and TN were considered as the two most important soil chemical parameters, because they markedly affected soil enzymes (urease, dehydrogenase, β-glucosidase, arylsulphatase and peroxidase), biological properties (total PLFA, gram-positive, gram-negative, fungi and AMF) and several other chemical properties (CEC, AZn). The results followed the observations of Bonanomi et al. (2014) [54], who found that high enzyme activities were mainly attributable to the relatively higher SOC and next to energy sources and nutrients that sustain growth and activity of soil microbes.

As presented in Table 4, total PLFAs was significantly correlated with most soil enzymes, and similar findings were also reported by Giacometti et al (2013) [9]. In addition, soil pH strongly affects soil enzyme activities (i.e. urease, β-glucosidase, arylsulphatase, acid phosphatase), and negative correlations were observed between pH and PLFA bioindicators. This is consistent with the findings of Zhou et al. (2014) [58], who concluded that soil pH might have an essential role in controlling soil microbial structure in purple soil. In addition, soil CEC was signficantly and positively correlated with pH, and the low CEC in HPPS may attribute to its soil acidification [34]. The negative correlations between rice yield and pH/CEC suggested that low levels of pH and CEC may be potential constrants limiting the productivity in HPPS.

Table 4. Correlation coefficients among rice yield and soil physico-chemical and biological properties.

	Yield	BD	pH	SOC	TN	AN	AP	AK	CEC	ASi	AZn	Urease	DH	BG	ARS	ACP	POD	PLFAs	G+	G-	TB	ACT	Fungi	AMF
Yield	1.00																							
BD	-0.83	1.00																						
pH	-0.60	0.53	1.00																					
SOC	0.47	-	-	1.00																				
TN	0.63	-0.55	-0.47	0.75	1.00																			
AN	-	-	-	-	-	1.00																		
AP	0.56	-0.39	-	0.34	0.36	-	1.00																	
AK	-	-	-	-	-	-	-	1.00																
CEC	-0.65	-	0.59	-0.74	-0.59	-	-0.50	-	1.00															
ASi	0.49	-0.38	-	-	-	-	0.45	-	-	1.00														
AZn	0.59	-0.59	-	0.62	0.72	-	0.61	-	-0.36	-	1.00													
Urease	0.69	-0.67	-0.48	0.39	0.48	-	0.63	-	-0.48	0.48	0.52	1.00												
DH	-	-	-	0.53	0.34	-	0.49	-	-0.64	0.42	-	-	1.00											
BG	0.65	-0.49	-0.69	0.64	0.61	-	-	-0.35	-0.72	0.37	0.39	0.42	0.44	1.00										
ARS	0.46	-	-0.48	0.75	0.53	-	0.52	-	-0.85	0.34	0.46	0.40	0.82	0.64	1.00									
ACP	0.46	-0.48	-0.34	-	-	-	-	0.44	-	0.35	-	-	-0.35	-	-	1.00								
POD	0.39	-	-	0.64	0.49	-	-	-	-0.56	-	-	0.33	0.55	0.55	0.61	-	1.00							
PLFAs	0.69	-0.68	-0.51	-	0.52	-	-	-	-0.37	0.39	0.35	0.49	-	0.58	0.34	0.43	0.47	1.00						
G+	0.65	-0.66	-0.46	-	0.43	-	-	-	-	0.47	-	0.39	-	0.62		0.47	0.45	0.95	1.00					
G-	0.65	-0.46	-0.48	0.82	0.77	-	0.49	-	-0.75	-	0.60	0.47	0.43	0.58	0.68	-	0.54	0.55	0.43	1.00				
TB	0.77	-0.63	-0.55	0.68	0.72	-	0.38	-	-0.67	0.41	0.52	0.49	0.39	0.73	0.60	-	0.59	0.86	0.81	0.87	1.00			
ACT	0.51	-0.59	-0.49	-	-	-	-	-	-	0.33	-	0.42	-	0.47	-	0.37	-	0.93	0.92	-	0.68	1.00		
Fungi	0.71	-0.51	-0.37	0.73	0.77	-	0.46	-	-0.69	-	0.63	0.35	0.42	0.69	0.58	-	0.59	0.56	0.57	0.76	0.82	0.33	1.00	
AMF	0.61	-0.41	-0.49	0.40	-	-	0.63	-	-0.65	0.72	-	0.58	0.67	0.55	0.77	-	0.47	0.66	0.62	0.58	0.69	0.58	0.43	1.00

Only the significant correlations are shown.
Abbreviations: BD, bulk density; SOC, soil organic carbon; TN, total N; AN, available N; AP, available P; AK, available K; CEC, cation exchange capacity; ASi, available Si; AZn, available Zn; DH, dehydrogenase; BG, β-glucosidase; ARS, arylsulfatase; ACP, acid phosphatase; POD, peroxidase; G+, gram-positive; G-, gram-negative; TB, total bacteria; ACT, actinomycetes; AMF, arbuscular mycorrhizal fungi.

Soil quality assessment

Based on correlation analysis results (Table 4), variables of AN, AK and dehydrogenase were not well correlated with rice yield, suggesting that those soil properties were not limiting or enhancing the rice growth. Therefore, AN, AK and dehydrogenase were excluded from soil quality evaluation in this study. Additionally, ASi and AZn were also omitted because of their concentrations typically higher than the critical values, although they presented significant correlations with rice yield. Ultimately, BD, pH, SOC, TN, AP, CEC, Urease, β-glucosidase, arylsulphatase, acid phosphatase, peroxidase, total PLFAs, G+, G-, TB, actinomycete, fungi and AMF were selected as optimal soil quality indicators for PCA.

The PCA showed that more than 82% of the variance in the data was explained by the first four PCs with Eigenvalues $\geqq 1$ (Table 5). The highly weighted variables under PC4, defined as those within 10% of the highest weight of the factor loading were total N and AMF. Both total N and AMF were selected for the MDS because they were not highly correlated ($r = 0.31$). In PC1, PC2 and PC3, only one highly weighted variable was observed. Therefore, total bacteria, acid phosphatase and AP were all retained in the MDS and the refined MDS was established including TN, AP, acid phosphatase, total bacteria and AMF.

Table 5. Results of principal components analysis (PCA) of statistically significant soil quality indicators.

Soil quality attribute	PC1	PC2	PC3	PC4
Eigenvalues	9.498	2.827	1.407	1.043
% of variance	52.77	15.71	7.81	5.79
Cumulative percent	52.77	68.47	76.29	82.08
		Factor loading		
Bulk density	-0.682	-0.438	0.181	0.402
pH	-0.661	-0.202	0.219	0.026
Soil organic C	0.730	-0.562	0.175	-0.159
Total N	0.769	-0.215	0.131	**-0.458**
Available P	0.526	-0.211	**-0.696**	0.006
CEC	-0.792	0.414	0.127	-0.125
Urease	0.658	0.095	-0.514	-0.106
Glucosidase	0.815	-0.059	0.207	0.051
Phosphatase	0.186	**0.736**	-0.225	-0.173
Arylsulfatase	0.733	-0.502	-0.097	0.339
Peroxidase	0.652	-0.289	0.312	0.207
PLFAs	0.801	0.514	0.185	0.099
Gram-positive	0.750	0.546	0.284	0.143
Gram-negative	0.846	-0.261	-0.001	-0.184
Total bacteria	**0.948**	0.103	0.173	-0.033
Actinomycetes	0.628	0.655	0.184	0.229
Fungi	0.818	-0.223	0.214	-0.239
AMF	0.769	0.08	-0.299	**0.504**

Boldface numbers are heavily weighted factors under each principal component (PC).

After the MDS selection process, an additional PCA was conducted with MDS indicators. The weighting factors for TN, AP, acid phosphatase, total bacteria and AMF were 0.186, 0.142, 0.253, 0.222 and 0.197, respectively. Each MDS variable was transformed to a value between 0 and 1 using their individual scoring functions. A standard scoring function was used to normalize TN and AP as described in Qi et al. (2009) [14]. The remaining MDS variables, phosphatase, total bacteria and AMF, were considered as "higher is better" indicators and were transformed using a linear scoring function [30]. Finally, SQI was calculated using the integrated quality index equation (Eq (1)). The soil quality indices of the selected waterlogged purple paddy fields varied from 0.295 to 0.794. The mean SQI was highest for HPPS (0.725±0.046) followed by MPPS (0.536±0.068) and LPPS (0.425±0.063) (Fig 3). While the HPPS represented the highest soil quality, the LPPS represented the poorest soil quality status, and low levels of TN and AP were found to be its major constraints limiting rice productivity. High soil quality can maintain high productivity without significant soil or environment degradation [59]. Mangers in our study area should pay more attention to the LPPS and particularly to any special limiting factor.

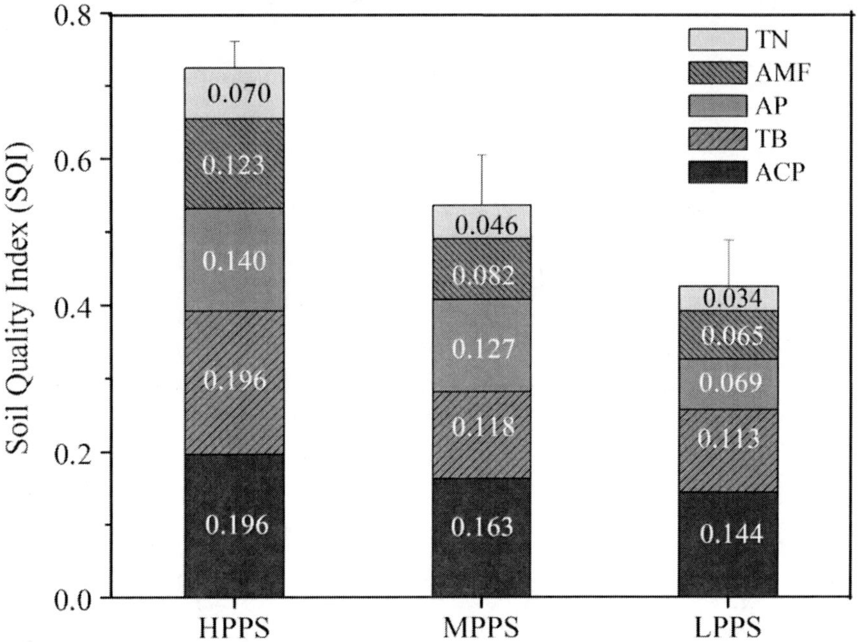

Figure 3. Soil quality indices (SQIs) of HPPS, MPPS and LPPS.

Correlation analysis indicated that SQI and rice yield were linearly related (Fig 4). The results followed the observations of Li et al. (2013) [17], who also reported significant positive correlation between SQI and ric yield. Our results also support previous conclusions that the selected MDS indicators could effectively evaluate soil quality status as a rice-production medium and the SQI approach is an appropriate way to develop a quantitative procedure to evaluate different land management practices [11,12].

CONCLUSION

An MDS including TN, AP, acid phosphatase, total bacteria and AMF was established for evaluating soil quality of waterlogged paddy soil in acidic purple soil region. SQI was also created based on the MDS method, and HPPS, MPPS and LPPS received mean SQI scores of 0.725, 0.536 and 0.425, respectively. LPPS presented a worst soil quality status, and low levels of TN, AP and soil microbial activities were considered to be the primary constraints limiting its

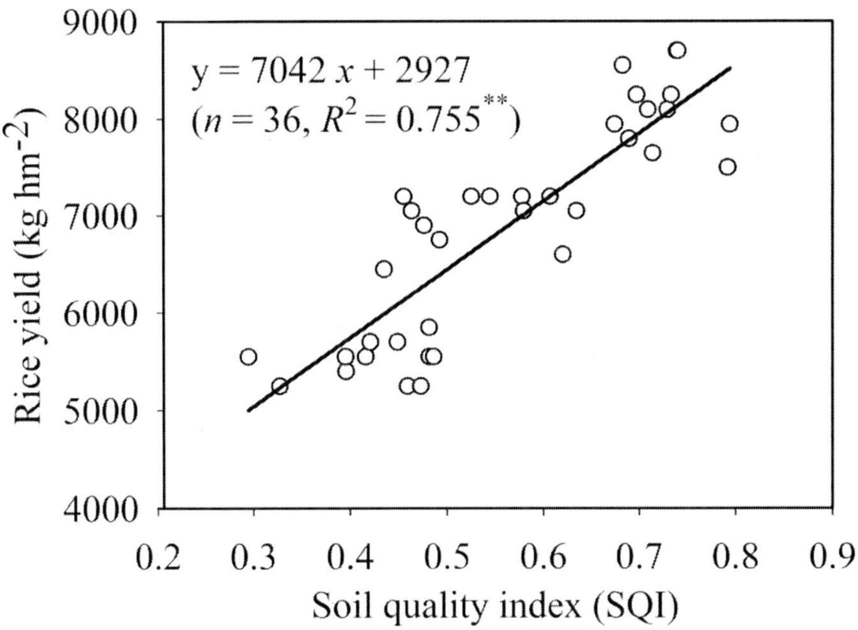

Figure 4. Correlation between soil quality index and rice yield of 36 paddy fields.

productivity compared to HPPS. All soil samples collected were rich in available N, K, Si and Zn, but deficient in available P, which may be the major constraint for the studied regions. Based on our results, a higher dose of phosphorus fertilizer should be considered to prevent severe deficiency of AP while the rate of basic N fertilizer should be decreased, with the saved N fertilizer to be applied as additional topdressing fertilizer. Other effective measures (i.e. soil-biochar additions, manure amendment) could be conducted because of their beneficial to counteract soil degradation [60] and improve soil biological quality [61], and future researches are resquired to confirm those hypothesises.

ACKNOWLEDGMENTS

This research was supported by the Special Fund for Agro-scientific Research in the Public Interest (201003016), the earmarked fund for China Agriculture Research System, and the National Basic Research Program of China (2013CB127405).

FUNDING STATEMENT

The Special Fund for Agro-scientific Research in the Public Interest (201003016); Funding institution: Ministry of Agriculture of the People's Republic of China; WZ received the funding. The earmarked fund for China Agriculture Research System, and the National Basic Research Program of China (2013CB127405); Funding institution: Ministry of Science and Technology of the People's Republic of China; WZ received the funding. The funders had no role in study design, data collection and analysis, decision to publish, or preparation of the manuscript.

DATA AVAILABILITY

All relevant data are within the paper and its Supporting Information files.

REFERENCES

1. Fu B, Wang YK, Xu P, Wang DJ. Changes in overland flow and sediment during simulated rainfall events on cropland in hilly areas of the Sichuan Basin, China. Prog Nat Sci. 2009; 19: 1613–1618.
2. Wang HJ, Shi XZ, Yu DS, Weidorf DC, Huang B, Sun WX, et al. Factors determining soil nutrient distribution in a small-scaled watershed in the purple soil region of Sichuan Province, China. Soil Till Res. 2009; 105: 300–306.
3. Gao Y, Zhu B, Yu GR, Chen WL, He NP, Wang T, et al. Coupled effects of biogeochemical and hydrological process on C, N, and P expert during extreme rainfall events in a purple soil watershed in southwestern China. J Hydrol. 2014; 511: 692–702.
4. Tesfahunegn GB, Tamene L, Vlek PLG. Evaluation of soil quality identified by local farmers in Mai-Negus catchment, northern Ethiopia. Geoderma 2011; 163: 209–218.
5. Doran JW. Soil health and global sustainability: translating science into practice. Agr Ecosyst Environ. 2002; 88: 119–127.
6. Bouma J. Land quality indicators of sustainable land management across scales. Agr Ecosyst Environ. 2002; 88: 129–136.
7. Doran JW, Zeiss MR. Soil health and sustainability: managing the biotic component of soil quality. Appl Soil Ecol. 2000; 15: 3–11.

8. Aparicio V, Costa JL. Soil quality indicators under continuous cropping systems in the Argentiean pampas. Soil Till Res. 2007; 96: 155–165.

9. Giacometti C, Demyan MS, Cavani L, Marzadori C, Ciavatta C, Kandeler E. Chemical and microbiological soil quality indicators and their potential to differentiate fertilization regimes in temperate agroecosystems. Appl Soil Ecol. 2013; 64: 32–48.

10. Bhardwaj AK, Jasrotia P, Hamiltona SK, Robertson GP. Ecological management of intensively cropped agro-ecosystems improves soil quality with sustained productivity.Agr Ecosyst Environ. 2011; 140: 419–429.

11. D'Hose T, Cougnon M, Vliegher AD, Vandecasteele B, Viaene N, Cornelis W, et al.The positive relationship between soil quality and crop production: A case study on the effect of farm compost application. Appl Soil Ecol. 2014; 75: 189–198.

12. Lima ACR, Brussaard L, Totola MR, Hoogmoed WB, de Goede RGM. A function evaluation of three indicator sets for assessing soil quality. Appl Soil Ecol. 2013; 64: 194–200.

13. Boluda R, Roca-Pérez L, Iranzo M, Gil C, Mormeneo S. Determination of enzymatic activities using a miniaturized system as a rapid method to assess soil quality. Eur J Soil Sci. 2014; 65: 286–294.

14. Qi YB, Darilek JL, Huang B, Zhao YC, Sun WX, Gu ZQ. Evaluating soil quality indices in an agricultural region of Jiangsu Province, China. Geoderma 2009; 149: 325–334.

15. Liu ZJ, Zhou W, Shen JB, Li ST, Ai C. Soil quality assessment of yellow clayey paddy soils with different productivity. Biol Fert Soils 2014; 140: 74–81.

16. Bonanomi G, D'Ascoli R, Antignani V, Capodilupo M, Cozzolino L, Marzaioli R, et al. Assessing soil quality under intensive cultivation and tree orchards in Southern Italy.Appl Soil Ecol. 2011; 47: 184–194.

17. Li P, Zhang TL, Wang XX, Yu DS. Development of biological soil quality indicator system for subtropical China. Soil Till Res. 2013; 126: 112–118.

18. Andrews SS, Mitchell JP, Mancinelli R, Karlen DL, Hartz TK, Horwath WR, et al.On-farm assessment of soil quality in California's central valley. Agron J. 2002; 94: 12–23.

19. Rahmanipour F, Marzaioli R, Bahrami HA, Feredouni Z, Bandarbadi SR. Assessment of soil quality indices in agricultural lands of Qzavin Province, Iran. Ecol Indic. 2014; 40: 19–26.

20. Beare MH, Bruce RR. A comparison of methods for measuring water-stable aggregates: implications for determining environmental effects on soil structure. Geoderma1993; 56: 87–104.

21. Page ALR, Miller H, Keeney DR. Methods of Soil Analysis. 2nd ed. Madison: American Society of Agronomy; 1982.

22. DeForest JL. The influence of time, storage temperature, and substrate age on potential soil enzyme activity in acidic forest soils using MUB-linked substrates and L-DOPA. Soil Biol Biochem. 2009; 41: 1180–1186.

23. Giacometti C, Cavani L, Baldoni G, Ciavatta C. Microplate-scale fluorometric soil enzyme assays as tools to assess soil quality in a long-term agricultural field experiment.Appl Soil Ecol. 2014; 75: 80–85.

24. Alef K, Nannipieri P. Methods in Applied Soil Microbiology and Biochemistry. London: Academic Press; 1995.

25. Wu YP, Ding N, Wang G, Xu JM, Wu JJ, Brookes PC. Effects of different soil weights, storage times and extraction methods on soil phospholipid fatty acid analyses.Geoderma 2009; 150: 171–178.

26. Moeskops B, Sukristiyonubowo, Buchan D, Sleutel S, Herawaty L, Husen E, et al.Soil microbial communities and activities under intensive organic and conventional vegetable farming in West Java, Indonesia. Appl Soil Ecol. 2010; 45: 112–120.

27. Brejda JJ, Moorman TB, Karlen DL, Dao TH. Identification of regional soil quality factors and indicators. Central and Southern High Plains. Soil Sci Soc Am J. 2000; 64: 2115–2124.

28. Andrews SS, Carroll C. Designing a soil quality assessment tool for sustainable agroecosystem management. Ecol Appl. 2001; 11: 1573–1585.

29. Shukla MK, Lal R, Ebinger M. Determining soil quality indicators by factor analysis.Soil Till Res. 2006; 87: 194–204.

30. Liebig MA, Varvel G, Doran JW. A simple performance-based index for assessing multiple agroecosystem functions. Agron J. 2001; 93: 313–318.

31. Doran JW, Parkin BT. Defining and assessing soil quality In: Doran JW, Coleman DC, Bezdicek DF, Stewart BA, editors. Defining Soil Quality for a Sustainable Environment. Madison: Soil Science Society of America; 1994. pp. 3–21.

32. Lu SG, Sun FF, Zong YT. Effect of rice husk biochar and coal fly ash on some physical properties of expansive clayey soil (Vertisol). Catena 2014; 114: 37–44.

33. Guo JH, Liu XJ, Zhang Y, Shen JL, Han WX, Zhang WF, et al. Significant acidification in major Chinese croplands. Science 2010; 327: 1008–1010.

34. Grehan PD, Scott BJ. Soil acidification—an agricultural and environmental problem In: Pratley JE, Robertson A, editors. Agriculture and the environmental imperative. Melbourne: CSIRO Publishing; 1998. pp. 98–128.

35. Wang YF, Tang CX, Wu JJ, Liu XM, Xu JM. Impact of organic matter addition on pH change of paddy soils. J Soil Sediment 2013; 13: 12–23.

36. Tong DL, Xu RK. Effects of urea and $(NH_4)_2SO_4$ on nitrification and acidification of Ultisols from Southern China. J Environ Sci. 2012; 24: 682–689.

37. Lal R. Soil carbon sequestration to mitigate climate change. Geoderma 2004; 123: 1–22.

38. State Soil Survey Service of China (SSSC). Dataset of National Soil Survey of China. Beijing: China Agriculture Press; 1996. Pp. 103–122 (in Chinese).

39. Hao QJ, Cheng BH, Jiang CS. Long-term tillage effects on soil organic carbon and dissolved carbon in a purple paddy soil of Southwest China. Acta Ecol Sin. 2013; 33: 260–265.

40. He YR. Purple soil in China Beijing: Science Press; 2003. (in Chinese).

41. Kamprath EJ, Watson M.E. Conventional soil and tissue tests for assessing the phosphorus status of soil In: Khasawneh FE, Sample EC, Hamprath EJ, editors. The Role of Phosphorus in Agriculture. Madison: Soil Science Society of America; 1980. pp. 443–469.

42. Bruun TB, Mertz O, Elberling B. Linking yields of upland rice in shifting cultivation to fallow length and soil properties. Agr Ecosyste Environ. 2006; 113: 139–149.

43. Jiang YH, Yu ZR, Ma YL. The effect of stubble return on agro-ecological system and crop growth. Soil Sci. 2001; 32: 209–213.

44. Li ZM. Purple soil in China Beijing: Science Press; 1991. (in Chinese).

45. Tang L, Zeng GM, Nourbakhsh F. Aftificial neural network approach for prdicting cation exchange capacity in soil based on physico-chemical properties. Environ Eng Sci. 2009; 26: 137–146.

46. Sinoga JR, Pariente S, Diaz AR, Murillo JFM. Variability of relationships between soil organic carbon and some soil properties in Mediterranean rangelands under different climatic conditions (South of Spain). Catena 2012; 94: 17–25.

47. Liang B, Lehmann J, Solomon D, Kinyangi J, Grossman J, O'Neill B, et al. Black carbon increases cation exchange capacity in soils. Soil Sci Soc Am J. 2006; 70: 1719–1730.

48. Ma JF, Yanmaji N. Silicon uptake and accumulation in higher plants. Trends Plant Sci. 2006; 11: 392–397.

49. Zhu QQ, Liu Z. The status of microelements in elation to crop production in paddy soils of China: III. Zinc. In: Proceeding of Symposium on Paddy Soils. New York: Springer Berlin Heidelberg; 1981. pp. 635–640.

50. Liu Z. Regularities of content and distribution of zinc in soils of China. Sci Agric Sin. 1994; 27: 30–37 (in Chinese).

51. Wang SY, Yu TQ, Wang JL, Yang L, Yang K, Lu P. Preliminary study on spatial variability and distribution of soil available microelements in Pinggu county, Beijing, China. Agr Sci China 2008; 7: 1235–1244.

52. Pajares S, Gallardo JF, Masciandaro G, Ceccanti B, Etchevers JD. Enzyme activities as an indicator of soil quality changes in degraded cultivated Acrisols in the Mexican Trans-volcanic belt. Land Degrad Dev. 2011; 22: 373–381.

53. Lee H, Ha HS, Lee CH, Lee YB, Kim PJ. Fly ash effect on improving soil properties and rice productivity in Korean paddy soils. Bioresource Technol. 2006; 97: 1490–1497.

54. Bonanomi G, D'Ascoli R, Scotti R, Gaglione SA, Caceres MG, Sultana S, et al. Soil quality recovery and crop yield enhancement by combined application of compost and wood to vegeables grown under plastic tunnels. Agr Ecosyst Environ. 2014; 192: 1–7.

55. Costa AL, Paixão SM, Caçador I, Carolino M. CLPP and EEA profiles of microbial communities in salt marsh sediments, J Soil Sediment 2007; 7: 418–425.

56. Ge TD, Nie SA, Qin HL, Xiao HA, Tong CL. Microcalorimetric studies for microbial activity of three vegetable soils, J Food Agric Environ. 2012; 10: 930–932.

57. Gu YF, Yun X, Zhang XP, Tu SH, Sun XF, Lindstrom K. Effect of different fertilizer treatments on quality of soil microbes and structure of ammonium oxidizing bacterial community in a calcareous purple paddy soil. Agr Sci China 2008; 7: 1481–1489.

58. Zhou ZF, Shi XJ, Zheng Y, Qin ZX, Xie DT, Li ZL, et al. Abundance and community structure of ammonia-oxidizing bacteria and archaea in purple soil under long-term fertilization. Eur J Soil Biol. 2014; 60: 24–33.

59. Griffiths BS, Ball BC, Daniell TJ, Hallett PD, Neilson R, Wheatley RE, et al.Integrating soil quality changes to arable agricultural systems following organic matter addition, or adoption of a ley-arable rotation. Appl Soil Ecol. 2010; 46: 43–53.

60. Willekens K, Vandecasteele B, Buchan D, Neve SD. Soil quality is positively affected by reduced tillage and compost in an intensive vegetable cropping system. Appl Soil Ecol. 2014; 82: 61–71.

61. Dong WY, Zhang XY, Dai XQ, Fu XL, Yang FT, Liu XY, et al. Changes in soil microbial community composition in response to fertilization of paddy soils in subtropical China. Appl Soil Ecol. 2014; 84: 140–147.

Chapter 7

Gross Alpha and Beta Activities and Trace Heavy Elemental Concentration Levels in Chemical Fertilizers and Agricultural Farm Soils in Nigeria

N. N. Jibiri[1*], K. P. Fasae[2]

[1]*Radiation and Health Physics Research Laboratory, Department of Physics, University of Ibadan, Ibadan, Nigeria*

[2]*Department of Science Technology, The Federal Polytechnic, Ado-Ekiti, Nigeria*

ABSTRACT

The gross alpha (α) and beta (β) activities and trace heavy elemental concentrations have been determined in Single Super Phosphate (SSP) and Nitrogen Phosphorous Potassium (NPK) fertilizers samples used for agricultural purposes in Nigeria. Cultivated farm soil samples from two regions (north and south) in the country and samples of phosphate rocks were also analyzed for gross α and β radioactivity and elemental concentrations. The beta activity concentration in the NPK (15-15-15) fertilizers was found to vary from 2410.0 ± 80.0 $Bq{\cdot}kg^{-1}$ to 4560 ± 140.0 $Bq{\cdot}kg^{-1}$, between 1340.0 ± 70.0 $Bq{\cdot}kg^{-1}$ and 1440.0 ± 70.0 $Bq{\cdot}kg^{-1}$ in the Single Super phosphate fertilizers while in the farm soil it ranged between 360.0 ± 40.0 $Bq{\cdot}kg^{-1}$ to 570.0 ± 50.0 for the north samples and 200.0 ± 50.0 to 230.0 ± 20 $Bq{\cdot}kg^{-1}$ for the south samples. The alpha activities in the NPK fertilizers was found to vary from 20.0 ± 10.0 to 90.0 ± 20.0 $Bq{\cdot}kg^{-1}$, in the SSP it varied from 60.0 ± 20.0 to 100.0 ± 20.0 $Bq{\cdot}kg^{-1}$ while in the soil samples it varied from 8.0 ± 6.0 to 40.0 ± 10.0 $Bq{\cdot}kg^{-1}$ (north) and 10.0 ± 10.0 to 20.0 ± 10.0 $Bq{\cdot}kg^{-1}$ (south). The trace heavy metal concentration in Phosphate rock samples, SSP and the NPK (15-15-15) fertilizers samples were lower than the safety limits, however, the heavy elements were all detected in NPK fertilizers samples including Hg.

Keywords: Chemical Fertilizers; Gross Alpha Activity; Gross Beta Activity; Trace Heavy Elements; Farm Soil; Nigeria

1. INTRODUCTION

The addition of inorganic phosphate fertilizers to soil for agricultural crop growth and production is a common practice worldwide because they are sources of macronutrient (N, P, K, Ca, Mg and S) and micronutrients (B, Cl, Co, Cu, Fe, Mn, Mo, Ni, Se, Si and Zn) to the crops especially when they are deficient in the soils [1]. Inorganic fertilizers may contain radionuclides, heavy metals and fluorine, heavy metal micro-nutrients (Co, Cu, Fe, Mn, Mo, Ni, Zn) and heavy metals (As, Al, Cd, Pb and Hg) considered toxic when their concentrations exceed safety limits and importantly when they find their way into human systems through eco-systemic mechanisms [2]. For a plant to grow properly, it needs close to 60 chemical elements, the key elements accounting for 90% of a plant's dry weight are carbon, oxygen and hydrogen while some 8% to 9% of nitrogen, phosphorus, magnesium, sulphur, calcium and potassium. The remaining elements amount to a mere 1% or 2%; these include the important elements; boron, iron, copper, manganese, zinc and cobalt [3]. Fertilizer effects on crops are classified as either direct or indirect based on the plant need for the fertilizer. They have direct effects if they supply nutrients for plants directly and indirect if they serve to improve the soil's physical chemical and biological properties so as to mobilize the nutrients present in the soil and to neutralize the soil's acidity. There are multi-nutrient fertilizers which contain two or more macronutrients such as N-P, P-K, or N-K (double-nutrients) while N-P-K is termed tripled-nutrient fertilizers referred to as complete fertilizers. Phosphorus is one of the limiting elements to plant growth and development because lack of available phosphorus in soils limits the growth of both cultivated and uncultivated plants. Today, the only important source of fertilizer and industrial phosphorus is phosphate rock [4]. The phosphate rocks are also rich in uranium, thorium and their daughters, deposited as calcium phosphate minerals [2]. Contamination of soil and water bodies by major and trace elements may be caused by natural processes or through human activities in the environment [5].

The long-continued application of fertilizers and their by products can redistribute and elevate heavy metal and radionuclide concentrations in soil profiles. Research has shown that uranium concentrations in phosphate rock samples (a raw material for phosphate fertilizer) are enhanced by a factor of more than 10 above background uranium in soil. The ingestion of radionuclides through food intake via the soil treated with fertilizer accounts for a substantial part of average radiation doses to various organs of the body and also represents one of the important pathways for long term health considerations [6-9]. The heavy use of NPK fertilizers by farmers to improve crop yield following impoverishment of the soil has been attributed to be responsible for their elevated concentrations in food crops [6]. Despite the wide application of fertilizers in farming in the country and the different brands used by farmers, the much needed research has not been carried out in brands of fertilizers available to farmers. Until recently the level of natural radionuclide concentration levels (^{226}Ra, ^{232}Th and ^{40}K) in brands of fertilizer used in some part of Nigeria was determined by Jibiri and Fasae [7] by gamma ray spectrometry. As a further step to this initial research effort, the gross alpha and beta radioactivity levels in both farm soil and fertilizer samples in the country are investigated in this study. In view of the contribution of fertilizer to natural radiation exposures through their applications in farming activities and by crop uptake to food ingestions, the present study aims are as follows: 1) determine the level of radioactivity in the brand of fertilizers sold in Nigeria through their alpha and beta emissions and also in different farm soils where they have been applied for agricultural purpose; 2) determine the trace heavy elemental compositions in the fertilizer brands; and finally, 3) provide requisite radiometric data that may be useful in policy formulation and implementation on fertilizer safety and standards in the agricultural sector in Nigeria.

2. MATERIALS AND METHODS

2.1. Sampling

In the collection of samples of fertilizers in this study consideration was given to the two fertilizer manufacturing companies in the country; the Federal Superphosphate Fertilizer Company Ltd. (FSFC), located in Kaduna State in the Northern part of the country and National Fertilizer Company of Nigeria (NAFCON) in PortHarcourt located in the southern part. Fertilizer samples from these two companies were collected and also the phosphate rocks, the raw materials (the Sokoto and Togo types) used by the companies in the manufacturing of fertilizers. Two samples of single super phosphate fertilizers were collected from FSFC marked SSP_1 and SSP_2 while NPK fertilizer brand were collected from markets outlets in six major commercial cities located in six states in southwestern Nigeria. This is because the markets represent the distribution channel of brands of fertilizers to the farmers and are centers of commerce in the region which are easily accessible by all strata of farmers. The markets in the city centers from where samples were collected in the southern states are: Oyo, Ogun, Osun, Ondo, Lagos and Ekiti and samples from each state was labeled NPK_{oy}, NPK_{og}, NPK_{os}, NPK_{od}, NPK_{La}, and NPK_{Ek}, respectively. Majorly, farmers from the southern part of the country use the NPK brand because of the fertile soil type while the north uses the SSP brand because it is an arid zone. Furthermore, samples of farm soil treated with SSP fertilizer were collected from School of Agriculture Kaduna farm and the Prison farm of the Nigerian Prison Services labeled $Soil_{North}$ while those treated with NPK were collected from farmlands around NAFCON labeled $Soil_{South}$. However, these sampling approaches were premised on the assumption that the farm soil samples from the farmlands in each of the region were treated with fertilizer types produced in that region. The validity of this assumption was however not investigated in this study.

2.2. Gross Alpha and Beta Counting Procedure

AN EURISYS MEASURE IN 20 a low background multiple (eight) channel alpha and beta counter was used for the measurements of the gross alpha and beta in the fertilizer and the farm soil samples. It is a gas flow proportional counter with 450 mg/cm^3 thick window of diameter 60 mm. The counting gas is an argon-methane mixture at the ratio of 90% and 10% respectively. The counting system incorporates an anti-coincident guard counters used to eliminate the influence of high energy cosmic radiation that would enter the measuring environment. The background radiation level within the measuring environment was estimated to be less than 1.0 mSv/ hr. The chambers were covered with 10 cm thick lead and the inside dimensions are 480 × 280 × 105 mm^3. The system was connected to a micro-processor IN-SYST, a spreadsheet programmed QUARTTRO-PRO and a graphic programmer. For the alpha activity measurements the standard used is ^{239}Pu α-sources with activity ranging from 133.3 Bq to 185.8 Bq for the eight channels at 2pstearadians. The radionuclide impurity in each of them varied from 0.74% - 0.82% [10]. For the beta activity measurements the standard used are ^{90}Sr β sources of diameter 38 mm and an active film of 12 $mg{\cdot}cm^{-3}$ thick. For the eight channels, eight sources of activities varied from 105.1 Bq to 117.7 Bq at 4p-stearadians. The radionuclide impurity in each of them was less than 0.1%. Cerca and LEA Laboratories in France certified all measurements with certificate No. CT001271/00/1778-1783 [10].

The beta and alpha specific activities were calculated using the expression in Eq.1. [10]:

2.3. Trace Heavy Metal Determination

A one gram aliquot of each sample of the fertilizers was transferred into a conical flask. A 10.0 ml of the digestion solution containing a mixture of concentrated HNO_3 and HCL mixed in ratio 3:1 was added to each sample in the flask for digestion procedure as described by Rowel [11]. The mixture was heated until digestion was complete, that is, when the samples became whitish and brown fumes of HNO_3 seized to appear. The residue was diluted with deionized water, filtered and the filtrate was further diluted to make 100.0 ml using deionized water in a standard flask. Series of standard solution of each metal were prepared and run along with each sample at the respective wavelength for maximum absorption of each metal. The metal concentrations were thereafter determined using Atomic Absorption Spectrophotometer (AAS) (Buck Scientific Model 200A (AOAC, 1990).

3. RESULTS AND DISCUSSION

3.1. Gross Alpha and Beta Activities in the Fertilizer and Farm Soil Samples

The gross alpha and beta activity concentrations in the single superphosphate and NPK fertilizers are presented in **Table 1** while the concentrations in the farm soil samples from the two regions is presented in **Table 2**. As could be seen from **Table 1** the gross beta activity concentration varied from 2410.0 $Bq{\cdot}kg^{-1}$ (Oy) to 4560 $Bq{\cdot}kg^{-1}$ (Od) in the NPK (15-15-15) while in the SSP fertilizers it varied from 1340.0 $Bq{\cdot}kg^{-1}$ to 1440.0 $Bq{\cdot}kg^{-1}$. This shows that there are more beta emitters in the NPK fertilizers than in the SSP fertilizers. This trend was a direct opposite on gross alpha activities in the sample. This shows that the gross alpha activities are higher in the SSP fertilizers (60.0 $Bq{\cdot}kg^{-1}$ and 100.0 $Bq{\cdot}kg^{-1}$) than in NPK fertilizers with the range of values (2.0 $Bq{\cdot}kg^{-1}$ to 90.0 $Bq{\cdot}kg^{-1}$). This observation was in agreement with the study by Jibiri and Fasae [7], that low activity concentration of ^{40}K is associated SSP and high values with NPK fertilizers. It could also be observed that the highest value of the gross beta activity concentrations in NPK fertilizers was from Ondo State (Od) samples (90.0 $Bq{\cdot}kg^{-1}$) and the lowest was obtained in Ogun state (Og) samples. Hence, from the previous work by Jibiri and Fasae [7] and confirmed in this study showed that there is a direct relationship between the gross beta and alpha activities with the radionuclide contents in the samples. High ^{226}Ra and ^{40}K concentrations will indicate high gross alpha and beta activities in the samples. This is in agreement with the study by Aytas et al., [12]. Generally we

3. RESULTS AND DISCUSSION

Table 1. The gross alpha and beta activity concentrations in super phosphate and NPK fertilizer samples.

Samples	Alpha Activity ($Bq \cdot kg^{-1}$)	Beta Activity ($Bq \cdot kg^{-1}$)
Super phosphate SSP_1	100.0 ± 20.0	1340.0 ± 70.0
Super phosphate SSP_2	60.0 ± 20.0	1440.0 ± 70.0
NPK_{Oy} (15-15-15)	3.0 ± 5.0	2410.0 ± 80.0
NPK_{Og} (15-15-15)	10.0 ± 10.0	2470.0 ± 80.0
NPK_{Os} (15-15-15)	10.0 ± 10.0	2420.0 ± 80.0
NPK_{Od} (15-15-15)	90.0 ± 20.0	4560.0 ± 140.0
NPK_{La} (15-15-15)	60.0 ± 20.0	3510.0 ± 100.0
NPK_{Ek} (15-15-15)	2.0 ± 10.0	2640.0 ± 80.0

Table 2. The Gross alpha and beta activity concentrations in farm soil samples from two regions of Nigeria.

Sample	Alpha Activity ($Bq \cdot kg^{-1}$)	Beta Activity ($Bq \cdot kg^{-1}$)
$Soil_{North\ 1}$	8.0 ± 6.0	360.0 ± 40.0
$Soil_{North\ 1}$	30.0 ± 10.0	450.0 ± 50.0
$Soil_{North\ 2}$	30.0 ± 20.0	570.0 ± 50.0
$Soil_{North\ 2}$	40.0 ± 10.0	450.0 ± 40.0
$Soil_{South\ 1}$	10.0 ± 10.0	200.0 ± 50.0
$Soil_{South\ 2}$	20.0 ± 10.0	230.0 ± 20.0

observed that there are mixes of fertilizer products sold in a particular State with varying radionuclide contents hence they will present different radioactivity scenario to natural radioactivity in farmlands in the States where they will be used for agricultural purposes. From**Table 2**, the gross beta activity concentration for farm soil samples from the northern part of Nigeria varied from 360.0 Bq·kg^{-1} to 570.0 Bq·kg^{-1} while for the alpha activities it varied from 8.0 Bq·kg^{-1} to 40.0 Bq·kg^{-1}. The gross beta activity concentrations for farm soil samples from the southern part of Nigeria varied from 200.0 Bq·kg^{-1} to 230.0 Bq·kg^{-1} while the gross alpha activity concentrations varied from 10.0 Bq·kg^{-1} to 20.0 Bq·kg^{-1}. The results show that the gross activity concentrations in farm soil samples in the northern part of the country are higher than the values obtained in the southern part of the coun-

$$\text{Specific Activity}(\alpha,\beta)\,\text{Bq/kg} = \frac{\text{Counting Rate}(\alpha,\beta) - \text{Background counting rate}(\alpha,\beta)}{\text{Sample Efficiency} \times \text{Channel Efficiency} \times \text{Weight of sample}} \quad (1)$$

try. It could also be observed that the introduction of fertilizer into the farm soil may have elevated the concentration of radionuclides in the northern part of the country more than in the southern part of the country, thereby elevating the concentration of radionuclides in food stuff possibly from the northern region of the country. As reported by Jibiri et al., [6], fertilizer largely used to improve soil fertility is seen to have enhanced the quantity of natural radionuclides in the soil and consequently through uptake are made available to the plants. However, the gross alpha and beta activity concentrations values obtained in this study were lower than the range of values 141 Bq·kg^{-1} to 2589 Bq·kg^{-1} for beta activity concentrations and 109 Bq·kg^{-1} to 660 Bq·kg^{-1} for gross alpha activity concentrations obtained in West Bengal [13,14].

3.2. Trace Heavy Elemental Concentration in Super Phosphate Fertilizer, N-P-K (15-15-15) and Phosphate Rock Samples

The results of the elemental analysis for the soil samples are presented in **Table 3**. It could be observed that the elemental concentrations were higher in phosphate rock samples followed by SSP. The values were relatively low in the NPK samples. However, while heavy elements were not detectable in the phosphate rocks of different origins and in the SSP they were detected in the NPK fertilizer samples. Micronutrients such as Cu, Ni and Zn are essential for plant life while Cd, Cr and Pb are not essential elements but can be classified as toxic to plants. When these heavy metal concentration is compared with the mean concentration in the Earth's crust (Cd, 0.5 ppm; Cr, 200 ppm; Cu, 100 ppm; Ni, 80 ppm; Pb,16 ppm; Zn, 50 ppm, as reported by Sauerbeck [15]it may be observed that the analysed samples are enriched in Cd, Ni, Pb and Zn since under natural conditions, Pb occurs in the soil at 5 to 20 ppm, cadmium at 0.05 to 0.5 ppm, and arsenic at 1 to 5 ppm. In highly urbanized areas, Pb occurs in the soil at 100 to 200 ppm, cadmium at 1 to 5 ppm, and arsenic at 10 to 20 ppm. Threshold levels where these elements could be harmful by direct ingestion are generally higher than levels commonly found in urban soils. For example, US government threshold levels for Pb in the soil are greater than 500 ppm and greater than 20 ppm of cadmium [16] if the comparison is performed with the world range for phosphate rocks, i.e. Cd, 1 - 100 ppm; Cr, 7 - 500 ppm; Cu, 1 - 1000 ppm; Ni, 0 - 100 ppm; Pb, 0 - 10 ppm; Zn, 4 - 1000 ppm [5] and lesser in the Tokyo region (0.05 ppm of cadmium and 0.5 ppm of lead and zinc) [17] and in Europe [18]. From the values obtained in **Table 3**, the concentrations of heavy metals considered generally to be toxic varied from BDL to 3.73 $mg \cdot kg^{-1}$. These values were less than the values obtained internationally for which Cd ranged from <1 to 36.8 $mg \cdot kg^{-1}$, with a median of 33.2 $mg \cdot kg^{-1}$ for phosphatic fertilizers [19]. This may be an indication that suggests that the addition of heavy metals into the soil through the application of inorganic fertilizer do not significantly raise their concentration in soils to harmful levels. This assertion was supported by various international studies that the entry of heavy metals, especially Cd, Pb and Ni, into the human food chain may not be associated with inorganic fertilizers but to other anthropogenic sources, domestic and industrial residues/wastes and pesticides [20]. It is likely that the continued long use of fertilizer may elevate the concentrations of these elements in the farm soil if no reduction and remediation procedures are considered.

Table 3. Elemental concentrations in Super Phosphate fertilizers, N-P-K (15-15-15) fertilizers and Phosphate rock samples.

ELEMENT	Super phosphate Fertilizer SSP_1	Super phosphate Fertilizer SSP_2	N-P-K* Fertilizers	Phosphate rock (Togo)	Phosphate rock (Sokoto, Nigeria)
Mg ($mg \cdot kg^{-1}$)	3.68	1.12	13.17 ± 0.882	0.39	3.70
Mn ($mg \cdot kg^{-1}$)	3.64	24.60	1.067 ± 0.145	1.28	19.44
Fe ($mg \cdot kg^{-1}$)	37.40	450.70	15.25 ± 4.394	BDL	450.70
Pb ($mg \cdot kg^{-1}$)	13.80	BDL	0.60 ± 0.010	18.14	20.20
As ($mg \cdot kg^{-1}$)	0.30	BDL	BDL	0.20	0.50
Ni ($mg \cdot kg^{-1}$)	0.70	0.30	0.62 ± 0.624	BDL	0.80
Co ($mg \cdot kg^{-1}$)	0.11	BDL	0.03 ± 0.013	0.07	0.20
Cu ($mg \cdot kg^{-1}$)	5.27	14.78	18.03 ± 1.55	16.93	11.96
Cd ($mg \cdot kg^{-1}$)	0.77	0.62	1.475 ± 0.079	0.86	0.94
Ca ($mg \cdot kg^{-1}$)	2.25	0.14	0.142 ± 0.042	3.67	8.98
Se ($mg \cdot kg^{-1}$)	1.95	BDL	BDL	2.95	3.73
Zn ($mg \cdot kg^{-1}$)	BDL	BDL	0.033 ± 0.024	BDL	BDL
Cr ($mg \cdot kg^{-1}$)	BDL	BDL	0.147 ± 0.046	BDL	0.10
Hg ($mg \cdot kg^{-1}$)	BDL	BDL	1.43 ± 1.264	BDL	BDL

*± in NPK indicates the standard deviation value of the six samples analyzed

4. CONCULSION

The gross beta and alpha activity concentrations and trace heavy elemental levels in the fertilizer samples used in Nigeria have been determined. NPK fertilizers were found to exhibit high gross beta activity concentrations while SSP exhibit high gross alpha activity concentrations. The farm soil samples from the northern part of the country exhibited high gross alpha and beta activities compared to farm soil samples from the Southern parts of Nigeria. The gross alpha and beta in the samples were good indicators for the presence of the natural radionuclide contents in the fertilizer samples. Though the concentrations of the heavy metals were very low in the fertilizer samples analyzed in this work compared to literature values it is anticipated that continually long use of the fertilizers in the farm soil may elevate their concentrations to toxic levels which may pose a health risk to the population if remedial actions are not considered.

5. ACKNOWLEDGEMENTS

We are grateful to the National Fertilizer Company, Onne, PortHarcourt, Rivers-State, for access into the company to take some samples. We are grateful to Mr. Hamidu of School of Agriculture Kaduna, Mr. Borisade the director of the Nigerian Prison farms and the staff of The National Superphosphate fertilizer, Kaduna for their cherished assistance. We are also grateful to the Center of Energy Research and Training (CERT) Ahmadu Bello University, Zaria, Nigeria and Mrs. Onoja R.A. for the access to the equipments.

REFERENCES

1. Abdel-Haleem, A.S., Sroor, A., El-Bahi, S.M. and Zohny, E. (2001) Heavy metals and rare earth elements in phosphate fertilizers components using instrumental neutron activation analysis. Applied Radiation and Isotopes, 55, 569-573.

2. Fabiano, T.C. and Daniel M.B. (2006) Radionuclides, heavy metals and fluorine incidence at Tapira phosphate rocks, Brazil, and their industrial (by) products. Environmental Pollution, 139, 232-243.

3. Pozin, M.E. (1986) Fertilizer manufacture. Mir Publishers, USSR 82 KVS, 234-236.

4. Tisdale, S.L., Nelson, W.L. and Beaton, J.D. (1985) Soil fertility and fertilizers. 4th Edition, Macmillan Publishing Company, New York, 188-239.

5. Abdulwahab, R.H. and Awadh, M. Al J. (1994) Elemental concentration of selected soil and water samples from AlMadinah Area, Saudi Arabia. Journal of King Saud University Science, 6, 127-136.

6. Jibiri, N.N., Farai, I.P. and Alausa, S.K. (2007) Estimation of annual effective dose due to natural radioactive elements in ingestion of foodstuffs in tin mining area of JosPlateau, Nigeria. Journal of Environmental Radioactivity, 94, 31-40.

7. Jibiri, N.N. and Fasae, P.K. (2011) Activity concentration of ^{226}Ra, ^{232}Th and ^{40}K in brands of fertilizers used in Nigeria. Radiation Protection Dosimetry, 148, 132-137.

8. McDonald, P., Jackson, D., Leonard, D.R.P. and McKay, K. (1999) An assessment of ^{210}Pb and ^{210}Po in terrestrial foodstuffs from regions of potential technological enhancement in England and Wales. Journal of Environmental Radioactivity, 43, 15-29.

9. Fernandez, G., Rodriquez, I.M., Castro, G.V., Carrazana, G. and Martizez, R.N. (2004) Radiological surveillance of foods and drinking water in the Cuban Republic. Proceedings of the 11th Conference of the International Radiation Protection Association (IRPA), Madrid, 23-28 March 2004.

10. Akpa, T.C., Mallam, S.P., Ibeanu, I.G.E. and Onoja, R.A. (2004) Characteristcs of gross alpha/beta proportional counter. Nigerian Journal of Physics, 16, 13-18.

11. Rowel, D.L. (1993) Soil science methods and applications. University of Reading, Longman.

12. Aytas, S., Yusan, S., Aslani, M.A.A., Karali, T., Turkozu, D.A., Gok, C., Erenturk, S., Gorkce, M. and Oguz, K. (2012) Determination and evaluation of natural radioactivity and heavy metal levels in surface sediments of Maritza and Tundja River. Scientific Technological Research Council of Turkey (TUBITAK) and Bulgarian Academy of Science (BAS), Contract ref No: 106Y201, INSINUME, Brussels, 12-15 June 2012.

13. Ghosh, D., Deb, A., Bera, S., Sengupta, R. and Patra, K.K. (2008) Measurement of natural radioactivity in chemical fertilizer and agricultural soil: Evidence of high alpha activity. Environmental Geochemistry and Health, 30, 79- 86.

14. Selcuk, Z.O., Cevlan, H. and Dogru, M. (2009) Gross alpha and beta radioactivity concentration in water, soil and sediment of the Bendimahi River and Van Lake (Turkey). Environmental Monit, 148, 39-46.

15. Sauerbeck, D. (1992) Conditions controlling the bioavailability of trace elements and heavy metals derived from phosphate fertilizers in soils.

Proceeding of IMPHOS Conference on Phosphorus, Life and Environment, Casablanca, 8-11 September 1992, 418-448.

16. Kitagishi, K. and Yamane, H. (1981) Heavy metal pollution in soils of Japan. CRC Press, Tokyo, 19-26.

17. Hodel, D.R. and Chang, A.C. (2009) Trace elements and urban gardens. University of Califonia Cooperative Extension Los Angeles County. http://celosangeles.ucdavis.ed

18. Kabata-Pendias, A. and Pendias, H. (2009) Trace elements in soils and plants. 3rd Edition, CRC Press, Boca Raton, 3-24.

19. Modaihsh, A.S., Al-Swailem, M.S. and Mahjoub, M.O. (2004) Metals content of commercial inorganic fertilizers used in the Kingdom of Saudi Arabia. Sultan Qaboos University Journal for Scientific. Research Agricultural and Marine Sciences, 9, 21-25.

20. Godson, R.E.E., Mynepali. A. and Sridhar, K.C. (2004) Soil quality near a chemical fertilizer industry at PortHarcourt Nigeria. African Journal of Environmental Assessment and Management, 8, 19-26.

Chapter 8

Effect of Drying on Heavy Metal Fraction Distribution in Rice Paddy Soil

Yanbing Qi[1,2], Biao Huang[3], Jeremy Landon Darilek[3]

[1] College of Resources and Environment, Northwest Agriculture and Forestry University, Yangling, Shaanxi, People's Republic of China, 2 Key Laboratory of Plant Nutrition and the Agri-environment in Northwest China, Ministry of Agriculture, Yangling, Shaanxi, People's Republic of China, 3 State Key Laboratory of Soil and Sustainable Agriculture, Institute of Soil Science, Chinese Academy of Sciences, Nanjing, Jiangsu, People's Republic of China

ABSTRACT

An understanding of how redox conditions affect soil heavy metal fractions in rice paddies is important due to its implications for heavy metal mobility and plant uptake. Rice paddy soil samples routinely undergo oxidation prior to heavy metal analysis. Fraction distribution of Cu, Pb, Ni, and Cd from paddy soil with a wide pH range was investigated. Samples were both dried according to standard protocols and also preserved under anaerobic conditions through the sampling and analysis process and heavy metals were then sequentially extracted for the exchangeable and carbonate bound fraction (acid soluble fraction), iron and manganese oxide bound fraction (reducible fraction), organic bound fraction (oxidizable fraction), and residual fraction. Fractions were affected by redox conditions across all pH ranges. Drying decreased reducible fraction of all heavy metals. $Cu_{residual\ fraction}$, $Pb_{oxidizable\ fraction}$, $Cd_{residual\ fraction}$, and $Ni_{residual\ fraction}$ increased by 25%, 33%, 35%, and >60%, respectively. $Pb_{residual\ fraction}$, $Ni_{acid\ soluble\ fraction}$, and $Cd_{oxidizable\ fraction}$ decreased 33%, 25%, and 15%, respectively. Drying paddy soil prior to heavy metal analysis overestimated Pb and underestimated Cu, Ni, and Cd. In future studies, samples should be stored after injecting N_2 gas to maintain the redox potential of soil prior to heavy metal analysis, and investigate the correlation between heavy metal fraction distribution under field conditions and air-dried samples.

INTRODUCTION

Soil heavy metals Cu, Pb, Ni, and Cd, are regarded as "chemical time bombs" because of their propensity for accumulation in the soil and uptake by crops. This ultimately causes human toxicity in both the short- and long-term [1]–[2], making farmland ecosystems dangerous to health [3].

Morphological characteristics and processes of heavy metals have been studied to better understand heavy metal occurrence in various environments, transport pathways, and crop uptake. There are a wide range of soil inorganic and organic substances affected by redox conditions. Redox changes the valence of ions, subsequently affecting the forms of various elements and compounds and their transport processes. Furthermore, heavy metal behavior is strongly correlated with redox potential (Eh) [4].

Many studies of river and lake sediments have reported the effect of redox conditions on the distribution of heavy metals. For example, such changes in redox conditions affect heavy metal association with organic matter (OM) [5], and iron (Fe) and manganese (Mn) oxidized fractions are unstable in reduced environments [3], [6]. Kelderman and Osman [2] reported that in river sediments, exchangeable and carbonate bound fractions of Cu, Zn, and Pb increased as Eh increased, and organic fractions, oxidizable Cu, and oxidizable Pb decreased. Lu et al. [7] found that in reduced Iraqi River sediment in Changchun City, Mn oxide fractions increased significantly, organic bound Cu decreased by 40%, but decrease in organic bound Pb, Zn, and Ni was less than Cu.

Soil heavy metal studies which include plant available indices, total concentrations, and fraction distributions have also been conducted on many agricultural systems, including rice (*Oryza sativa*) production. Rice has a high water requirement compared to other crops and during the growing season, Eh can reach roughly 300–200 mV [4]. Soil samples, including from rice paddies under submerged conditions, are routinely air-dried prior to heavy metal analysis [8]–[12]. Zheng and Zhang [13] studied the effect of moisture regimes on paddy soil heavy metals and found that soil moisture did not affect the direction or pathways of fractionation distribution (from active to stable fractions), but did affect the transformation rate. Zheng and Zhang [13] dried rice paddy soil samples after collection and then reconstituted three moisture regimes under controlled conditions in the lab. It does not appear that sample preparation was conducted in an anaerobic environment, therefore this result does not represent in situ soil moisture regime that controls the distribution of heavy metals.

Redox conditions are known to have a significant effect on heavy metal speciation in sediments [14]. As indicated by Calmano et al. [5], if anoxic sediments are exposed to atmosphere, redox condition change and redistribution and transformation of heavy metal fractions in the sediments takes place. A few studies show the effect of redox conditions on heavy metal availability [15]–[17]. Paddy soil has the anoxic condition during rice growing season similar to the river sediment. Thus, we hypothesize that drying prior to analysis for paddy soil changes the anoxic condition to redox condition as a consequence mobility and uptake of heavy metal by crop are changes. To date, the effect of drying on soil heavy metal fraction distribution in rice paddies is largely unknown.

In order to fill this knowledge gap and work towards a better methodology for sampling and handling paddy soil prior to heavy metal analysis, we conducted a study in Zhangjiagang County, a typical rice production region of China, to 1) understand heavy metal distribution in paddy soil under various redox and pH conditions, and 2) determine the effect of soil air-drying on heavy metal fractions. This study intends to provide useful information for heavy metal assessment in situ. It is hoped that through a better understanding of heavy metal fractions in situ, a better understanding of correlation of heavy metal between soils and crops can be drawn, and a suitable set of corrective measures to prevent heavy metal pollution can be put in place.

MATERIALS AND METHODS

Ethics Statement

All the sample sites were distributed in the private land and permission were approved by the land owner in each site (Zhigang Wang can be contacted for the future permissions). The field studies did not involve endangered or protected species because all the sample sites were in the farmland with rice planted. The coordinates of sample sites ranged from 120°35′ to 120°42′E and from 31°44′ to 31°51′N.

Sample Collection

Zhangjiagang County was selected as the research area because of two reasons, firstly, the soil is typical paddy soil with rice planting history of more than 100 years, secondly, heavy metals have been accumulated slightly in this area because of sparkly distributed chemical factories. Zhangjiagang County has a humid monsoon climate in the north subtropical zone, with four distinct seasons, and average annual temperature and precipitation is 15.0°C and 1045.9 mm, respectively.

An investigation of soil physical and chemical characteristics of Zhangjiagang County, Jiangsu, was conducted in 2004 [18]–[19]. In the current study, ten locations from the 2004 investigation, each with a long history of rice production, were selected to represent a wide pH range (4.81–8.16).

Topsoil (0–20 cm) was collected in October, 2007 from rice paddies before the harvest and after a full season of submersion. At the time of collection, there was still a 1–2 cm water layer covering the soil. Each soil sample was a composite of sub-samples taken from 6–8 points within 50 m^2 with a stainless steel soil probe. Half of the sample was quickly placed in polyethylene plastic tubes and the headspace was immediately purged with nitrogen (N) gas [20], pre-treated with acid, and weighed. Once transported to the laboratory, samples were stored under refrigeration at 4°C for the determination of heavy metals fractions under reduction conditions. The remaining sample was placed in polyethylene bags.

Chemical Analysis

Soil moisture was measured using soil from the polyethylene bags by drying at 105°C for 24 h. The remaining soil from the polyethylene bags were air-dried and sieved through a 2 mm sieve. Soil pH was measured with a glass electrode in a 1:2.5 soil: water suspension [21]. Soil OM was determined using the dichromate-wet combustion method [22]. The samples were digested using HNO_3-$HClO_4$-HF for measurement of the total heavy metals [19].

The Community Bureau of Reference (BCR) sequential extraction procedure described by Ure et al. [23] was used to extract heavy metal fractions. For the exchangeable and carbonate bound fraction (acid soluble fraction), acetic acid (20 ml, 0.11 mol l^{-1}) was added to a 50 ml polypropylene centrifuge tube containing 0.5 g sample and shaken for 16 h at ambient temperature (20±1°C) on an end-over mechanical shaker operating at 40 rpm. The extract was centrifuged at 4000 rpm, decanted into a polyethylene container, and stored at −4°C until analysis. For the Fe and Mn oxide bound fraction (reducible fraction), the residue from the step above was shaken with hydroxylamine hydrochloride (20 ml, 0.1 mol l^{-1}, acidified to pH 2 with nitric acid). This was also centrifuged, decanted, and stored as described above. For the organic bound fraction (oxidizable fraction), the residue from reducible fraction was shaken with 30% hydrogen peroxide (20 ml, acidified to pH 2 with nitric acid) and then ammonium acetate (25 ml, 1 mol l^{-1}, acidified to pH 2 with nitric acid). This was also centrifuged, decanted, and stored as above. For the residual fraction, the residue from oxidizable fraction was digested using the method described above for the total digestion of heavy metals.

The entire extraction procedure for soil samples in polyethylene tubes was the same as described for the air-dried samples except that samples were handled in an anaerobic chamber, using N as the purge gas. All analysis of heavy metal concentrations was determined using inductively coupled plasma-atomic emission spectrometry (ICP-AES).

Statistical Analysis

Analysis of variance was performed using the GLM procedure. Mean separations were performed using Duncan's Multiple Range to test soil pH and redox on soil heavy metal fractions. Linear relationship (correlation) within fractions and regression within the percentage distribution of each metal fraction in oxidized and reduced condition were conducted. All statistical analysis was conducted using SPSS (version 13.0).

RESULTS

Soil Physicochemical Properties

Soil physicochemical properties of rice paddies in Zhangjiagang are shown in Table 1. Average soil moisture of submerged soil was 30.2% and ranged from 24.6% to 33.9%. Soil OM ranged from 19.40 g kg^{-1} to 31.30 g kg^{-1} and had a mean of 25.47 g kg^{-1}. The most abundant metal was Pb, which had a mean concentration of 43.27 mg kg^{-1}. The least abundant metal was Cd, which had a mean concentration of 0.20 mg kg^{-1}. The coefficient of variation (CV) for Cd (41.3%) was higher than Cu, Pb, and Ni (25.6%, 8.8% and 9.3%, respectively).

Soil Heavy Metal Fraction Distribution

Heavy metal fraction distribution varied among elements (Fig. 1 and 2). Redox conditions had a significant effect on all Cu fractions (Fig. 1, Table 2), soil pH did not affect Cu fractions, and there was only a significant interaction effect of redox conditions and pH with reducible fraction. The predominant fractions were reducible fraction and oxidizable fraction (35–40%), followed by residual fraction (15–18%) and acid soluble fraction (<1%). Air-dried samples had significantly different Cu fraction distribution. Compared with reduced soil, acid soluble fraction was still the smallest fraction accounting for 5% of the total Cu. Reducible fraction and oxidizable fraction had significantly lower proportions while residual fraction had significantly higher proportion and was the predominant fraction.

Table 1. Soil chemical properties of rice paddies in Zhangjiagang County, Jiangsu.

Samples	pH	Moisture (%)	Soil Organic Matter (g kg^{-1})	Cu	Pb	Ni	Cd
				(mg kg^{-1})			
1	4.81	33.93	31.30	34.87	48.96	28.82	0.236
2	5.54	30.75	26.50	28.35	38.10	26.41	0.087
3	5.87	26.74	26.80	28.57	41.64	24.66	0.160
4	6.12	31.21	28.70	26.64	40.10	25.70	0.190
5	6.45	24.62	23.70	46.71	42.92	30.15	0.107
6	6.94	28.10	29.30	25.73	39.13	24.61	0.135
7	7.17	29.77	19.40	32.56	43.46	31.74	0.180
8	7.61	31.79	21.10	30.99	44.81	28.80	0.327
9	8.02	32.66	26.70	53.46	44.26	28.54	0.300
10	8.16	32.53	21.20	30.26	49.34	24.62	0.281
Average	6.67	30.2	25.47	33.82	43.27	27.41	0.200
St dev	1.10	2.9	3.95	8.66	3.79	2.55	0.083
Cv (%)	16.49	9.7	15.53	25.62	8.76	9.31	41.28

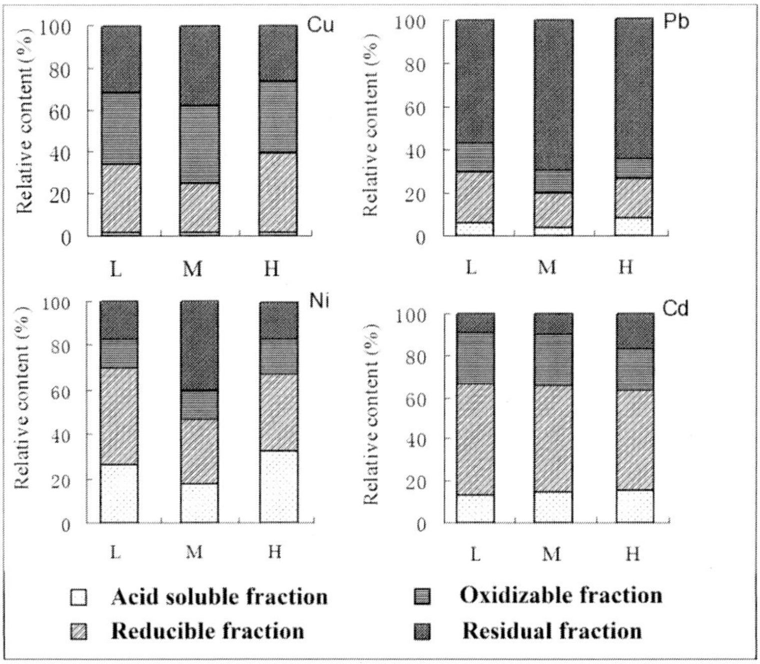

Figure 1. The percentage distribution of each metal fraction in reduced condition in paddy soil samples in Zhangjiagang County. L means lower pH (<6),

M means moderate pH (6–7), H means higher pH (>7). Note: significant interactions within and between factors are presented in Table 2.

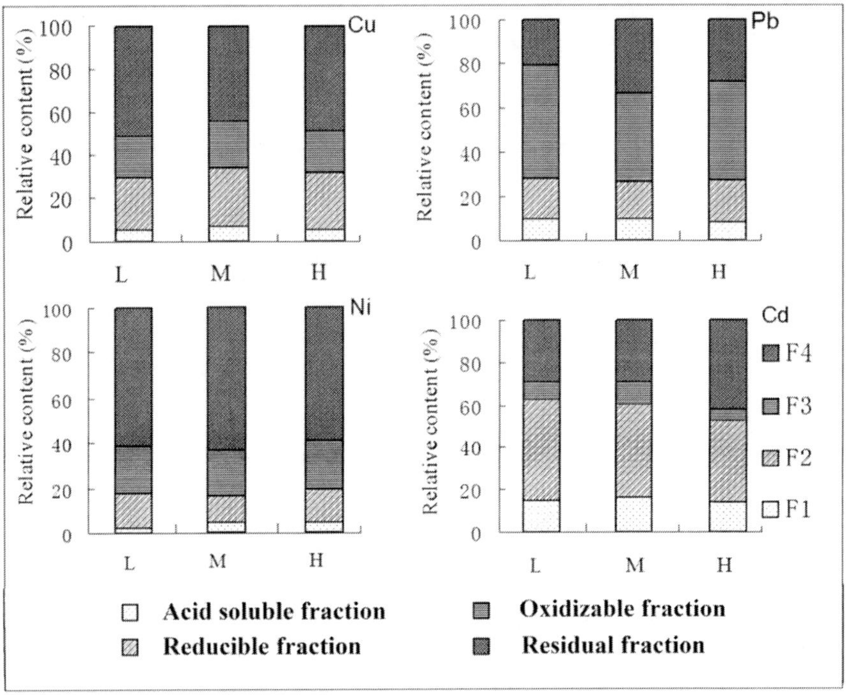

Figure 2. The percentage distribution of each metal fraction in oxidized condition in paddy soil samples in Zhangjiagang County. L means lower pH (<6), M means moderate pH (6–7), H means higher pH (>7). Note: significant interactions within and between factors are presented in

Table 2. P-values from the analysis of variance of redox conditions and pH for heavy metal fraction distributions.

Heavy Metal	Factors	P value			
		Acid soluble fraction	Reducible fraction	Oxidizable fraction	Residual fraction
Cu	Redox	<0.001	0.050	<0.001	<0.001
	pH	0.685	0.109	0.325	0.555
	Interaction	0.308	0.034	0.873	0.111
Pb	Redox	0.001	0.901	<0.001	0.000
	pH	0.230	0.544	0.124	0.068
	Interaction	0.123	0.092	0.345	0.970
Ni	Redox	0.000	0.000	0.000	0.000
	pH	0.023	0.009	0.314	0.009
	Interaction	0.040	0.304	0.356	0.164
Cd	Redox	0.08	0.316	0.000	0.024
	pH	0.810	0.261	0.015	0.142
	Interaction	0.089	0.359	0.277	0.413

P-values <0.05 show significant difference within factors and significant interaction between factors for each heavy metal fraction, separately.

Redox conditions affected all Pb fraction distributions except reducible fraction, and pH did not affect fraction distribution ($p>0.05$) (Table 2). There was also no interaction effect of redox conditions and pH with Pb fractions. In reduced soil, the predominant fraction was residual fraction (60%), followed by reducible fraction (20%), oxidizable fraction and acid soluble fraction. In oxidized soil, acid soluble fraction was 3% higher than in the reduced soil, oxidizable fraction was 20% higher and residual fraction was 33% lower, respectively, than in reduced soil, and reducible fraction was not significantly different.

Redox conditions affected all Ni fraction distributions and pH affected all fraction distributions except oxidizable fraction (Table 2). In reduced soil with pH 6 to 7, $Ni_{\text{oxidizable fraction}}$ had the lowest proportion, and $Ni_{\text{residual fraction}}$ had the highest (Fig. 1). There was significant interaction of redox conditions and pH with acid soluble fraction. In oxidized soil, acid soluble fraction decreased from 30% to 5% and reducible fraction decreased from 35% to 15%. The residual fraction increased by more than 60% and oxidizable fraction increased less than 5%.

Redox conditions and pH affected $Cd_{\text{oxidizable fraction}}$ and $Cd_{\text{residual fraction}}$ (Table 2). In reduced soil the predominant fraction was reducible fraction (55%), followed by oxidizable fraction (25%), acid soluble fraction and residual fraction. In oxidized soil, residual fraction increased by 15% and oxidizable fraction decreased by 15%. Treatments had no significant effect on acid soluble fraction or reducible fraction.

Regression of Reduced and Oxidized Samples

Regression analysis did not reveal significant relationships between soil heavy metal distribution of oxidized and reduced soil. None of the Cu fractions had significant regression results. For Pb, only residual fraction was significant, and for Ni, only reducible fraction was significant (Fig. 3). For Cd fractions, regression of oxidized and reduced samples had R^2 values of 0.26, 0.69, 0.48, and 0.66 for acid soluble fraction, reducible fraction, oxidizable fraction, and residual fraction, respectively (Fig. 4). Regression analysis between relative concentration in oxidized soil and relative concentration in reduced heavy soil showed significant correlation for reducible fraction, oxidizable fraction and residual fraction (Fig. 4). Regression analysis separately by pH and OM did not improve correlation between the dependent and independent variable in Fig. 4.

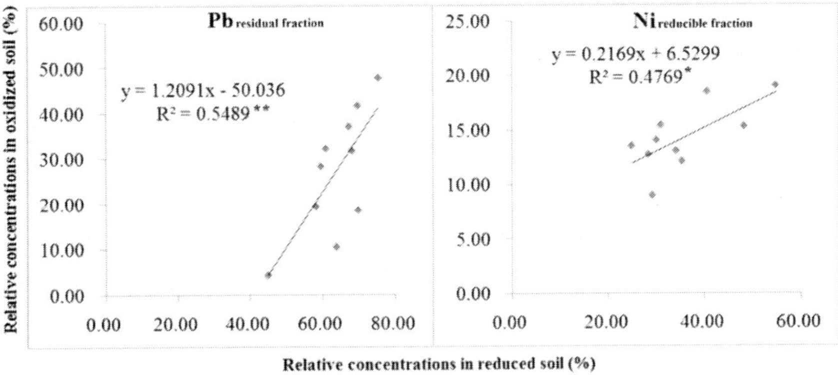

Figure 3. Correlation between the percentage distribution of iron and manganese bound nickel ($Ni_{reducible\ fraction}$) and residual lead ($Pb_{residual\ fraction}$) in oxidized and reduced paddy soil samples in Zhangjiagang County.

*, and ** denote the significant correlation at $p \leq 0.05$ and 0.01, respectively.

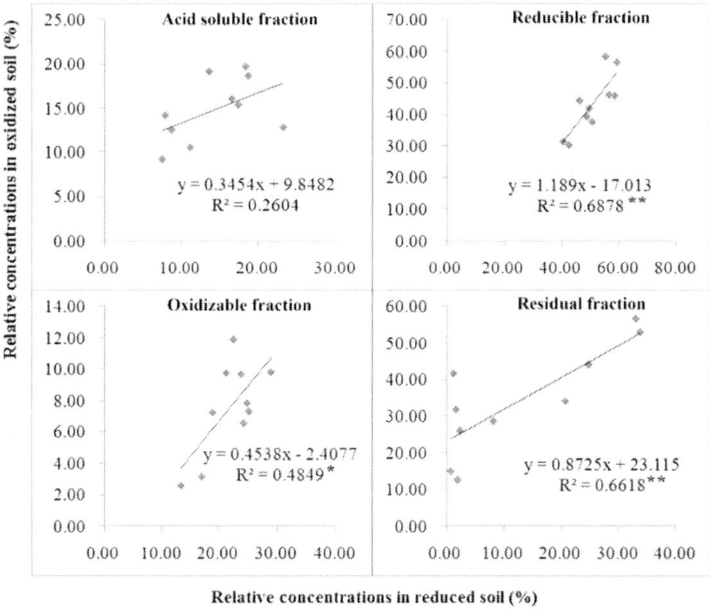

Figure 4. Correlation between the percentage distribution of cadmium in oxidized and reduced paddy soil samples in Zhangjiagang County.

*, and ** denote the significant correlation at $p \leq 0.05$ and 0.01, respectively.

DISCUSSION

Total Heavy Metal Concentrations in Rice Paddies

Soil heavy metals originate from both the environment and anthropogenic sources. Although all heavy metal concentrations in this study, except three Cd samples, were higher than typical background concentrations [24], they are lower than the critical limits established by the Soil Environmental Quality Standard of China (GB 15618–1995) [25]. The results of our study are consistent with previous research in the area [25]. In rice paddies of China, there is some evidence that this accumulation is a result of heavy metal rich irrigation water [9]. However, low concentrations of heavy metals suggest that although heavy metal concentrations in Zhangjiagang have experienced accumulation from anthropogenic sources, it is not yet a pressing environmental concern.

Heavy Metal Fractionation in Reduced Rice Paddies

Generally, rice paddies are submerged during rice cultivation and have lower Eh values during this part of the year. Reduced soil had lower $Cu_{acid\ soluble\ fraction}$ and $Pb_{acid\ soluble\ fraction}$ and higher $Ni_{acid\ soluble\ fraction}$, $Ni_{reducible\ fraction}$, and $Cd_{reducible\ fraction}$ than air-dried soil. This is similar to heavy metal fractionation changes that occur in various redox states of sewage sludge [26]–[28]. Förstner and Wittmann [3] explain that for Cu and Pb the most unstable fraction, acid soluble fraction, is replaced by reducible fraction and oxidizable fraction under reduced conditions.

For Ni and Cd, acid soluble fraction accounted for a relatively high proportion (15%–50%), which is not consistent with heavy metal fractionation ranges in sewage sludge. This may be due to Eh differences in rice paddies and sewage sludge, interactions in the rhizosphere of rice paddies, or both. Several studies address changes in soil heavy metal fraction distribution in the rhizosphere [29]–[31]. High $Ni_{acid\ soluble\ fraction}$ and $Ni_{reducible\ fraction}$ indicated that Ni is adsorbed to solid materials or combined with weak acid and Fe and Mn oxides in reduced soil [27]. High $Cd_{reducible\ fraction}$ was also found in dredged sediments [32].

High $Cu_{reducible\ fraction}$ and $Cu_{oxidizable\ fraction}$ in reduced soil indicates that Fe and Mn oxides play a crucial role in Cu processes, and that Fe and Mn oxides cause this fraction to be relatively stable [33]–[34]. This result is consistent with many studies which show dominant oxidizable fraction in reduced sediments [26], [34]–[37].

We found the predominant fraction for Pb in rice paddies to be residual fraction. This fraction is generally considered to be the most stable heavy metal fraction in the soil and unavailable to plants because they are found in the lattice of clay mineral [35], [38]. In addition to residual fraction, sediment Pb in many studies has been found to have large percentages of reducible fraction [2]. Lower reducible fraction in our study might be because of the higher Eh values, interactions in the rhizosphere of rice paddies, or both.

Effect of Drying on Heavy Metal Fractionation

When soils are routinely dried in the lab prior to chemical analysis, Eh rapidly increases and metals are oxidized. While this may not have a significant effect on heavy metal fractions in agriculture soil which is not submerged, this procedure can drastically change the heavy metal chemistry of reduced soils.

Generally, heavy metal acid soluble fraction increases as Eh increases, however, in our study, Ni and Cd did not respond accordingly. The reason for this is not clear and further study is needed. As Eh increases, reducible fraction generally decreases, most likely because of heavy metal bonding with Fe and Mn oxides [2], [4]–[5], [39]. Our study bears this out as well.

Previous research showed that the relationship of Eh with heavy metal oxidizable fraction and residual fraction is not clear [2]–[5], [7], [39]. In our study, increasing Eh led to increased $Pb_{oxidizable\ fraction}$ and decreased $Cu_{oxidizable\ fraction}$. Collavini et al. [40] suggested that decreased $Cu_{oxidizable\ fraction}$ as soil oxidizes occurs due to the interaction of copper sulfide oxidation and Fe and Mn oxides. Released Cu^{2+} is then redistributed to other fractions such as acid soluble fraction and residual fraction. Part of the obscurity may be because of the complex nature of bonding in copper sulfides, which have both covalent and ionic bonding characteristics and a high occurrence of delocalized electrons [41]–[42].

In this study, increasing Eh led to increased $Cu_{reducible\ fraction}$, which is in agreement with Kelderman and Osman [2], who reported a decrease in reducible fraction for river sediments under submerged conditions, and with Saeki et al. [14] who reported an increase in $Cu_{reducible\ fraction}$ after drying of lake sediments.

Drying caused an increase in $Pb_{oxidizable\ fraction}$, which contradicts previous literature [2], [7]. As Eh increases, $Pb_{reducible\ fraction}$ slightly decreases, mainly due to bonding of Pb with Fe and Mn oxides [7]. Dried soil had lower $Ni_{reducible\ fraction}$ and $Cd_{oxidizable\ fraction}$ and higher $Ni_{residual\ fraction}$ and $Cd_{residual\ fraction}$ than reduced soil. This was consistent with the results of other studies [2], [7], [40].

There is strong evidence that drying significantly affects soil heavy metal fractionation of submerged rice paddy soil samples but effects are different than lake sediment, river sediment, and sewage sludge. One of the main reasons most likely is the differences in Eh of the sample sources. Generally, sediment Eh is <−100mv, while rice paddy soil Eh ranged from 200–300 mv [4]. Abundant roots in rice paddies may be another important factor. Root exudates can affect many soil properties such as pH and microbial activity. Changes in these properties can then indirectly affect soil heavy metal fraction distribution. Chen et al.[43] reported that abundant rhizosphere can significantly increase $Cu_{acid\ soluble\ fraction}$ and significantly decrease $Cd_{acid\ soluble\ fraction}$. Shuman and Wang [44] reported that rice rhizosphere significantly increased $Zn_{reducible\ fraction}$ and $Cu_{oxidizable\ fraction}$. Similarly, Lin et al. [24] reported that rice rhizosphere significantly increased $Cd_{reducible\ fraction}$ and $Cd_{oxidizable\ fraction}$.

Effects of Drying on Heavy Metal Availability

The distribution, migration, and plant availability of soil heavy metals are reflected not in the total concentrations, but in specific fractions. With the BCR sequential extraction method, heavy metal bioavailability decreases with each progressively stronger extraction [45]. acid soluble fraction has the highest bioavailability and residual fraction has the lowest. Acid soluble fraction, reducible fraction, and oxidizable fraction are considered active fractions and each contain some bioavailable heavy metals [46]. In this study, Ni and Cd have large active fractions (80%), followed by Cu (>60%), and Pb (<40%). Drying submerged paddy soil samples caused decreases in the active fractions of Cu, Ni, and Cd and increased active fractions of Pb. Zheng and Zhang [13] found an increased bioavailability with increased Eh across all heavy metals measured (Cu, Pb, Cd, and Hg). Our results suggest that when samples are dried prior to heavy metal bioavailability analysis, Pb is overestimated and Cu, Ni, and Cd are underestimated.

Regression analysis results showed that samples which have been dried prior to analysis are not suitable indicators of reduced soil conditions and cannot be used to predict field conditions. However, adding pH and OM did not improve the predictive power, particularly for Cd (data not shown).

CONCLUSIONS

Almost all heavy metal fractions were significantly affected by redox conditions, and few heavy metal fractions were affected by pH. Drying the soil prior to analysis caused metal ions to change valence and a redistribution of heavy metal fractions. Some redistribution occurred as metals moving from active fractions to stable fractions. Therefore, drying decreases the representativeness of *in situ* conditions and the availability of some heavy metals are misestimated. We suggest that soil heavy metal fractionation procedures for rice paddies ensure anaerobic conditions from the time of sampling until analysis. Methodology studies to improve sampling and lab handling techniques would be beneficial, as would further investigation of possible correlation of non-dried sampling and analysis using bioavailable extractions with air-dried samples. For some heavy metals, the processes driving redistribution trends are still unclear. Because the relationship between soil Eh and heavy metal fractionation distribution directly affects heavy metal mobility and bioavailability, a better understanding of the soil heavy metal chemistry under various redox conditions is imperative.

AUTHOR CONTRIBUTIONS

Conceived and designed the experiments: YQ. Performed the experiments: YQ JLD. Analyzed the data: YQ BH. Contributed reagents/materials/analysis tools: BH. Wrote the paper: YQ.

REFERENCES

1. Gong ZT, Huang B (1998) Studies on potential "chemical timer bombs" and their igniting factors in soils. Adv. Earth Sci. 23: 184–191.

2. 2Kelderman P, Osman AA (2007) Effect of redox potential on heavy metal binding forms in polluted canal sediments in Delft (the Netherlands). Water Res. 41: 4251–4261.

3. Förstner U, Wittmann GTW (1983) Metal pollution in the aquatic environment, second ed. Springer, Berlin.

4. Li YG, Xue SG, Wu XY (2004) Transport and transformation of heavy metals in the soil-paddy plants system. Geo. China.31(supp.): 87–92.

5. Calmano W, Hong J, Förstner U (1993) Binding and mobilization of heavy metals in contaminated sediments affected by pH and redox potential. Water Sci. Tech. 28: 223–235.

6. Stumm W, Morgan JJ (1996) Aquatic chemistry: an introduction emphasizing chemical equilibria in natural water, third ed. Wiley, New York.

7. Lu YZ, Dong DM, Fu Y (2006) Comparison of speciation patterns of heavy metals in Yitong River sediment under reduced and oxidized conditions. Chem. J. Chinese. Uni. 27: 449–453.

8. Andreu V, Gimeno-García E (1999) Evolution of heavy metals in marsh areas under rice farming. Environ. Pollution 104: 271–282.

9. Luo YM, Jiang XJ, Wu LH, Song J, Wu SC, Lu RH, Christie P (2003) Accumulation and chemical fractionation of Cu in a paddy soil irrigated with Cu-rich wastewater. Geoderma 115: 113–120.

10. Li JX, Yang XE, He ZL, Jilani G, Sun CY, Chen SM (2007) Fractionation of lead in paddy soils and its bioavailability to rice plants. Geoderma 141: 174–180.

11. Zhao KL, Liu XM, Xu JM, Selim HM (2010) Heavy metal contaminations in a soil-rice system: Identification of spatial dependence in relation to soil properties of paddy fields. J. Hazard. Mater. 181: 778–787.

12. Ngoc MN, Dultz S, Kasbohm J (2009) Simulation of retention and transport of copper, lead and zinc in a paddy soil of the Red River Delta, Vietnam, Agr. Ecosyst. Environ. 129: 8–16.

13. Zheng SA, Zhang MK (2011) Effect of moisture regime on the redistribution of heavy metals in paddy soil. J. Environ. Sci. 23: 434–443.

14. Saeki K, Okazaki M, Matsumoto S (1993) The chemical phase changes in heavy metals with drying and oxidation of the lake sediments. Water Res. 27: 1243–1251.

15. Reddy CN, Patrick WHJ (1977) Effect of redox potential and pH on the uptake of cadmium and lead by rice plants. J. Environ. Qual. 6: 259–262.

16. Sajwan KS, Lindsay WL (1986) Effects of redox on zinc deficiency in paddy rice. Soil Sci. Soc. Am J. 50: 1264–1269.

17. Kashem MA, Singh BR (2001) Metal availability in contaminated soils: I. Effects of flooding and organic matter on changes in Eh, pH and solubility of Cd, Mi and Zn. Nutr. Cycl. Agroecosys. 61: 247–255.

18. Qi YB, Huang B, Gu ZQ, Sun WX, Zhao YC (2008) Spatial and temporal variation of C/N ratio of agricultural soils in typical area of Yangtze Delta region and its environmental significance. B. Miner. Petro. Geochem. 27: 50–56.

19. Shao XX, Huang B, Gu ZQ, Sun WX, Zhao YC (2006) Spatial-temporal variation of pH values of soil in a rapid economic developing area in the Yangtze Delta Region and their causing factors. B. Miner. Petro. Geochem. 25: 35–38.

20. Moore P, Coale F (2000) Phosphorus fractionation in flooded soils and sediments. In: G. M. Pierzenski, (Eds.), Methods of phosphorus analysis for soils, sediments, residuals, and waters. Available online athttp://www.soil.ncsu.edu/sera17/publications/sera17-2/pm_cover.htm.

21. Lu RK (2004) Soil and agro-chemical analytical methods, China Agricultural Science and Technology Press, Beijing.

22. Nelson DW, Sommers LE (1996) Total carbon, organic carbon, and organic matter, In: D. L. Sparks (Eds.), Methods of soil analysis, part 3: Chemical methods, Madison, Wisc.: SSSA and ASA, 961–1010.

23. Ure AM, Quevauviller PH, Muntau H (1993) Speciation of heavy metals in soils and sediments-an account of the improvement and harmonization of extraction techniques undertaken under the auspices of the BCR of the Comission of European Communities. Int. J. Environ. Anal. Chem. 51: 135–151.

24. Lin Q, Zheng CR, Chen HM (1998) Transformation of cadmium species in rhizosphere. Acta Pedologica Sinica 35: 461–467.

25. Shao XX, Huang B, Zhao YC (2008) Sun (2008) WX Pollution assessment of soil heavy metals in a representative area of the Yangtze River Delta region. Environ. Chem. 27: 218–221.

26. Tsai LJ, Yu KC, Chang JS (1998) Fractionation of heavy metals in sediment cores from the Ell-Ren River, Taiwan. Water Sci. Tech. 37: 217–224.

27. Tessier A, Campbell PGC, Bisson M (1979) Sequential extraction procedure for the speciation of particulate trace metals. Anal. Chem. 51: 844–851.

28. Argese E, Bettiol C (2001) Heavy metal partitioning in sediments from the lagoon of Venice (Italy). Toxicol. Environ. Chem. 79: 157–170.

29. Hinsinger P, Plassard C, Tang C, Jaillard B (2003) Origins of root-mediated pH changes in the rhizosphere and their responses to environmental constraints: a review. Plant Soil. 248: 43–59.

30. Tao S, Chen YJ, Xu FL, Cao J, Li BG (2003) Changes in copper speciation in maize rhizosphere soil. Environ Pollution. 122: 447–454.

31. Martínez-Alcalá I, Walker DJ, Bernal MP (2010) Chemical and biological properties in the rhizosphere of Lupinus albus alter soil heavy metal fractionation. Ecotox. Environ. Safe. 73: 595–602.

32. Singh SP, Ma LQ, Tack FMG (2000) Trace metal leach ability of land-disposed dredged sediments. J. Environ. Qual. 29: 1124–1132.

33. 33.Wu HS, Zhang AQ, Wang LS (2004) Immobilization study of biosorption of heavy metal ions onto activated sludge. J. Environ. Sci. 16: 640–645.

34. 3Wang C, Li XC, Ma HT, Qian J, Zhai JB (2006) Distribution of extractable fractions of heavy metals in sludge during the wastewater treatment process. J. Hazard. Mater. 137: 1277–1283.

35. Šurija B, Branica M (1995) Distribution of Cd, Pb, Cu and Zn in carbonate sediments from the Krka River estuary obtained by sequential extraction. Sci. Total Environ. 170: 101–118.

36. Li RY, Yang H, Zhou ZG, Lv JJ, Shao XH, Jin F (2007) Fractionation of heavy metals in sediments from Dianchi lake, China. Pedosphere 17: 265–272.

37. Shrivastaia SK, Banerjee DK (2004) Speciation of metals in sewage sludge and sludge-amended soils. Water, Air, Soil Pollut. 152: 219–232.

38. Teixeira E, Ortiz L, Alves M, Sanchez J (2001) Distribution of selected heavy metals in fluvial sediments of the coal mining region of Baixo Jacuí, RS, Brazil. Environ Geol. 41: 145–154.

39. Kazi TG, Jamali MK, Kazi GH, Arain MB, Afridi HI, et al. (2005) Evaluating the mobility of toxic metals in untreated industrial wastewater sludge using a BCR sequential extraction procedure and a leaching test. Anal. Bioanal. Chem. 383: 297–304.

40. Collavini F, Zonta R, Novelli AA, Zaggua L (2000) Heavy metals behavior during resuspension of the contaminated reduced sludge of the Venice canals. Toxicol. Environ. Chem. 77: 171–187.

41. Zheng SA, Zheng XQ, Chen C (2012) Leaching Behavior of Heavy Metals and Transformation of Their Speciation in Polluted Soil Receiving Simulated Acid Rain. PLoS ONE 7(11): e49664

42. Goh SW, Buckley AN, Lamb RN, Rosenberg RA, Moran D (2006) The oxidation states of copper and iron in mineral sulfides, and the oxides formed on initial exposure of chalcopyrite and bornite to air. Geochim. Cosmochim. Ac. 70: 2210–2228.

43. Chen YJ, Tao S, Deng BS, Zhang XQ, Huang Y (2001) Effect of root system on metal fractionation in rhizosphere of contaminated soil. Acta Pedologica Sinica 38: 54–59.

44. Shuman LW, Wang J (1997) Effect of rice variety on zinc, cadmium, iron and manganese contents in rhizosphere and non-rhizosphere soil fractions. Commun. Soil Sci. Plant 28: 23–36.

45. Kashem MA, Singh BR, Kswai S (2007) Mobility and distribution of Cadmium, Nickel, and Zinc in contaminated soil profiles. Earth Environ. Sci. 77: 187–198.

46. Wang CQ, Dai TF, Li B (2007) The speciation and bioavailability of heavy metals in paddy soils under the rice-wheat cultivation rotation. Acta Ecologica Sinica 27: 889–897.

Chapter 9

Soil Oxidation-Reduction in Wetlands and Its Impact on Plant Functioning

S. R. Pezeshki [1,*] and R. D. DeLaune [2]

[1]*Department of Biological Sciences, University of Memphis, Memphis, TN 38152, USA*

[2]*Department of Oceanography of Coastal Sciences, School of Coast & Environment, Louisiana State University, Baton Rouge, LA 70803, USA*

ABSTRACT

Soil flooding in wetlands is accompanied by changes in soil physical and chemical characteristics. These changes include the lowering of soil redox potential (Eh) leading to increasing demand for oxygen within the soil profile as well as production of soil phytotoxins that are by-products of soil reduction and thus, imposing potentially severe stress on plant roots. Various methods are utilized for quantifying plant responses to reducing soil conditions that include measurement of radial oxygen transport, plant enzymatic responses, and assessment of anatomical/morphological changes. However, the chemical properties and reducing nature of soil environment in which plant roots are grown, including oxygen demand, and other associated processes that occur in wetland soils, pose a challenge to evaluation and comparison of plant responses that are reported in the literature. This review emphasizes soil-plant interactions in wetlands, drawing attention to the importance of quantifying the intensity and capacity of soil reduction for proper evaluation of wetland plant responses, particularly at the process and whole-plant levels. Furthermore, while root oxygen-deficiency may partially account for plant stress responses, the importance of soil phytotoxins, produced as by-products of low soil Eh conditions, is discussed and the need for development of methods to allow differentiation of plant responses to reduced or anaerobic soil conditions *vs.* soil phytotoxins is emphasized.

Keywords: flooding; photosynthesis; plant stress; soil redox potential; soil phytotoxins; wetlands

1. INTRODUCTION

Excess water in wetland soils is a major factor affecting plant survival and functioning. In saturated soils, the supply of atmospheric oxygen into the soil is curtailed and various facultative and obligate anaerobic microorganisms use oxidized compounds as electron acceptors for respiration, thus converting them to reduced forms. The reduction and the associated processes influence plant survival, growth and functioning in wetlands. This review focuses on the literature relevant to this topic and the need for development of research methods associated with properly quantifying of soil reduction processes that can aid in the evaluation of plant responses in wetland environments.

2. SOIL OXIDATION-REDUCTION

A chain of reactions is initiated upon soil flooding leading to reduced (low) soil redox potential (Eh, mV) conditions. These reactions include physical, chemical and biological processes that have significant implications for wetland plants [1,2,3]. Physical processes include restriction of atmospheric gas diffusion in the soil leading to depletion of soil oxygen and accumulation of carbon dioxide [4,5]. Shortly after flooding, the limited supply of oxygen in soil pore spaces is depleted rapidly by roots, microorganisms, and soil reductants [6]. This process leads to oxygen depletion and reduction in soil oxidation reduction potential (Eh) followed by a chain of soil chemical changes. The processes that follow include denitrification, reduction of iron, manganese and sulfate, and changing soil pH and Eh [3]. For example, in a typical series of reductions NO_3^- is reduced to N2, Mn^{+4} to Mn^{+2}, Fe^{+3} to Fe^{+2}, SO_4^{2-} to H_2S, S^{2+} or HS^- (depending upon pH) and accumulations of acetic and butyric acids that are produced by microbial metabolism [1,2].

3. QUANTIFYING SOIL REDOX POTENTIAL CONDITIONS

Review of literature reveals that many terms such as "flooded", "saturated", "waterlogged" are utilized to describe oxygen-deficient root medium. Obviously, these terms are qualitative and do not quantitatively define the rhizosphere [7]. Furthermore, methods used to quantify oxygen content and oxygen diffusion rate in drained soils cannot be employed effectively in wet soils [3]. In contrast, measurement of soil Eh is a valuable tool because it can be measured in laboratory and in the field (for details see [8,9] and the references cited therein). Furthermore, quantifying soil Eh is particularly advantageous in periodically flooded soils since the range of Eh is much wider, between approximately −300 to +700 mV, than either aerated (Eh > +400 mV) or permanently waterlogged (Eh < +350 mV) soils [10,11,12], Thus, soil Eh or redox measurements represent an excellent quantifying tool for defining the presence or absence of oxygen and the soil chemical status in wetland soils.

4. INTENSITY AND CAPACITY OF SOIL REDUCTION

In wetland soils, the oxidation-reduction is used to signify the intensity of reduction. However, the reduction of inorganic redox systems including oxygen following flooding can be described in "intensity" and "capacity" terms [7]. Although critical Eh values (intensity of reduction) at which the inorganic redox systems become unstable do provide valuable information, they do not provide any indication of the total capacity of the system to accept electrons, supporting respiration or oxygen demand in the root zone [9]. In contrast, redox capacity represents the amount of electrons accepted by oxidants (such as oxygen, nitrate, manganese, iron, and sulfate) supporting respiratory of microorganism decomposing organic matter. The oxidants may have a relative high redox buffering capacity which explains why some wetland soil with relatively large amount of bioreducible iron do not undergo a rapid decrease in redox potential (intensity) following flooding. High soil reduction capacity at any intensity of reduction in the root zone can compete for oxygen transported thru aerenchyma into the root rhizosphere and potentially can increase plant stress above that caused by intensity of reduction alone (Figure 1).

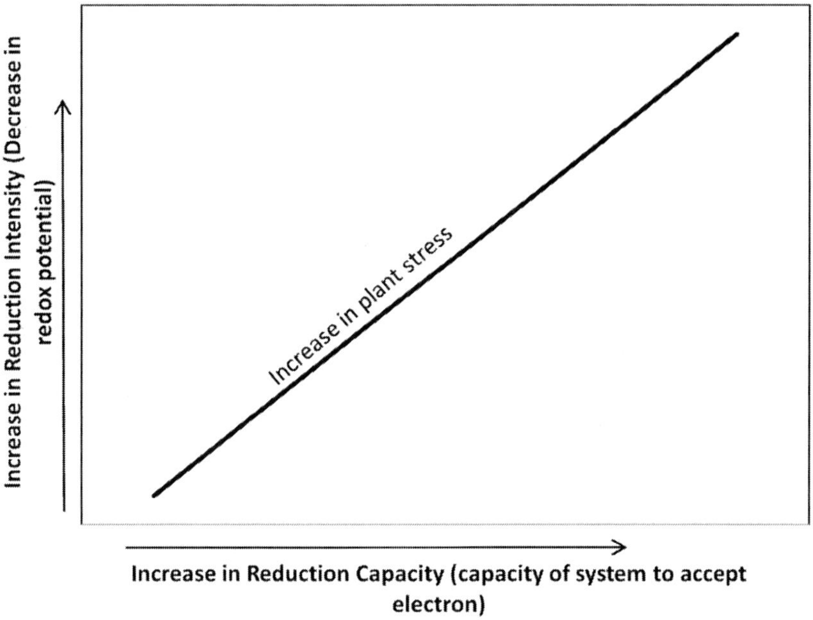

Figure 1. Oxidation-reduction potential in wetland soils: relationship of intensity and capacity of reduction in soil to wetland plant stress.

The intensity factor determines the relative ease of the reduction, while the capacity factor denotes the amount of redox system undergoing reduction, e.g., oxygen consumption at root interface [7,13,14]. This relationship may be expressed as follows: Eh = Eo + 2.3 [RT/nF] $\log_e$ [Ox/Red]; where Ox and Red are the concentrations of the oxidized and reduced forms, respectively, of the substance under consideration, Eo is the electrode potential of the 50% oxidized system specific to that substance, n is the number of moles of electrons transferred, F is the Faraday constant, R is the gas constant, and T is the temperature (K) [12]. It then follows that Eh is dependent on the ratio of oxidized and reduced forms and not on their absolute quantities. Therefore, a 90% oxidized system has the same electrode potential no matter whether the total concentration is 0.01% or 10%, but the poising (capacity) of the latter will be 1000 times greater. Thus, a system of Eo + 0.1 volt will oxidize a system of Eo − 0.1 volt, but will be oxidized by a system of Eo + 0.3 volts; however, the extent to which the reactions will take place depend upon the capacity of the systems.

The redox capacity factor is important although less is known about whether it is the intensity or the capacity that is most affecting plant functioning. As noted above, two soils with the same level of intensity of reduction may differ substantially in the capacity for reduction. The reduction capacity is estimated using measurements of soil respiration and calculating oxygen equivalent by stoichiometry [14]. However, part of the oxygen consumption is due to microbial respiration and partly due to direct oxidation of the accumulated reductants. Development of methodology for distinction of these two components therefore, is needed. In the laboratory, levels of soil redox capacity may be created and/or manipulated by adding extra carbon (organic matter) as energy source to the soil while maintaining the redox intensity level. In an experimental set up, reduction capacity may be controlled by adding granular D-glucose to the growth medium that is maintained under set redox intensity conditions [15].

To properly evaluate wetland plant responses to soil flooding, both intensity and capacity of soil reduction must be quantified because these two components influence oxygen demand in the soil [7,14]. Kludze *et al.* [16,17] and Sorrell *et al.* [18] using titanium citrate solution to create a high O_2-demand in root environment, reported that root oxygen transport and release were affected by such conditions in several wetland species. However, such a solution, while a significant improvement over de-oxygenated solution, at best only mimics wet soil conditions [14] and does not represent other important characteristics of wet soils, such as soil capacity for phytotoxin production that may have significant adverse effects on many wetland plants.

5. SOIL REDUCTION AND WETLAND PLANT FUNCTIONING

Responses of plants to low oxygen in the root zone have been typically assessed by growing plants hydroponically, then introducing pressurized nitrogen through the solution to remove oxygen [7]. In such a system, roots are exposed to Eh only slightly below values where oxygen disappears on the redox scale (*i.e.*, +350 to +400 mV; [7]). However, low oxygen conditions are represented by Eh values ranging between +400 and −300 mV. The Eh around −300 mV may occur in highly reduced soils. Since oxygen is absent at Eh values beginning at or below +350 mV, the absence of oxygen alone does not provide much information on the intensity of reduction. Furthermore, studies designed to evaluate responses of plants grown at the upper portion of the anaerobic range of the redox scale may yield results that are not typically the same as those displayed by plants growing in a more reducing environments.

Based on the considerations outlined above, it is argued that the common laboratory approaches of pressurized N_2 may be sufficient for evaluation of responses of flood-sensitive plants, such as many crop species, to low oxygen conditions. However, evaluation of responses of wetland plants to oxygen-deficient soils require methods that represent root zone reduction at levels at which flood-tolerant wetland species are subjected to in natural wetland soil enviornment. These plants can survive low to extreme reducing conditions covering a significant portion along the redox scale in their natural environments.

Clearly, in wetland soils, plants are faced with not only the lack of oxygen but a substantial demand for oxygen in the sediment due to plant roots and microbial demand [7,9,13,19,20,21,22,23,24,25,26]) that is represented by low soil Eh. Extensive field data on the relationship between Eh and plant community distribution in saltmarshes were presented by Armstrong *et al.* [27]. Pennington and Walters [28] and Davy *et al.* [29] reported sediment Eh as a major factor affecting survival and growth of their study species. It is known that in sediments characterized by weak redox capacity, certain wetland plants are capable of increasing the Eh of the bulk sediment [30,31,32].

In a typical flooded soil, plants may respond to soil physicochemical changes by displaying a wide range of stress symptoms that have been reported in numerous publications (for reviews see [19,22,23,33,34,35,36]. However, due to lack of quantifying Eh conditions in many reports, the relationship between soil Eh and wetland plant functioning is less understood. From a physiological-ecology standpoint, soil Eh data provide insights into the status of various soil compounds, many of them important to wetland plant functioning. For example, a soil Eh of zero mV indicates that oxygen and nitrate are not likely to be present while the bioreducible iron and manganese compounds are in a reduced state. Eh reading of +400 mV indicates that oxygen may be present despite the presence of excess water in the soil [7,37,38].

A major consequence of soil flooding and the subsequent reducing conditions is development of high competitive oxygen between root and soil microbial demand that may affect internal plant tissue oxygen concentrations thus, its critical metabolic processes [7,19,21,26,39,40,41]. Other consequences of soil reduction processes include changes in availability and/or concentrations of various nutrients that are essential for plant functioning, and production of a host of compounds known to be phytotoxic [3,8,42,43]. These compounds include: reduced forms of Fe and Mn, ethanol, lactic acid, acetaldehyde and aliphatic acids such as formic, acetic, butyric acids, and cyanogenic compounds [2,6,44]. The accumulation of these compounds in flooded soils may reach levels that can cause injury to plants [1,9,45]. For instance, excess soil sulfide is known to inhibit growth of various marsh macrophytes [20,21,46,47,48]. The soluble sulfide species including H_2S can be toxic to plant roots [38,49,50]. The organic acids such as acetic, butyric, propionic and caproic acids have a variety of adverse effects on plants [20,21,38,51,52].

In the following sections, the effects of soil flooding, and the concomitant drop in soil Eh, on certain plant metabolic processes, physiological functions, anatomical and morphological features, growth and survival is discussed.

6. METABOLIC RESPONSES TO FLOODING

Metabolic adaptations are complex mechanisms that allow wetland plants to survive anoxic soil conditions [8,38]. Normal growth and functioning of roots require more oxygen than is needed for root respiration processes alone [4,53]. Under aerated conditions, oxygen primarily diffuses into the roots from soil air spaces via the root epidermis. However, when roots are flooded, the presence of water column in soil creates a severe barrier to gas diffusion, thus oxygen must reach the roots through internal paths from aerial parts [38,54]. Such conditions lead to root energy deficiency because of the inhibition of normal (aerobic) root respiration. Therefore, much of the immediate flood-injury to roots is attributed to their oxygen status [4,28,55]. Under root oxygen stress, mitochondrial respiration is compromised while metabolism is shifted to fermentation. The process converts pyruvate to ethanol (Figure 2) via activity of enzyme alcohol dehydrogenase (ADH) or to lactate via activity of lactate dehydrogenase (LDH) [56].

Increased ADH activity in response to flooding has been reported for many plants as an indicator of oxygen deficiency [57,58,59]. The response has been reported for flood-sensitive as well as flood-tolerant plants [60,61]. The role of ADH in flood-tolerance has been known for sometime and may include maintaining intracellular pH. The ability to survive is related to the extent of cytosolic acidosis due to accumulation of organic acids from fermentation, such as lactate. The changes in cell pH have considerable impact on cell metabolism due to reduction in activities of several enzymes that require pH close to 7 for optimum activity [62]. The ADH activity appears to be inversely related to soil Eh due to anaerobic respiration [9]. Nonetheless, it is not known if the induction of ADH activity contributes to the plant survival or signals the harmful effects of root anoxia [54].

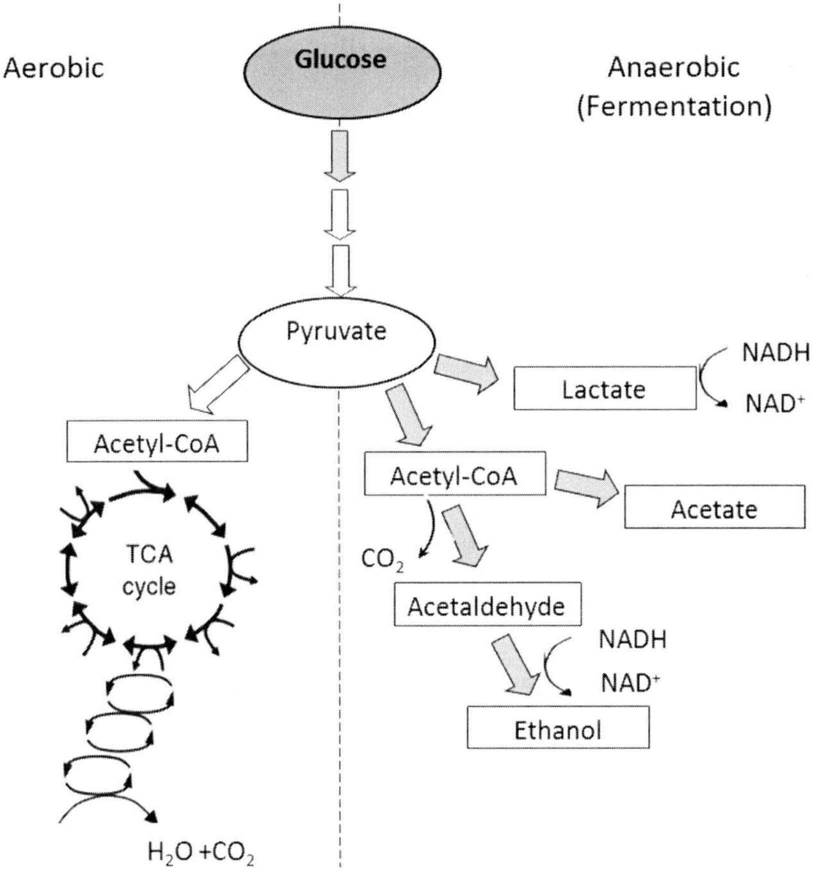

Figure 2. Metabolic pathways under aerobic and anaerobic conditions (from Reddy and DeLaune [9]).

Major products of fermentation are ethanol and CO_2. Anaerobic respiration, however, is rather an inefficient pathway of energy production because it produces 2 moles of ATP per each glucose molecule broken down to pyruvic acid compared to 36–38 ATPs generated from each glucose in aerobic respiration. The low ATP production creates an energy deficiency that affect many plant metabolic functions including water and nutrient uptake, internal solute transport, and photosynthetic carbon fixation. Additionally, the increased glucose consumption during fermentation may lead to carbohydrate depletion in plants leading to low growth. The metabolic processes affected may include reduction in activity of photosynthetic enzymes. The activity of these enzymes is highly sensitive to changes in environmental conditions [63,64,65,66]. In addition, decrease in leaf tissue chlorophyll content has been reported in response to environmental stresses including soil flooding [66,67]. Many species rely on anaerobic metabolism as a means of surviving anaerobic root conditions yielding very few molecules of ATP [4,68]. Wetland plants maintain carbohydrate reserves to support sustained ATP generation [69,70]. Thus, the carbohydrate stored in the roots may potentially play an important role in plant survival since presumably the carbohydrates can be used to feed anaerobic respiration under reduced soil Eh conditions.

The ability to survive flooding is also related to avoidance of ethanol production in glycolytic pathway. Ethanol accumulation in flood-sensitive species may lead to death of root cells, therefore, to avoid excess accumulation of ethanol, it may be leaked out of plant via transpiration stream and/or leakage as well as production of alternate end products such as malate and lactate. Reduction of water and nutrient uptake, disturbance of hormonal balance such as a decrease in gibberellin and cytokinin and an increase in abscisic acid and ethylene [71] due to flooding has been reported in the literature [72,73].

7. THE INTERNAL OXYGEN TRANSPORT SYSTEM

Root and rhizomes in flooded wetland plants obtain oxygen via gas-phase transport from the shoot system, internal photosynthetic oxygen production, or atmospheric oxygen through an extensive oxygen transport system (aerenchyma tissue) in roots, stems, and leaves [17]. This system allows a plant to transport the needed oxygen to the roots for maintaining aerobic respiration and to oxidize reducing compounds in the rhizosphere, thus establishing a gradient of soil oxygen availability. The role that roots of certain wetland woody species have on establishing such radial gradients in a soil profile that influences soil redox status has been reported by several researchers [74,75,76]. However, the development of morphological and anatomical features in response to flooding in most plants is time dependent. Thus, during the initial period of stress, the required energy for survival is mostly generated through anaerobic metabolism [4,63].

The internal system of large gas spaces reduces internal volume of respiring tissues and oxygen consumption, thus, enhancing the potential for oxygen reaching the distant underground portions of the plant [17,20,21]. Because of the advantages, the oxygen transport system has been considered as a major mechanism critical to plant's ability to cope with soil anoxia [17,19,20,21,28,29,30,50,77,78,79,80,81,82]. Diffusion, while the major pathway of root aeration in wetland plants, is not the only means, as rhizome ventilation due to pressurized throughflows of gases has been demonstrated for some species [20,21,83,84].

The effectiveness of the gas transport is primarily dependent on two factors: (1) the resistance to diffusion that is proportional to root length and inversely proportional to root porosity; and (2) the oxygen demand along the diffusion path resulting from respiratory needs as well as oxygen leakage from the roots into the rhizosphere [49,85]. Indeed, oxygen demands of roots and rhizosphere, which may include large communities of facultative anaerobes, are competitive for oxygen molecules because in reduced soils these systems compete for the plant pool of oxygen simultaneously [19,79,84]. As soil reduction continues, there is a progressively greater demand imposed upon roots for oxygen [7].

It is known that Eh conditions and microbial oxygen demand in the soil are among major factors affecting root oxygen release to the rhizosphere [26]. However, literature concerning the relationship between functional aspects of gas transport within plants and soil Eh conditions is limited. In a few studies that evaluated the relationship, it is evident that intense soil reduction promotes oxygen loss from root to the rhizosphere. For instance, high correlation was found between radial oxygen loss rates from roots and soil Eh intensity; there was an increasingly higher oxygen loss rates as soil Eh became more reduced [14]. However, the enhancement may require soil Eh below some threshold levels [86]. It is reported that in*Spartina patens*, a dominant US Gulf coastal brackish marsh species, root porosity increased as sediment Eh decreased, leading to root porosity of 22% in plants grown at +200 mV while porosity was 45% in plants grown at −300 mV Eh. Radial oxygen loss was significantly greater for plants grown in −300 mV Eh compared to plants grown at +200 mV Eh [87]. Other studies have shown similar patterns of responses for root porosity-soil Eh intensity relationship in wetland plants including swamp and bottomland woody species [17,88]. In contrast, Brix and Sorrell [89] reported that root porosity in two wetland species, *Phalaris arundinacea* and *Glyceria maxima*, did not change in response to a short-term (7–12 days) reducing Eh in root medium.

Despite the reported increase in aerenchyma tissue formation (and hence porosity) in wetland species in response to reducing Eh conditions, this enhancement may not be sufficient to satisfy the root respiratory needs for oxygen due to the greater radial oxygen loss rates in response to high intensity of reduction. Pezeshki *et al.* [90,91] concluded that despite a substantial enhancement of aerenchyma tissue formation in*Spartina patens*, alcohol dehydrogenase (ADH) activity continued to be higher in flooded than control plants indicating continued oxygen stress in the roots of flooded plants.

In addition to the effects of the intensity of reduction, differences in Eh capacity among wetland soils may influence many plant functions including oxygen transport, rhizosphere oxygenation, and, thus, many aspects of plant functioning [13,14]. Studies showed that increased Eh capacity under a constant Eh intensity did not have any significant effect on root porosity in *S. patens* but oxygen release from roots to the rhizosphere was increased in response to the increasing Eh capacity [14]. However, the authors reported that there was a threshold Eh capacity beyond which oxygen release remained constant and/or decreased in this species. The response was attributed to the potential effects of several factors such as soil phytotoxins as well as stomatal closure. Nonetheless, the reason(s) for such a response remains unknown.

8. PLANT NUTRITION

Plant nutrition is influenced by soil flooding and the processes associated with reducing soil Eh conditions. Many factors including soil physicochemical characteristics, nutrient pools, plant developmental and physiological status, and flood-tolerance capabilities are important [22,78,92]. Despite the adaptations reported for wetland plants, various nutritional deficiencies and toxicities may occur. Reduced soil conditions may lead to inhibition of nutrient uptake and transport due to root dysfunction, death [22,61,93,94] and blockages in the vascular and aerenchyma systems resulting from phytotoxin damages [20]. Oxygen supplies to the roots are critical for nutrient uptake and ion transport. Flood-induced stress include reduced water uptake due to root dysfunction that results from soil O2-deficiency including altered cation and anion concentration in xylem sap as well as decreased hydraulic conductance of roots has been reported in the literature for many species [95,96]. The generation of adequate ATP is an important requirement for active uptake that occur via the H+ translocating ATPase in the plasma membrane [43]. Jackson *et al.* [96] reported that flooding degraded root integrity within 10 hours of initiating treatment in tomato plants and solute uptake was deregulated. DeLaune *et al.* [93] using add labeled N-15 measured uptake in cherrybark oak (*Quercus falcata* var. *pagodaefolia*) and overcup oak (*Q. lyrata*) and reported decreased N-15 uptake under moderate reducing soil conditions as compared to oxidized conditons

There are ample reports of reduction of water and nutrient uptake, disturbance of hormonal balance such as a decrease in gibberellin and cytokinin and an increase in abscisic acid and ethylene in response to flooding [71,72,73]. Under reduced soil conditions, some wetland plants may continue ion uptake partly because of the internal O_2 supply system but partial anoxia in roots can reduce solute intake [43,61,97]. Nutrient concentrations at toxic levels may be accumulated in tissues under reduced conditions due to higher availability of certain nutrients and root dysfunction [9,98,99]. During prolonged flooding, as soil Eh reduction continues, pH decreases while zinc availability increases leading to high tissue zinc concentrations [100] and reduced ferric and manganic forms that are soluble [6]. Thus, tissue Mn and Fe concentrations are greater than found in plants under aerated conditions [101,102,103]. Leaf discoloration (bronzing) due to high soluble ferrous iron has been reported in some species [49].

Further soil Eh reduction leads to reduction of sulfate to sulfide by anaerobic microorganisms [104]. Although sulfide is phytotoxic, in most cases wetland plants have the capability to oxidize sulfide in the rhizosphere thus, avoiding or minimizing injury [49]. However, there are several reports confirming that excess soil sulfide may inhibit plant growth [20,21,46,47,48]. The soluble sulfide species including H_2S are toxic to the roots [49,50]. For instance, sulfate uptake, translocation and accumulation in foliage is documented for several species [105,106]. The inhibitory effect of H_2S on cytochrome oxidase is disruptive for aerobic respiration and excess cytosolic Fe and Mn is harmful to enzymatic structures [33]. The inhibitory effects of elevated sulfide concentrations on leaf photosynthetic capacity have been demonstrated in several wetland species including *Panicum hemitomom* and *Spartina alterniflora* (Figure 3, Pezeshki *et al.* [99,107]. Such photosynthetic response has been attributed to disruption of light reactions [108] and/or photophosphorylation [109], and alterations in activity of photosynthetic enzymes [110,111,112]. Sulfide utilization and injuries have been reported in hypoxic roots and rhizomes of *Phragmites australis* [113]. Sulfide has been implicated as a factor responsible for decreased plant growth and productivity in several wetland species [20,21,99,114].

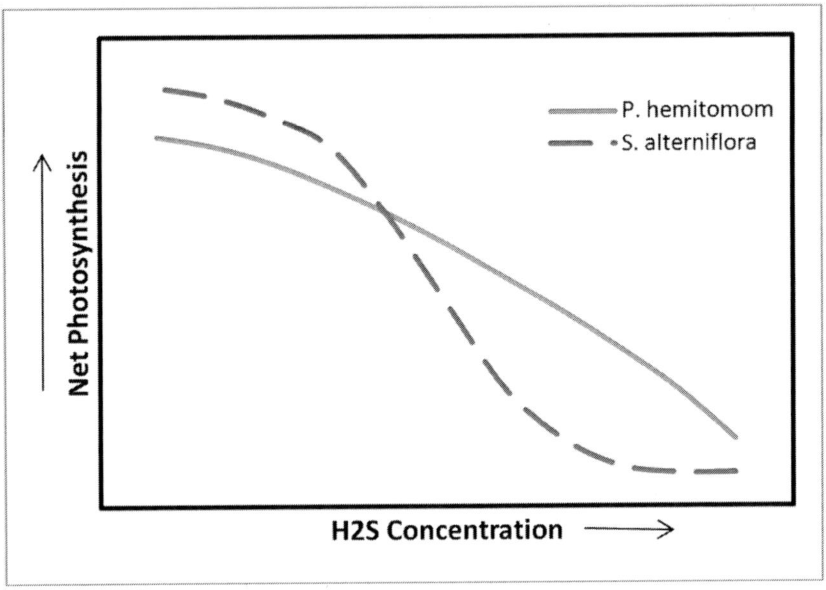

Figure 3. The relationship between net photosynthesis and H_2S concentration in the sediment for two wetland species, *Panicum hemitomom* and *Spartina alterniflora* (from Pezeshki *et al.* [99,107].

9. PLANT WATER RELATIONS AND GAS EXCHANGE

Factors associated with low soil Eh conditions can influence plant water relations through stomatal closure and slower water uptake than under aerated conditions [23,95,96,115,116,117,118]. Increased internal water stress and leaf dehydration leading to stomatal closure have been reported in some species due to a decrease in root permeability when roots were flooded [23,78,119]. Reduced water uptake due to root dysfunction may include altered cation and anion concentration in xylem sap. Decreased hydraulic conductance of roots has been reported in the literature for many species [95,96].

The extent of development of internal water stress reported for some species shows a wide range, however, in most cases the initial stomatal closure occurs in the absence of significant changes in plant water status [120,121,122]. The rapid stomatal closure and maintenance of a favorable water status thus, is likely due to low transpiration rates for which a slow water absorption rate by roots may sufficiently compensate rather than a sustained root conductivity [123,124]. Else *et al.* [116] found that in some species, flooded roots led to rapid stomatal closure via a root-shoot signaling mechanism, thus avoiding the possibility of dehydration. Many wetland species initially close stomata in response to soil flooding, however, exceptions have been reported in the literature. For instance, stomata in certain mangroves did not respond to Eh as low as −180 mV over short-term experimental exposures [125,126]. The stomatal closure is concomitant with a reduction in photosynthesis. However, under prolonged soil reduction stomatal reopening may occur leading to photosynthetic recovery [127]. The degree of resumption of stomatal functioning appears to be dependent on species, duration of reducing conditions, and the intensity of soil reduction.

Decreases in photosynthetic rates have been reported for some species in response to reducing soil Eh conditions; Figure 4, Figure 5 [23,28,81,82,117,128]. For example, DeLaune *et al.* [7] studied photosynthetic responses to soil Eh using Titanium-citrate solution. The authors noted that the intensity of reduction in growth medium had significant influence on plant phtosynthetic activity in marsh grass *Spartina patens* as Eh below −200 mV led to decreased photosynthesis. Krauss *et al.* [128] noted decreased stomatal conductance and net photosynthesis for baldcypress (*Taxodium distichum*) plants flooded in a greenhouse. As soil Eh dropped to −40 to −70 mV range, stomatal conductance reduced 36% and net photosynthesis dropped 40% compared to controls. In general, reducing soil Eh conditions adversely impact photosynthetic rates in wetland plants (Figure 6) although the extent of the impact varies across species as shown in Table 1. Additional studies by Pennington and Walters [28], Gravatt and Kirby [129], and Krauss *et al.* [128] have further confirmed the inhibitory effects of low soil Eh on photosynthesis of woody species. The photosynthetic decline may be attributed to a combination of diffusional limitations on gas exchange due to stomatal closure and metabolic (non-stomatal) inhibition [127]. However, the contribution of each component requires further assessment. In response to low soil Eh in range of −200 to +100 mV, a shift in light response curves was noted in some woody species suggesting the adverse effects of low Eh on the photosynthetic capacity of the leaves [88].

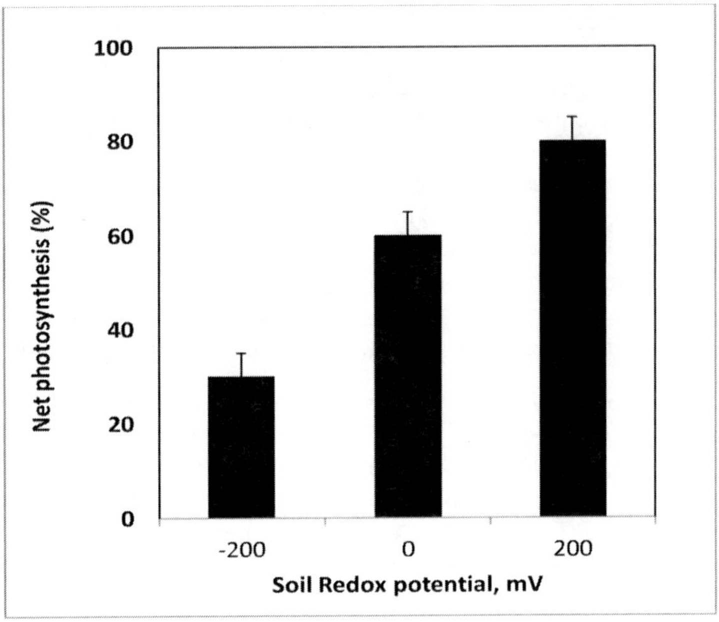

Figure 4. Changes in net photosynthesis of *Typha domingensis* in response to soil redox conditions. Values are presented as percent of controls (from Pezeshki *et al.* [157]).

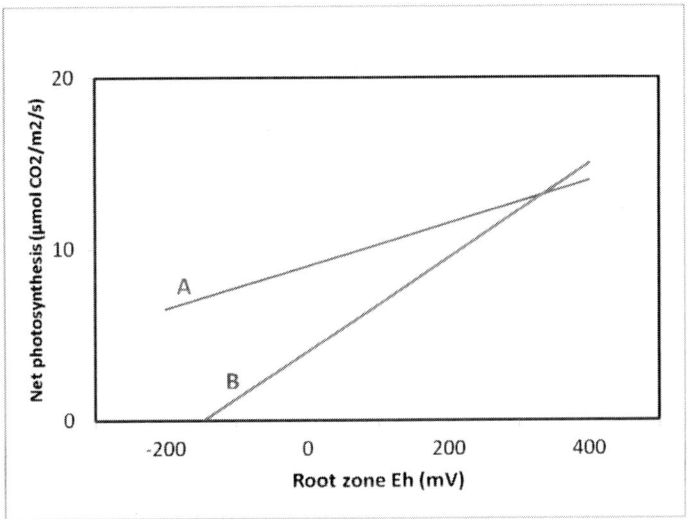

Figure 5. Response of light saturated leaf photosynthesis to root-zone redox potential (Eh) for 5-year-old (A) *Fraxinus pennsylvanica* and (B)*Quercus bicolor* in a created wetland (from Pennington and Walters [28]).

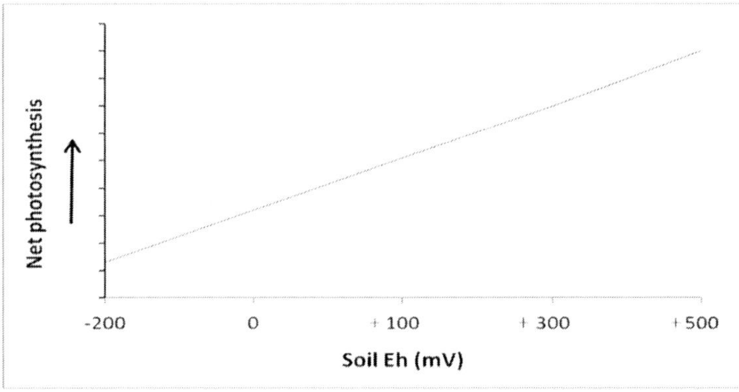

Figure 6. A generalized relationship between photosynthetic activity in wetland plants and soil redox potential (from Pezeshki [22,23]).

Table 1. Photosynthetic responses of selected marsh plants and wetland/bottomland woody species to the intensity of soil reduction. Values represent percent reduction in net photosynthetic rates compared to plants grown under oxidized (Eh > +400 mV) conditions or percent of pre-stress levels.

Species	Soil Eh (mV)	Photosyn. Response (%)	Reference
Marsh plants			
Typha domingensis	+200	18	[157]
	0	49	[157]
	−200	75	[157]
Cladium Jamaicense	+200	24	[157]
	0	46	[157]
	−200	100	[157]
Spartina alterniflora	<−200	15–21	[131]
Spartina patens	+230	7	[91]
	−110	18	[91]
Woody speccies			
Quercus lyrata	+340	54	[93]
	+175	66	[93]
Ligustrum sinese	+200/−200	86	[132]
Taxodium distichum	−70	5	[158]
	−160	22	[158]
Taxodium distichum	+350/+175	5	[159]
Taxodium distichum	−40/−70	36	[128]
	+18/+172	Decreased/recovered	[160]
Salix nigra	−130	53	[158]
	+50 to −80	5	[86]
Quercus nuttallii	+100/−220	35–68	[88]
Quercus nuttallii	+350/+175	21	[159]
Quercus falcata	+100/−220	65–87	[88]
Quercus michauxii	+350/+175	58	[159]

Chlorophyll fluorescence has been used to assess intact plants for detecting stress effects such as flooding effects [130] particularly the effects on PSII functioning. Utilizing this technique, Pociecha *et al.* [117] reported that in flooded *Vicia faba* L. plants, there was an apparent damage to photosynthetic apparatus as indicated by lower Fv/Fm compared to control plants. However, in wetland plants such as *Salix nigra*, Eh drop to the range of −50 to −100 mV did not have any significant effect on Fv/Fm [86]. This Eh range represents a moderately reduced soil conditions thus may explain the lack of PSII response in a wetland species such as *S. nigra*. Additional research is needed to assess PSII response to low soil Eh in wetland plants.

Photosynthetic recovery is reported for some wetland species despite the continuation of low soil Eh conditions provided that the soil is moderately reduced [88,127,131,132]. Soil Eh below +150 mV led to decreased net photosynthesis in *Alnus japonica* seedlings and the rate of recovery was correlated with the development of adventitious roots [81,82]. Even if a substantial recovery of photosynthetic rates occur, the impact of initial reduction and the delayed and/or slow recovery on plant survival and growth may be substantial due to the disruption of carbon fixation and decrease in photosynthate production [28,61,88,128,129].

What are the mechanisms through which the low soil Eh condition reduces leaf photosynthetic activity in many wetland plants? There are numerous factors that may be responsible for the decrease of net photosynthesis. For instance, ethylene accumulation has been implicated and the effects may be due to loss of photosynthetic capacity of mesophyll [133]. Low soil Eh conditions could lead to reductions of net photosynthesis due to decreased leaf water potential, reduced rubisco activity [66], disruption in photosynthate transport [61], alterations in source-sink relationship or reduced sink demand [23,33,134,135]. Other factors that may contribute to the reduction in photosynthetic capacity include low leaf chlorophyll content and/or leaf chlorophyll degeneration, and dysfunctioning of PSII. Recent reports showed that flooding led to limitations on the quantum efficiency of PSII in certain crop plants [116,136]. Leaf chlorophyll content decreases in some wetland plants that are subjected to reduced soil conditions [137].

Rubisco is an important enzyme in photosynthesis that catalyzes carboxylation yielding two molecules of 3-phosphoglycerate and oxygenation that produces one molecule of 3-phosphoglycerate and one molecule of 2-phosphoglycolate [138]. Because activity of this enzyme is critical for photosynthesis, flood-induced reductions of photosynthesis may be due, at least partially, to decreases in rubisco activity. In addition, the recovery of rubisco activity may contribute to the observed recovery of net photosynthesis in some wetland plants [22,66,127]. Indeed changes in rubisco (EC 4.1.1.39) activity may be an early signal contributing to the reduction of leaf photosynthesis in flooded plants as reported in the literature [66,139,140]. In some crop species, stress leads to reduced rubisco activity and enhanced rubisco degradation [65,141,142]. Li *et al.* [143] showed that decreased gene expression of oxygen evolving complex, large subunits of Rubisco and ferredoxin in response to flooding may have contributed to the observed photosynthetic reduction.

Low soil Eh conditions influences translocation of various photosynthetic products as reported for some [61,144,145]. The effects include low ATP production resulting from disruption of the oxidative phosphorylation [56,146,147], carbohydrate synthesis, transport, allocation, and utilization [78,148,149,150]. Carbohydrate allocation patterns and translocation rates appear to be critical for withstanding hypoxic condition [149,150,151]. However, the specifics of wetland plant responses remain to be evaluated.

10. GROWTH AND BIOMASS PRODUCTION

Decrease in survival and biomass accumulation in response to low soil Eh conditions is reported in certain wetland species. For example, while *T. distichum* remained relatively unaffected by Eh as low as −160 mV (Figure 7a–c), *Carya illinoesis* showed dramatic response at Eh around −70 mV (Kludze and DeLaune [13,14,23,61,86,87]. Significant changes are also reported in root to shoot ratio in some species as the effects of soil reduction are usually more drastic on root systems than shoots [86,114,152]. While a range of responses to low soil Eh conditions may occur in a given species, these changes, nevertheless, occur within the context of an overall reduction in biomass accumulation as reported for flooded plants of many species (Pezeshki [23] and the references cited therein). Root and shoot dry weights in *S. patens* decreased by 40% and 25% as soil Eh dropped from +200 mV to −300 mV, respectively. Results clearly indicated the influence of soil Eh intensity on growth of this dominant U.S. Gulf coast marsh species. It also was demonstrated that roots were more sensitive to Eh intensity than shoots (also data by Kludze and DeLaune [87]). In general, Eh reduction has pronounced effects on root elongation (Figure 8). Similar conclusions were drawn from a study on a flood-tolerant woody species *Taxodium distichum* [17]. Pezeshki and DeLaune [153] reported cessation of root growth in *S. patens* at soil Eh of −100 mV. In addition, Pezeshki *et al.* [90] noted smaller root system in *S. patens* under reducing soil conditions and concluded that such reduction in sink size, may in part, be responsible for a negative feed-back inhibition of photosynthesis thus, reduction in productivity of this species.

The increase in oxygen loss rates reported under reducing soil Eh [14,87] may explain the reductions in root growth of several wetland species under low soil Eh conditions. *Cladium jamaicense* and *Typha domingensis* produced significantly less biomass at Eh of −150 mV compared to plants grown at oxidized Eh of +600 mV [155]. Inhibitory effects of low soil Eh on photosynthesis and growth of woody species has also been reported by Pennington and Walters [28]. Root growth is an energy dependent process requiring oxygen, thus, under flooded conditions, root functioning is affected rapidly because molecular oxygen is required as an electron acceptor for oxidative phosphorylation [33,154]. Root elongation was inhibited in some bottomland woody species when soil Eh fell below +350 mV [152]. Interestingly, the +350 mV is the approximate Eh level that signifies the onset of oxygen disappearance from the soil system [7]. Root penetration depth was also adversely affected under low soil Eh treatment resulting in development of shallow root system different in architecture than those of plants growing in aerated (Eh > +400 mV) conditions [152,153]. The critical threshold Eh that inhibited root elongation differed among wetland species ranging from +300 mV to −200 mV [152,156].

Soil Eh capacity also influences wetland plant gas exchange and growth [13,14]. In the study, plants were grown under controlled Eh to examine the effect of Eh on CO_2 fixation. Soil redox capacity was imposed by application of different levels of extra energy source, by adding granular D-glucose to the growth medium, while maintaining Eh intensity at −200 mV. Increases in soil Eh capacity led to decreased photosynthesis, root growth, and biomass in *Oryza sativa* while enhanced root porosity and radial oxygen loss rate (Figure 9). Other studies demonstrated that plant carbon fixation, root, and shoot growth were significantly inhibited in *S. patens* under increasing soil reduction capacity. Root and shoot dry weights decreased by 70 and 37% in high reduction capacity conditions compared to control plants, respectively [14].

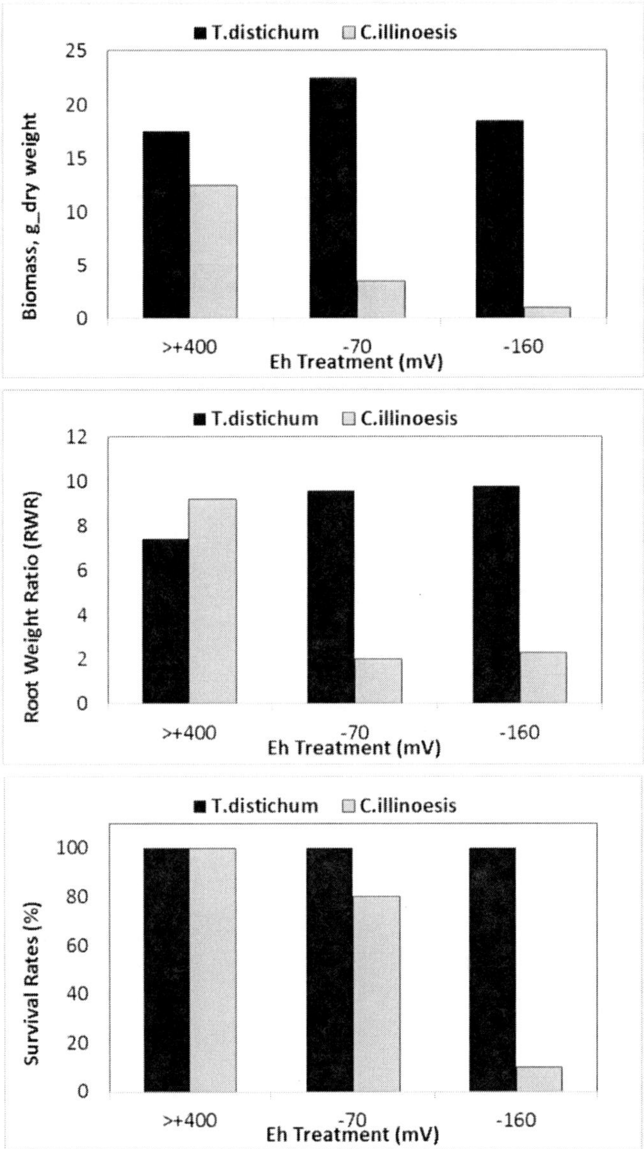

Figure 7. Biomass accumulation, root/weight ratio, and survival of two woody species *Taxodium distichum* and *Carya illinoesis* in response to changes in soil Eh conditions (from Pezeshki and Delaune [158]).

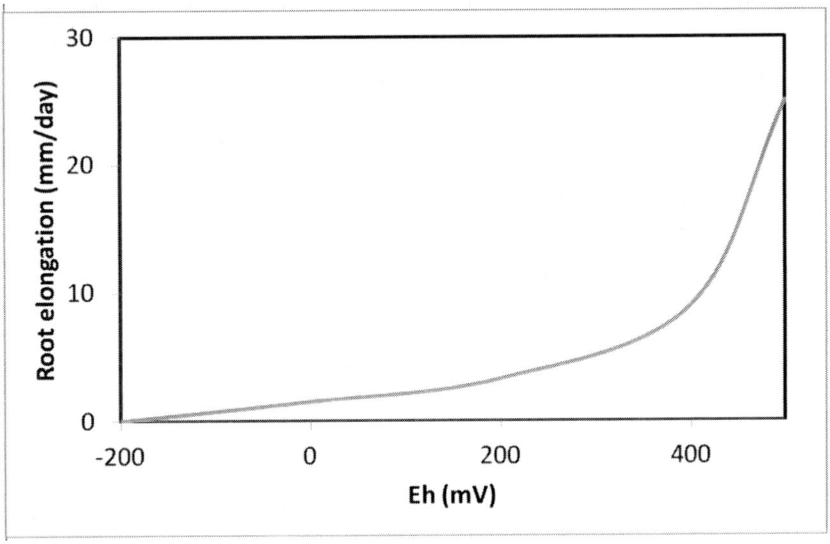

Figure 8. Responses of root elongation in *Spartina patens*, a brackish marsh species, to soil redox potential (Eh) (from Pezeshki and DeLaune [153]).

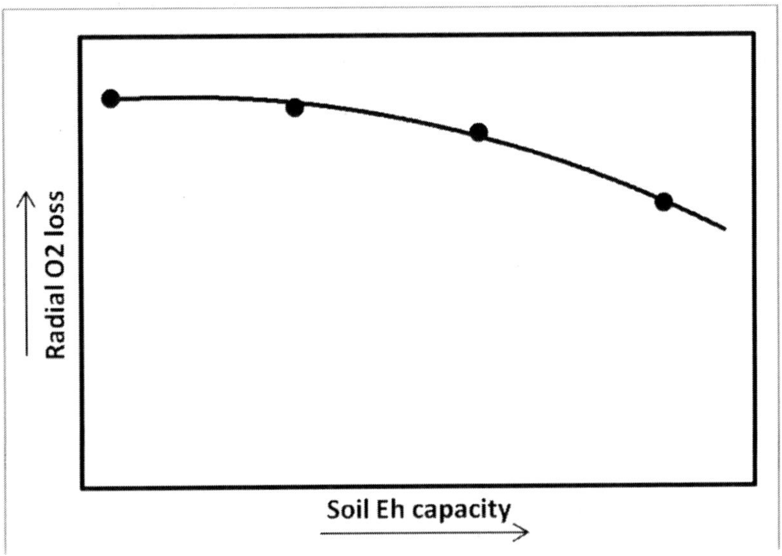

Figure 9. Radial O_2 loss (ROL) in *Spartina patens* in response to changes in soil reduction capacity while soil reduction intensity was maintained at −200 mV (Kludze and DeLaune [14]).

11. CONCLUDING REMARKS

There is a sufficient body of data to show that many plant functions are influenced by low soil Eh conditions. Wetland plant response is dependent on several factors including the species, duration of soil reduction, the timing, the intensity, and the capacity of soil reduction. The reducing condition of the soil is a major factor in wetland ecosystems that influences plant survival, growth, and productivity. Thus, quantifying soil reduction is critical to the understanding and interpretation of wetland plant responses to such conditions. In addition, plant response to low soil Eh conditions also reflects species' ability to respond to such conditions by utilizing a variety of morphological, anatomical and metabolic defense mechanisms. Nevertheless, many wetland species including those that possess a wide range of tolerance/avoidance capabilities to cope with low soil redox conditions are impacted negatively. The impact is a reflection of the fact that reducing soil conditions encompass not only soil oxygen deprivation but also production of various compounds in the soil, many of which considered highly phytotoxic. Thus, soil reducing conditions exert substantial influence on critical plant processes including gas exchange, water relations, photosynthate partitioning, translocation, hormonal balance, nutrition, growth, and biomass production.

Based on the limited available data, both intensity and capacity of reduction appear to influence plant functioning in wetland ecosystems although the roles of both factors need further investigations. In wetland soils, plants are faced with a substantial demand for oxygen from roots and soil microbial populations, the potential for loss of oxygen to soil through root radial oxygen loss that could improve bulk soil Eh conditions but nevertheless may impose plant internal O_2 deficiency, and the adverse effects of soil phytotoxins that are by-products of soil reduction. The severity of oxygen loss and the effects of reduction intensity and capacity on plant functioning are clearly common but vary across species. The need for additional data on various aspects of plant functioning and growth in wetland ecosystems in response to soil redox conditions, in terms of the intensity and the capacity, as well as the specific effects of soil phytotoxins is clear.

REFERENCES

1. Gambrell, R.P.; Patrick, W.H. Chemical and Microbiological Properties of Anaerobic Soils nad Sediments. In Plant Life in Anaerobic Environments; Hook, D.D., Crawford, R.M.M., Eds.; Ann Arbor Science: Ann Arbor, MI, USA, 1978; pp. 375–423.

2. Ponnamperuma, F.N. Effects of Flooding on Soils. In Flooding and Plant Growth; Kozlowski, T.T., Ed.; Academic Press Inc.: Orlando, FL, USA, 1984; pp. 1–44.

3. Gambrell, R.P.; DeLaune, R.D.; Patrick, W.H. Redox Processes in Soils Following Oxygen Depletion. InPlant Life Under Oxygen Deprivation: Ecology, Physiology, and Biochemistry; Jackson, M.B., Davies, D.D., Lambers, H., Eds.; SPB Academic Publishing BV: The Hague, The Netherlands, 1991; pp. 101–117.

4. Jackson, M.B.; Drew, M.C. Effects of Flooding on Growth and Metabolism of Herbaceous Plants. InFlooding and Plant Growth; Kozlowski, T.T., Ed.; Academic Press, Inc: Waltham, MA, USA, 1984; pp. 47–128.

5. Greenway, H.; Armstrong, W.; Colmer, T.D. Coditions leading to high CO2 (>5 kPa) in waterlogged flooded soils and possible effects on root growth and metabolism. Ann. Bot. 2006, 98, 9–32.

6. Ponnamperuma, F.N. The chemistry of submerged soil. Adv. Agron. 1972, 24, 29–96.

7. DeLaune, R.D.; Pezeshki, S.R.; Pardue, J.H. An oxidation-reduction buffer for evaluating physiological response of plants to root oxygen stress. Environ. Exp. Bot. 1990, 30, 243–247.

8. Patrick, W.H.; DeLaune, R.D. Chemical and biological redox systems affecting nutrient availability in the coastal wetlands. Geosci. Man 1977, XVIII, 131–137.

9. Reddy, K.R.; DeLaune, R.D. Biogeochemistry of Wetalnds: Science and Applications; CRC Press: Boca Raton, FL, USA, 2008; p. 774.

10. Pearsall, W.H.; Mortimer, C.H. Oxidation-reduction potentials in waterlogged soils, natural waters and muds. J. Ecol. 1939, 27, 483–501.

11. Mortimer, C.H. The exchange of dissolved substances between mud and water in lakes. J. Ecol. 1941,29, 280–329.

12. Mitsch, W.J.; Gosselink, J.G. Wetlands, 4th ed; John Wiley & Sons, Inc.: Hoboken, NJ, USA, 2007; p. 582.

13. Kludze, H.K.; DeLaune, R.D. Straw application effects on Methane and oxygen exchange and growth in rice. Soil Sci. Soc. Am. J. 1995, 59, 824–830.

14. Kludze, H.K.; DeLaune, R.D. Gaseous exchange and wetland plant response to soil redox intensity and capacity. Soil Sci. Soc. Am. J. 1995, 59, 939–945.

15. Kludze, H.K. Gaseous exchange and wetland plant response to soil redox conditions. Ph.D. Dissertation, Louisiana State University, Baton Rouge, LA, USA, 1994.

16. Kludze, H.K.; DeLaune, R.D.; Patrick, W.H. Aerenchyma formation and methane and oxygen exchange in rice. Soil. Sci. Soc. Am. J. 1993, 51, 386–391.

17. Kludze, H.K.; Pezeshki, S.R.; DeLaune, R.D. Evaluation of root oxygenation and growth in baldcypress in response to short-term soil hypoxia. Can. J. For. Res. 1994, 24, 804–809.

18. Sorrell, B.K.; Brix, H.; Orr, P.T. Oxygen exchange by entire root system of Cyperus involucratus and Eleocharis sphacelata. J. Aquat. Plant Manage. 1993, 31, 24–28.

19. Armstrong, W.; Brandle, R.; Jackson, M.B. Mechanisms of flood tolerance in plants. Acta Bot. Neerl.1994, 43, 307–358.

20. Armstrong, J.; Armstrong, W.; Beckett, P.M.; Halder, J.E.; Lythe, S.; Holt, R.; Sinclair, A. Pathways of aeration and the mechanisms and beneficial effects of humidity- and venturi-induced convections inPhragmites australis. Aquat. Bot. 1996, 54, 177–197.

21. Armstrong, J.; Armstrong, W.; van Der Putten, W.H. Phragmites die-back: Bud and root death, blockage within the aeration and vascular systems and the possible role of phytotoxins. New Phytol.1996, 133, 399–414.

22. Pezeshki, S.R. Plant Responses to Flooding. In Plant-Environment Interactions; Wilkinson, R.E., Ed.; Marcel Dekker, Inc.: New York, NY, USA, 1994; pp. 289–321.

23. Pezeshki, S.R. Wetland plant responses to flooding. Environ. Exp. Bot. 2001, 46, 299–312.

24. Sorrell, B.K. Effect of external oxygen demand on radia oxygen loss by Juncus roots in titanium citrate solutions. PlantCell Environ. 1999, 22, 1587–1593.

25. Van Wijck, C.; de Groot, C.J.; Grillas, P. The effect of anaerobic sediment on the growth of Potamogeton pectinatus: The role of

organic matter, sulphide, and ferrous iron. Aquat. Bot. 1992, 44, 31–49.

26. Laskov, C.; Horn, O.; Hupfer, M. Environmental factors regulating the radial oxygen loss from roots of Myriophyllum spicatum and Potamogeton crispus. Aquat. Bot. 2006, 84, 333–340.

27. Armstrong, W.; Wright, E.J.; Lythe, S.; Gaynard, T.J. Plant zonation and the effects of the spring-neap tidal cycle on soil aeration in a humber salt marsh. J. Ecol. 1985, 73, 323–339.

28. Pennington, M.R.; Walters, M.B. The response of vegetation zonation and soil redox potential in created wetlands. For. Ecol. Manag. 2006, 233, 1–10.

29. Davy, A.J.; Brown, M.J.H.; Mossman, H.L.; Grant, A. Colonization of a newly developing saltmarsh: Disentangling independent effects of elevation and redox potential on halophytes. J. Ecol. 2001, 99, 1350–1357.

30. Justin, S.H.F.W.; Armstrong, W. The anatomical characteristics of roots and plant response to soil flooding. New Phytol. 1987, 106, 465–495.

31. Pedersen, O.; Sand-Jensen, K. Adaptation of submerged Lobelia dortmanna to aerial life form: Morphology, carbon sources, and oxygen dynamics. Oikos 1992, 65, 89–96.

32. Pedersen, O.; Sand-Jensen, K.; Revsbech, N.P. Diel pulses of O2 and CO2 in sandy lake sediments inhabited by Lobelia dortmanna. Ecology 1995, 76, 1536–1545.

33. Drew, M.C. Sensing soil oxygen. Plant Cell Environ. 1990, 13, 681–693.

34. Drew, M.C. Oxygen deficiency and root metabolism: Injury and acclimation under hypoxia and anoxia. Ann. Rev. Plant Physiol. Plant Mol. Biol. 1997, 48, 223–250.

35. Hook, D.D.; Crawford, R.M.M. Plant Life in Anaerobic Environments; Ann Arbor Science: Woburn, MA, USA, 1978; p. 564.

36. Vartapetian, B.B.; Jackson, M.B. Plant adaptations to anaerobic stress. Ann. Bot. 1997, 79, 3–20.

37. Armstrong, W. The relationship between oxidation-reduction potentials and oxygen diffusion levels in some waterlogged organic soils. J. Soil Sci. 1967, 18, 27–34.

38. Turner, F.T.; Patrick, W.H. Chemical changes in waterlogged soils as a result of oxygen depletion.Trans. 9th Int. Congr. Soil Sci. 1968, 4, 53–56.

39. Armstrong, W.; Armstrong, J.; Beckett, P.M. Pressurized aeration in wetland macrophytes: Some theoretical aspects of humidity-induced convection and thermal transpiration. Folia Geobot. Phytotax.1996, 31, 25–36.

40. Greenway, H.; Gibbs, J. Mechanisms of anoxia tolerance in plants: II> Energy requirements for maintenance and energy distribution to essential processes. Funct. Plant Biol. 2003, 30, 999–1036.

41. Voesenek, L.A.C.J.; Colmer, T.D.; Pierik, R.; Millenaar, F.F.; Peeters, A.J.M. How plants cope with submergence. New Phytol. 2006, 170, 213–226.

42. DeLaune, R.D.; Smith, C.J.; Patrick, W.H. Nitrogen losses from a Louisiana Gulf Coast salt marsh. Est. Coast. Shelf Sci. 1983, 17, 133–141.

43. Armstrong, J.; Drew, M.C. Root Growth and Metabolism under Oxygen Deficiency. In Plant Roots: The Hidden Half; Waisel, Y., Eshel, A., Kafkaf, U., Eds.; Marcel Dekker: New York, NY, USA, 2002; pp. 729–761.

44. DeLaune, R.D.; Patrick, W.H.; Buresh, R.J. Sedimentation rates determined by Cs-137 dating in a rapidly accreting salt marsh. Nature 1978, 275, 532–533.

45. Drew, M.C.; Lynch, J.M. Soil anaerobiosis, micro-organisms and root functions. Ann. Rev. Phytopathol.1980, 18, 37–66.

46. King, G.M.; Klug, M.J.; Wiegert, R.G.; Chalmers, A.G. Relation of soil water movement and sulfide concentration to Spartina alterniflora production in a Georgia satlmarsh. Science 1982, 218, 61–63.

47. Ingold, A.; Havill, D.C. The influence of sulfide on the distribution of higher plants in salt marshes. J. Ecol. 1984, 72, 1043–1054.

48. Havill, D.C.; Ingold, A.; Pearson, J. Sulfide tolerance in coastal halophytes. Vegetatio 1985, 62, 279–285.

49. Tanaka, A.; Mulleriyawa, R.P.; Yasu, T. Possibility of hydrogen sulfide induced iron toxicity of the rice plant. Soil Sci. Plant Nutr. 1968, 4, 1–6.

50. Allam, A.I.; Hollis, J.P. Sulfide inhibition of oxidases in rice roots. Phytopathology 1972, 62, 634–639.

51. Rao, D.N.; Mikkelsen, D.S. Effects of acidic, propionic, and butyric acids on rice seedling growth and nutrition. Plant Soil 1977, 47, 323–334.

52. Armstrong, J.; Armstrong, W. Phragmites dieback: Toxic effects of propionic, butyric and caproic acids in relation to pH. New Phytol. 1999, 142, 201–217.

53. Atwell, B.J.; Greenway, H. The relationship between growth and oxygen uptake in hypoxic rice seedlings. J. Exp. Bot. 1987, 38, 454–465.

54. Armstrong, W. Aeration in higher plants. Adv. Bot. Res. 1979, 7, 225–332.

55. Drew, M.C. Soil aeration and plant root metabolism. Soil Sci. 1992, 154, 259–268.

56. Hopkins, W.G.; Huner, N.P. Introduction to Plant Physiology, 4th ed; John Wiley & Sons, Inc.: New York, NY, USA, 2009; p. 503.

57. Keeley, J.E. Population differentation along a flood frequency gradient: Physiological adaptation to flooding in Nyssa sylvatica. Ecol. Monogr. 1979, 49, 89–108.

58. Mendelssohn, I.A.; McKee, K.L.; Patrick, W.H. Oxygen deficiency in Spartina alterniflora roots: Metabolic adaptation to anoxia. Science 1981, 214, 439–441.

59. Crawford, R.M.M. Oxygen availability as an ecological limit to plant distribution. Adv. Ecol. Res. 1992,23, 93–285.

60. Kennedy, R.A.; Rumpho, M.E.; Fox, T.C. Aerobic metabolism in plants. Plant Phys. 1992, 100, 1–6.

61. Kogawara, S.; Yamanoshita, T.; Norisada, M.; Masumori, M.; Kojima, K. Photosynthesis and photoassimilate transport during root hypoxia in Melaleuca cajuputi, a flood-tolerant species, and in Eucalyptus camadulensis, a moderately floo-tolerant species. Tree Physiol. 2006, 26, 1413–1423.

62. Roberts, J.K.M.; Andrade, F.H.; Anderson, J.C. Further evidence that cytoplasmic acidosis is a determinant of flooding intolerance in plants. Plant Physiol. 1985, 77, 492–494.

63. Barmore, C.R. Effect of ethylene on chlorophyllase activity and chlorophyll content in calamond in rind tissue. HortScience 1975, 10, 595–596.

64. Berry, J.; Bjorkman, O. Photosynthetic and adaptation to temperature in higher plants. Annu. Rev. Plant Physiol. 1980, 31, 491–543.

65. Majumdar, S.; Ghosh, S.; Glick, B.R.; Dumbroff, E.B. Activities of chlorophyllase, phosphoenolpyruvate carboxylase and ribulose-1, 5-bisphosphate carboxylase in the primary leavaes of soybean during senescence and drought. Physiol. Plant 1991, 81, 473–480.

66. Pezeshki, S.R. Response of baldcypress seedlings to hypoxia: Leaf protein content, ribulose-1, 5-bisphosphate carboxylase/oxygenase activity and photosynthesis. Photosynthetica 1994, 30, 59–68.

67. Li, S.; Pezeshki, S.R.; Goodwin, S.; Shields, F.D. Physiological responses of black willow (Salix nigra) cuttings to a range of soil moisture regimes. Photosynthetica 2004, 42, 585–590.

68. Saglio, P.H.; Drew, M.C.; Pradet, A. Metabolic acclimation to anoxia induced by low (2–4 KPa partial pressure) oxygen pretreatment (hypoxia) in root tips of Zea mays. Plant Physiol. 1988, 86, 61–66.

69. Bertani, A.; Brambilla, I.; Menegus, F. Effect of anaerobiosis on rice seedlings: Growth, metabolic rate and fate of fermentation products. J. Exp. Bot. 1980, 31, 325–331.

70. MacDonald, R.C.; Kimmerer, T.W. Metabolism of transpired ethanol by eastern cottonwood (Populus deltoides). Plant Physiol. 1993, 102, 173–179.

71. Jackson, M.B. Long-distance signaling from roots to shoots assessed: The flooding story. J. Exp. Bot.2002, 53, 175–181.

72. Dat, J.F.; Capelli, N.; Folzer, H.; Bourgeade, P.; Badot, P.-M. Sensing and signaling during plant flooding. Plant Physiol. Biochem. 2004, 42, 273–282.

73. Hattori, Y.; Nagai, K.; Furukawa, S.; Song, X.; Kawano, R.; Sakakibara, H.; Wu, J.; Matsumoto, T.; Yoshimura, A.; Kitano, H.; et al. The ethylene response factors SNORKEL1 and SNORKEL2 allow rice to adapt to deep water. Nature 2009, 460, 1026–1030.

74. McKee, K.L. Growth and physiological responses of neotropical mangrove seedlings to root zone hypoxia. Tree Physiol. 1996, 16, 883–889.

75. Youssef, T.; Saenger, P. Anatomical adaptive strategies to flooding and rhizophere oxidation in mangrove seedlings. Aust. J. Bot. 1996, 44, 297–313.

76. Youssef, T.; Saenger, P. Photosynthetic gas exchange and accumulation of phytotoxins in mangrove seedlings in response to soil physico-chemical characteristics associated with waterlogging. Tree Physiol. 1998, 18, 317–324.

77. Teal, J.M.; Kanwisher, J.W. Gas transport in the marsh grass Spartina alterniflora. J. Exp. Bot. 1966, 17, 355–361.

78. Kozlowski, T.T. Responses of woody plants to flooding and salinity. Tree Physiol. Monogr. 1997, 1, 1–29.

79. Armstrong, W.; Justin, S.H.F.W.; Beckett, P.M.; Lythe, S. Root adaptation to soil waterlogging. Aquat. Bot. 1991, 39, 57–73.

80. Perata, P.; Alpi, A. Plant responses to anaerobiosis. Plant Sci. 1993, 93, 1–17.

81. Iwanaga, F.; Yamamoto, F. Growth, morphology and photosynthetic activity in flooded Alnus japonica seedlings. J. For. Res. 2007, 12, 243–246.

82. Iwanaga, F.; Yamamoto, F. Effects of flooding depth on growth, morphology and photosynthesis inAlnus japonica species. New For. 2008, 35, 1–14.

83. Dacey, J.W.H. Pressurized ventilation in the yellow water lily. Ecology 1981, 62, 1137–1147.

84. Armstrong, W.; Beckett, P.M. Internal aeration and the development of stelar anoxia in submerged roots: A multi-shelled mathematical model combining axial diffusion of oxygen in the cortex with radial losses to the stele, the wall layers, and the rhizosphere. New Phytol. 1987, 105, 221–245.

85. Luxmoore, R.J.; Sojka, R.E.; Stolzy, L.H. Root porosity and growth responses of wheat to aeration and light intensity. Soil Sci. 1972, 113, 354–357.

86. Li, S.; Pezeshki, S.R.; Shields, F.D. Partial flooding enhances aeration in adventitious roots of black willow (Salix nigra) cuttings. J. Plant Physiol. 2006, 163, 619–628.

87. Kludze, H.K.; DeLaune, R.D. Methane emission and growth of Spartina patens in response to soil redox intensity. Soil Sci. Soc. Am. J. 1994, 58, 1838–1845.

88. Pezeshki, S.R.; Anderson, P.A. Responses of three bottomland woody species with different flood-tolerance capabilities to various flooding regimes. Wetl. Ecol. Manag. 1997, 4, 245–256.

89. Brix, H.; Sorrell, B.K. Oxygen stress in wetland plants: Comparison of de-oxygenated and reducing root environments. Funct. Ecol. 1996, 10, 521–526.

90. Pezeshki, S.R.; Matthews, S.W.; DeLaune, R.D. Root cortex structure and metabolic response ofSpartina patens to soil redox conditions. Environ. Exp. Bot. 1991, 31, 91–97.

91. Pezeshki, S.R.; Pardue, J.H.; DeLaune, R.D. The influence of oxygen deficiency and redox potential on alcohol dehydrogenase activity, root porosity, ethylene production and photosynthesis in Spartinapatens. Environ. Exp. Bot. 1993, 33, 565–573.

92. Pezeshki, S.R.; DeLaune, R.D.; Anderson, P.H. Effect of flooding on elemental uptake and biomass allocation in seedlings of three bottomland tree species. J. Plant Nutr. 1999, 22, 1481–1494.

93. DeLaune, R.D.; Pezeshki, S.R.; Lindau, C.W. Influence of soil redox potential on nitrogen uptake and growth of wetland oak seedlings. J. Plant Nutr. 1998, 21, 757–768.

94. DeLaune, R.D.; Jugsujinda, A.; Reddy, K.R. Effect of root oxygen stress on phosphorus uptake by cattail. J. Plant Nutr. 1999, 22, 459–466.

95. Else, M.; Coupland, A.D.; Dutton, L.; Jackson, M.B. Decreased root hydraulic conductivity reduces leaf water potential, initiates stomatal closure, and slows leaf expansion in flooded plantsof castor oil (Ricinus communis) despite diminished delivery of ABA from roots to shoots in xylem sap. Physiol. Plant 2001, 111, 46–54.

96. Jackson, M.B.; Saker, L.R.; Crisp, C.M.; Else, M.A.; Janowiak, F. Ionic and pH signaling from roots to shoots of flooded tomato plants in relation to stomatal closure. Plant Soil 2003, 253, 103–113.

97. Gibbs, J.; Turner, D.W.; Armstrong, W.; Darwent, M.J.; Greenway, H. Response to oxygen deficiency in primary maize roots. I. Development of oxygen deficiency in the stele reduces radial solute transport to the xylem. Aust. J. Plant Physiol. 1998, 25, 745–758.

98. Hook, D.D.; DeBell, D.S.; McKee, W.H.; Askew, J.L. Responses of loblolly pine (mesophyte) and swamp tupelo (hydrophyte) seedlings to soil flooding and phosphorous. Plant Soil 1983, 71, 387–394.

99. Pezeshki, S.R.; Pan, S.Z.; DeLaune, R.D.; Patrick, W.H. Sulfide-induced toxicity: Inhibition of carbon assimilation in Spartina alterniflora. Photosynthetica 1988, 22, 437–442.

100. Pavanasasivam, V.; Axley, J.H. Influence of flooding on the availability of soil zinc. Commun. Soil Sci. Plant Anal. 1980, 11, 163–174.

101. Good, B.G.; Patrick, W.H. Gas composition and respiration of water oak (Quercus nigra) and green ash (Fraxinus pennsylvanica) roots after prolonged flooding. Plant Soil 1987, 97, 419–427.

102. McKevlin, M.R.; Hook, D.D.; McKee, W.H.; Wallace, S.U.; Woodruff, J.R. Loblolly pine seedling root anatomy and iron accumulation as affected by soil waterlogging. Can. J. For. Res. 1987, 17, 1257–1264.

103. Gries, C.; Kappen, L.; Losch, R. Mechanism of flood tolerance in reed (Phragmites australis). New Phytol. 1990, 114, 589–593.

104. Armstrong, W. Waterlogged Soils. In Environment and Plant Ecology; Etherington, J.R., Ed.; Wiley: New York, NY, USA, 1975; p. 181.

105. Carlson, P.R.; Forresst, J. Uptake of dissolved sulfide by Spartina alterniflora: Evidence from natural sulfur isotope abundance ratios. Science 1982, 216, 633–635.

106. Pearson, J.; Havill, D.C. The effect of hypoxia and sulfide on culture grown wetland an non-wetland plants. J. Exp. Bot. 1988, 39, 363–370.

107. Pezeshki, S.R.; DeLaune, R.D.; Pan, S.Z. Relationship of soil hydrogen sulfide level to net carbon assimilation of Panicum hemitomon and Spartina patens. Vegetatio 1991, 95, 159–166.

108. Shimazaki, K.; Sugahara, K. Inhibition site of the electron transport system in lettuce chloroplasts by fumigation of leaves with SO2. Plant Cell Physiol. 1980, 21, 125–135.

109. Wellburn, A.R.; Higginson, C.; Robinson, D.; Walmsley, C. Biochemical explanations of more than additive inhibitory effects of low atmospheric levels of sulphur dioxide plus nitrogen dioxide upon plants. New Phystol. 1981, 88, 223–237.

110. Garsed, S.G. The use of sulfite solution for studying the effects of SO2 on higher plants. Environ. Pollut. Ser. 1981, 24, 883–886.

111. Khan, A.A.; Malhotra, S.S. Ribulose bisphosphate carboxylase and glycollate oxidase from jack pine: Effects of sulphur dioxide fumigation. Phytochemistry 1982, 21, 2607–2612.

112. Dropff, M.J. Physiological effects of sulfide dioxide: 1. The effect of SO2 on photosynthesis and stomatal regulation of Vicia faba L. Plant Cell Environ. 1987, 10, 753–760.

113. Furtig, K.; Ruegsegger, A.; Brunhold, C.; Brandle, R. Sulfide utilization and injuries in hypoxic roots and rhizomes of Common Reed (Phragmites australis). Folia Geobot. Phytotax. 1996, 31, 143–151.

114. DeLaune, R.D.; Smith, C.J.; Patrick, W.H. Relationship of marsh elevation, redox potential and sulfide to Spartina alterniflora productivity. Soil Sci. Soc. Am. J. 1983, 47, 930–935.

115. Everard, J.D.; Drew, M.C. Mechanisms controlling changes in water movement through the roots ofHelianthus annuus L. during continuous exposure to oxygen deficiency. J. Exp. Bot. 1989, 40, 95–103.

116. Else, M.A.; Janowiak, F.; Atkinson, C.J.; Jackson, M.B. Root signals and stomatal closure in relation to photosynthesis, chlorophyll a fluorescence and adventitious rooting of flooded tomato plants. Ann. Bot. 2009, 103, 313–323.

117. Pociecha, E.; Koscielniak, J.; Filek, W. Effects of root flooding and stage of development on the growth and photosynthesis of field bean (Vicia faba L. minor). Acta Physiol. Plant. 2008, 30, 529–535.

118. Naumann, J.C.; Young, D.R.; Anderson, J.E. Leaf chlorophyll fluorescence and physiological response to freshwater and saltwater flooding in the evergreen shrub, Myrica cerifera. Environ. Exp. Bot. 2008, 63, 402–409.

119. Hiron, R.W.P.; Wright, S.T.C. The role of endogenous abscisic acid in the response of plants to stress.J. Exp. Bot. 1973, 24, 769–781.

120. Pereira, J.S.; Kozlowski, T.T. Variation among woody angiosperms in response to flooding. Physiol. Plant. 1977, 41, 184–192.

121. Sena Gomes, A.R.; Kozlowski, T.T. Responses of Melaleuca quinquenervia seedlings to flooding.Physiol. Plant 1980, 49, 373–377.

122. Tang, Z.C.; Kozlowski, T.T. Some physiological and morphological responses of Quercus macrocarpaseedlings to flooding. Can. J. For. Res. 1982, 12, 196–202.

123. Pezeshki, S.R.; Chambers, J.L. Stomatal and photosynthetic response of sweetgum (Liquidambarstyraciflua L.) to flooding. Can. J. For. Res. 1985, 15, 371–375.

124. Pezeshki, S.R.; Chambers, J.L. Response of cherrybark oak (Quercus falcata var. pagodaefolia) seedlings to short-term flooding. For. Sci. 1985, 31, 760–771.

125. Pezeshki, S.R.; DeLaune, R.D.; Patrick, W.H. Differential Response of selected mangroves to soil flooding and salinity: Gas exchange and biomass partitioning. Can. J. For. Res. 1990, 20, 869–874.

126. Pezeshki, S.R.; DeLaune, R.D.; Meeder, J.F. Carbon assimilation and biomass partitioning in Avicennia germinans and Rhizophora mangle seedlings in response to soil redox conditions. Environ. Exp. Bot.1997, 37, 161–171.

127. Pezeshki, S.R. Differences in patterns of photosynthetic responses to hypoxia in flood-tolerant and flood-sensitive tree species. Phtosynthetica 1993, 28, 423–430.

128. Krauss, K.W.; Doyle, T.W.; Howard, R.J. Is there evidence of adaptation to tidal flooding in saplings of baldcypress subjected to different salinity regimes? Environ. Exp. Bot. 2009, 67, 118–126.

129. Gravatt, D.A.; Kirby, C.J. Patterns of photosynthesis and starch allocation in seedlings of four bottomland hardwood tree species subjected to flooding. Tree Physiol. 1998, 18, 411–417.

130. Mielke, M.S.; Shaffer, B. Leaf gas exchange, chlorophyll fluorescence and pigment indexes of Eugenia uniflora L. in response to changes in light intensity and soil flooding. Tree Physiol. 2010, 30, 45–55.

131. Pezeshki, S.R.; DeLaune, R.D.; Patrick, W.H. Effect of fluctuating rhizosphere redox potential on carbon assimilation of Spartina alterniflora. Oecologia 1989, 80, 132–135.

132. Brown, C.E.; Pezeshki, S.R. A study on waterlogging as a potential silvicultural tool to controlLigustrum sinense populations in western Tennessee. Wetlands 2000, 20, 429–437.

133. Taylor, G.E.; Gunderson, C.A. Physiological site of ethylene effects on carbon dioxide assimilation inGlycine max. Plant Physiol. 1988, 86, 85–92.

134. Wample, R.L.; Thornton, R.K. Differences in the response of sunflower (Helanthus annuus) subjected to flooding and drought stress. Physiol. Plant 1984, 61, 611–616.

135. Pezeshki, S.R.; Santos, M.I. Relationship among rhizosphere oxygen deficiency, root restriction, photosynthesis and growth in baldcypress (Taxodium distichum) seedlings. Photosynthetica 1998, 35, 381–390.

136. Ahmed, S.; Nawata, E.; Sakuratania, T. Changes of endogenous ABA and ACC and their correlation to photosynthesis and water relations in mugbean (Vigna radiate) during waterlogging. Environ. Exp. Bot. 2006, 57, 278–284.

137. McKevlin, M.R.; Hook, D.D.; McKee, W.H. Growth and nutrient use efficiency of water tupelo seedlings in flooded and well drained soil. Tree Physiol. 1995, 15, 753–758.

138. Bowes, G. Growth at elevated CO2: Photosynthetic responses mediated through rubisco. Plant Cell Environ. 1991, 14, 795–806.

139. Vu, J.C.V.; Yelenosky, G. Photosynthetic responses of rough lemon and sour orange to soil flooding, chilling and short-term temperature fluctuation during growth. Environ. Exp. Bot. 1992, 32, 471–477.

140. Liao, C.T.; Lin, C.H. Effect of flooding stress on photosynthetic activities of Momordica charantia.Plant Physiol. Biochem. 1994, 32, 479–485.

141. Dann, M.S.; Pell, E.J. Decline of activity and quantity of ribulose bisphosphate carboxylase/owygenase and net photosynthesis in ozone-treated potato foliage. Plant Physiol. 1989, 91, 427–432.

142. Lauer, M.J.; Pallardy, S.G.; Belvins, D.C.; Randall, D.D. Whole leaf carbon exchange characteristics of phosphate deficient soybeans (Glycine max L.). Plant Physiol. 1989, 91, 848–854.

143. Li, S.; Goodwin, S.; Pezeshki, S.R. Photosynthetic gene expression in black willow under various soil moisture regimes. Biol. Plant 2007, 51, 593–596.

144. Sij, J.W.; Swanson, C.A. Effect of petiole anoxia on phloem transport in squash. Plant Physiol. 1973, 51, 368–371.

145. Qureshi, F.A.; Spanner, D.C. The effect of nitrogen on the movement of tracers down the stolon of Saxifraga sarmentosa with some observations on the influence of light. Planta 1973, 110, 131–144.

146. Vartapetian, B.B. Flood-Sensitive Plants under Primary and Secondary Anoxia: Ultrastructural and Metabolic Responses. In Plant Life Under Oxygen Deprivation; Jackson, M.B., Davies, D.D., Lambers, H., Eds.; SPB Academic Publishing: The Hague, The Netherlands, 1991; pp. 201–216.

147. Kennedy, R.A.; Fox, T.C.; Everard, J.D.; Rumpho, M.E. Biochemical Adaptations to Anoxia: Potential Role of Mitochondrial Metabolism to Flood Tolerance in Echinochloa Phyllopogon (Barnyard Grass). InPlant Life Under Oxygen Deprivation; Jackson, M.B., Davies, D.D., Lambers, H., Eds.; SPB Academic Publishing: The Hague, The Netherlands, 1991; pp. 217–227.

148. Yamamoto, F.; Kozlowski, T.T. Effect of flooding, tilting of stem, and ethrel application on growth, stem anatomy, and ethylene production of Acer platanoides seedlings. Scand. J. For. Res. 1987, 2, 141–156.

149. Yamamoto, F.; Sakata, T.; Terazawa, K. Growth, morphology, stem anatomy and ethylene production in flooded Alnus japonica seedlings. IAWA J. 1995, 16, 47–59.

150. Yamamoto, F.; Sakata, T.; Terazawa, K. Physiological, anatomical and morphological responses ofFraxinus mandshurica seedlings to flooding. Tree Physiol. 1995, 15, 713–719.

151. Webb, T.; Armstrong, W. The effects of anoxia and carbohydrates on the growth and viability of rice, pea and pumpkin roots. J. Exp. Bot. 1983, 34, 579–603.

152. Pezeshki, S.R. Root responses of flood-tolerant and flood-sensitive tree species to soil redox conditions. Trees 1991, 5, 180–186.

153. Pezeshki, S.R.; DeLaune, R.D. Influence of sediment oxidation-reduction potential on root elongation in Spartina patens. Acta Oecologia 1990, 11, 377–383.

154. Bertani, A.; Brambilla, I. Effect of decreasing oxygen concentration on wheat roots: Growth and induction of anaerobic metabolism. Z. Pflanzenphysiol. 1982, 108, 283–288.

155. Lissner, J.; Mendelssohn, I.A.; Anastasiou, C.J. A method for cultivating plants under controlled redox intensities in hydroponics. Aquat. Bot. 2003, 76, 93–108.

156. Will, R.E.; Seiler, J.R.; Feret, P.P.; Aust, W.M. Effects of rhizosphere inundation on the growth and physiology of wet and dry -site Acer rubrum (red maple) populations. Am. Midl. Nat. 1995, 134, 127–139.

157. Pezeshki, S.R.; DeLaune, R.D.; Kludze, H.K.; Choi, H.S. A comparative study of gas exchange characteristics of cattail (Typha domingensis) and sawgrass (Cladium jamaicense) to soil redox conditions. Aquat. Bot. 1996, 54, 25–35.

158. Pezeshki, S.R.; DeLaune, R.D. Responses of seedlings of selected woody species to soil oxidation-reduction conditions. Environ. Exp. Bot. 1998, 40, 123–133.

159. Anderson, P.H.; Pezeshki, S.R. The effects of intermittent flooding on seedlings of three forest species.Photosynthetica 1999, 37, 543–552.

160. Elcan, J.M.; Pezeshki, S.R. Effects of flooding on susceptibility of Taxodium distichum seedlings to drought. Photosynthetica 2002, 40, 177–182.

Chapter 10

Temporal Variability in Trace Metal Solubility in a Paddy Soil not Reflected in Uptake by Rice (*Oryza Sativa* L.)

Yunyu Pan [1,2], Gerwin F. Koopmans [2], Luc T. C. Bonten[3], Jing Song[1], Yongming Luo[4], Erwin J. M. Temminghoff [2] Rob N. J. Comans[2]

[1]*Key Laboratory of Soil Environment and Pollution Remediation, Institute of Soil ScienceChinese Academy of SciencesNanjingPeople's Republic of China*

[2]*Department of Soil QualityWageningen UniversityWageningenThe Netherlands*

[3]*AlterraWageningenThe Netherlands*

[4]*Yantai Institute of Costal Zone ResearchChinese Academy of SciencesYantaiPeople's Republic of China*

ABSTRACT

Alternating flooding and drainage conditions have a strong influence on redox chemistry and the solubility of trace metals in paddy soils. However, current knowledge of how the effects of water management on trace metal solubility are linked to trace metal uptake by rice plants over time is still limited. Here, a field-contaminated paddy soil was subjected to two flooding and drainage cycles in a pot experiment with two rice plant cultivars, exhibiting either high or low Cd accumulation characteristics. Flooding led to a strong vertical gradient in the redox potential (Eh). The pH and Mn, Fe, and dissolved organic carbon concentrations increased with decreasing Eh and vice versa. During flooding, trace metal solubility decreased markedly, probably due to sulfide mineral precipitation. Despite its low solubility, the Cd content in rice grains exceeded the food quality standards for both cultivars. Trace metal contents in different rice plant tissues (roots, stem, and leaves) increased at a constant rate during the first flooding and drainage cycle but decreased after reaching a maximum during the second cycle. As such, the high temporal variability in trace metal solubility was not reflected in trace metal uptake by rice plants over time. This might be due to the presence of aerobic conditions and a consequent higher trace metal solubility near the root surface, even during flooding. Trace metal solubility in the rhizosphere should be considered when linking water management to trace metal uptake by rice over time.

Keywords: Paddy soils Trace metal contamination Water management Redox potential Bioavailability Uptake

INTRODUCTION

Contamination of paddy soils with trace metals due to mining and smelting activities, application of fertilizers and sewage sludge, and wastewater irrigation is a widespread problem (Kuo et al. 2006; Römkens et al. 2009a; Zeng et al. 2011). Trace metal contamination of paddy soils can lead to the accumulation of trace metals in rice grains (Römkens et al. 2009b; Zeng et al. 2011). Since rice is one of the most important staple foods for human beings in South and Southeast Asia (Römkens et al. 2009a), consumption of rice with elevated trace metal levels is a serious threat to food safety and human health (Römkens et al. 2009b; Meharg et al. 2013). For typical rice-eating populations, the dietary intake of Cd via rice consumption is the major exposure pathway for human beings to Cd (Tsukahara et al. 2003; Meharg et al. 2013). For minimizing human health risks, it is important to understand the processes controlling the uptake of trace metals by rice plants.

The uptake of trace metals by rice plants depends on the bioavailability of trace metals in paddy soils (Simmons et al. 2008; Römkens et al. 2009b). For trace metals in paddy soils, three pools with a different bioavailability can be distinguished: the total, reactive, and directly available trace metal pools (Römkens et al. 2009a). The reactive pool represents trace metals adsorbed to reactive surfaces of soil organic matter (SOM), short-range ordered metal-(hydr)oxides, and clay, and it controls the trace metal concentrations in soil solution (Weng et al. 2001; Tipping et al. 2003). The size of this pool can be determined by an extraction of soil with 0.05 M EDTA, 0.1 M HCl, or 0.43 M HNO_3 (Römkens et al. 2009a). The directly available pool represents the free or total dissolved trace metal concentration in soil solution which can be directly related to the uptake of trace metals by plants (Lofts et al. 2004; Peijnenburg et al. 2007). The size of this pool can be determined by either sampling of the soil solution with soil moisture samplers or lysimeters (Reynolds et al. 2004; Shen and Hoffland 2007) or it can be mimicked by an extraction of soil with weak salt extracts such as $CaCl_2$, $Ca(NO_3)_2$, or $NaNO_3$ (Houba et al. 2000; Peijnenburg et al. 2007). The size of the total pool is normally larger than the size of the reactive pool, and the difference between both pools is interpreted to be nonreactive on a time scale relevant to the duration of a cropping season for rice (Römkens et al. 2009a). For paddy soils varying widely in soil properties such as pH and cation exchange capacity (CEC), regression-based log–log relationships have been used to link the directly available Cd pool as determined by an 0.01 M $CaCl_2$ extraction of soil to the Cd content of rice grains (Simmons et al. 2008; Römkens et al. 2009b). Other studies linked the Cd content in rice grains to the reactive Cd pool (0.43 M HNO_3) in combination with soil properties including pH, SOM, and clay (Brus et al.2009) or pH and CEC (Römkens et al. 2009b). These soil properties were included because they affect the solid–solution partitioning of trace metals (Weng et al. 2001). However, most of the relationships in the aforementioned studies were determined for ripened and harvested rice grains.

Rice is usually cropped in lowland paddy fields. Flooding conditions prevail during almost the entire cropping period, because lowland rice is extremely sensitive to water shortage (Kögel-Knabner et al. 2010). Flooded paddy fields are usually drained during the late tillering stage to control ineffective tillering of the rice plants and shortly before harvest to enable faster rice grain ripening and to facilitate harvesting. Consequently, lowland paddy fields undergo at least two flooding and draining cycles, which are known to lead to fluctuations in the redox conditions (Kögel-Knabner et al. 2010; Pan et al. 2014). Upon flooding, the redox potential (Eh) will decrease and the reduction of Mn- and Fe-(hydr)oxides will lead to their dissolution while organic matter will be released, resulting in elevated levels of Mn^{2+} and Fe^{2+} and dissolved organic matter (DOM) in soil solution (Grybos et al. 2007; Pan et al. 2014). Since SOM and metal-(hydr)oxides play an important role in the adsorption of trace metals in soils (Weng et al.2001), a decrease in the reactive surfaces of these soil constituents can lead to an increase in the solubility of trace metals. The mobilization of trace metals might be even further amplified by the release of DOM after flooding (Pan et al. 2014), because trace metals including Cu and Pb can bind strongly to DOM (Amery et al. 2007; Koopmans and Groenenberg 2011). On the other hand, reduction of SO_4^{2-} to S^{2-} can cause trace metals to precipitate as poorly soluble minerals (de Livera et al. 2011; Fulda et al. 2013). Additionally, the pH of acidic soils can increase during flooding (Kögel-Knabner et al. 2010), which enhances the adsorption of trace metals by SOM and metal-(hydr)oxides (Weng et al. 2001). During drainage and subsequent oxidation of paddy soils, the above-mentioned reduction reactions can reverse, leading either to the retention of trace metals via adsorption to freshly formed Mn (III/IV)- and Fe(III)-(hydr)oxides or to their release via dissolution of trace metal sulfide precipitates. Also, the pH can return to its antecedent value, thereby influencing the adsorption of trace metals by SOM and metal-(hydr)oxides.

These complex and dynamic effects of flooding and drainage on the Eh and redox chemistry of paddy soils can lead to a strong temporal variability in trace metal solubility (Du Laing et al. 2009; Schulz-Zunkel and Krueger 2009; Shaheen et al. 2014), which, in turn, might affect trace metal uptake by rice plants. However, current knowledge of how the effects of water management of paddy fields on trace metal solubility are linked to uptake of trace metals by rice plants over time is still limited.

In our previous work, we investigated how alternating flooding and draining conditions affected redox chemistry and solubility of heavy metals in two field-contaminated soils with similar properties but a contrasting pH (Pan et al. 2014, 2016). The primary objective of our current study is to investigate the relationship between the solubility of trace metals and their uptake by rice plants over time. To realize this objective, we carefully simulated two consecutive flooding and drainage cycles during rice growth using an acidic field-contaminated paddy soil in a pot experiment. Besides trace metal bioavailability in paddy soils, rice plant-specific factors can have a profound influence on the accumulation of trace metals in rice grains (Yu et al. 2006; Römkens et al. 2009b). For example, Indica and Japonica rice plant cultivars have been demonstrated to differ substantially in their ability to translocate Cd from the roots into their shoots, with a higher ratio of Cd in rice grains to Cd in the roots for Indica than for Japonica (Römkens et al. 2009b). Additionally, considerable variation in Cd accumulation in rice grains has been identified among various Indica varieties (Römkens et al. 2009b; Hu et al. 2013). Therefore, the secondary objective of our study was to quantify the differences in Cd content in rice grains of two Indica rice plant cultivars with different Cd accumulation characteristics. The Eh was monitored continuously at five depths along the soil profile during the two successive flooding and drainage cycles in the pot experiment. Soil solution was sampled using soil moisture samplers during the major rice growth stages at the same depths to determine its chemical composition including pH and the concentrations of total dissolved Mn, Fe, and trace metals and dissolved organic carbon (DOC). The trace metal contents in different rice plant tissues including the roots, stem, leaves, husks, and brown rice were measured during the major rice growth stages. This experimental approach allowed us to link the solubility of trace metals to their uptake by rice plants over time.

MATERIALS AND METHODS

Soil collection

Soil with a clay loam texture was sampled from a paddy field (0–20 cm) nearby a former transformer and electronic waste stripping and recycling factory in Taizhou city, Zhejiang Province, P.R. China (Pan et al. 2014). This soil was severely contaminated with trace metals because of industrial activities in the proximity of the sampling site (Sun et al. 2013). The soil was air-dried, passed through a 2-mm sieve, and stored at room temperature before analyzing the soil characteristics and preparing the pot experiment.

Experimental setup of the pot experiment

For the pot experiment, pots made of PVC material with a height of 35 cm and a diameter of 18 cm were used (Fig. 1). Quartz sand, pretreated and cleaned with 0.14 M HNO_3 and deionized water, was placed at the bottom of the pot in a layer of 3 cm (Fig. 1). A piece of 1-mm nylon sieve material was put on top of the quartz sand. Each pot was filled with 4.5 kg of air-dried paddy soil. Before transferring the soil to each pot, it was thoroughly mixed with basal N, P, and K fertilizer including 0.83 g N as urea ($CO[NH_2]_2$), 0.23 g P as KH_2PO_4, and 0.52 g K as KCl. During mixing, the soil moisture content was adjusted to about 30 % (w:w) to facilitate water penetration in the potted soil during the first flooding period. The thickness of the potted soil layer was 25 cm. In the pot experiment, two different rice cultivars were used (see "Management and harvesting of the rice plants" section). The pot experiment was set up with 12 replicates for each of the two rice cultivars (24 pots in total) in a greenhouse. The temperature and light intensity in the greenhouse were regulated in accordance with the ambient weather conditions, because an optimal rice growth requires a high light intensity and temperature. The pot experiment was subjected to two successive flooding and drainage cycles, which is in accordance with the water management of lowland paddy rice fields in practice: one drainage period at the late tillering growth stage and one drainage period shortly before the harvest (Fig. 2). The first flooding period lasted for 58 days until the late tillering stage and was followed by a drainage period of 8 days aiming to control ineffective tillering. The second flooding period lasted for 46 days until the ripening stage and was followed by a drainage period of 16 days to enable faster rice ripening and to facilitate harvesting. For flooding of the pots, three liter of deionized water was added to each pot in a step-wise manner leaving a final water layer on top of the soil layer of 5 cm (Fig. 1). The flooding procedure was completed within 3 h for all 24 pots. For each rice plant cultivar, three pots were randomly chosen to monitor the Eh and chemical composition of the soil solution. For Eh monitoring, a combined Eh-temperature glass fiber probe (Paleoterra Products, Amsterdam, The Netherlands) was inserted vertically in each of the six pots. The probes were tailor-made with five platinum Eh measurement points enabling us to monitor the Eh 3 cm above the soil–water interface (+3 cm), at the interface (0 cm), and at depths of 5 (−5 cm), 10 (−10 cm), and 20 cm (−20 cm) below the interface (Fig. 1). The Ag–AgCl reference electrodes were positioned in the water layer. The Eh signal was automatically logged every 15 min. The Eh relative to the standard hydrogen electrode was calculated from the measured potential with a temperature-dependent correction for the potential of the Ag–AgCl reference electrode. The temperature was measured using the combined Eh-temperature probe by platinum electrodes at five positions within the pots: +3, 0, −5, −10, and −20 cm (Fig. 1). For collecting soil solution samples from the pots, five soil moisture samplers (SMS) (Rhizosphere Research Products, Wageningen, The Netherlands) with a diameter of 2.5 mm, a mean pore size of 0.15 µm, and a porous section of 10 cm were placed horizontally at the same depths (+3, 0, −5, −10, and −20 cm) in the pot at which the Eh was measured. The silicone connector of the SMS tightly sealed off the holes of the pots, preventing air from penetrating the pots. Soil solution samples were taken with a vacuumed syringe at 11 occasions during the major rice growth stages (Fig. 2). This syringe was connected to the female luer lock of the SMS, which was closed again immediately after sampling. The soil solution samples were split in two subsamples: In one subsample, metal and sulfur concentrations were measured after acidification by adding concentrated HNO_3 to a level of 1.2 % (v:v), whereas the other subsample was used for the measurement of pH and DOC. The samples were stored in polypropylene tubes at 4 °C until chemical analysis as further detailed in the “Chemical analysis” section.

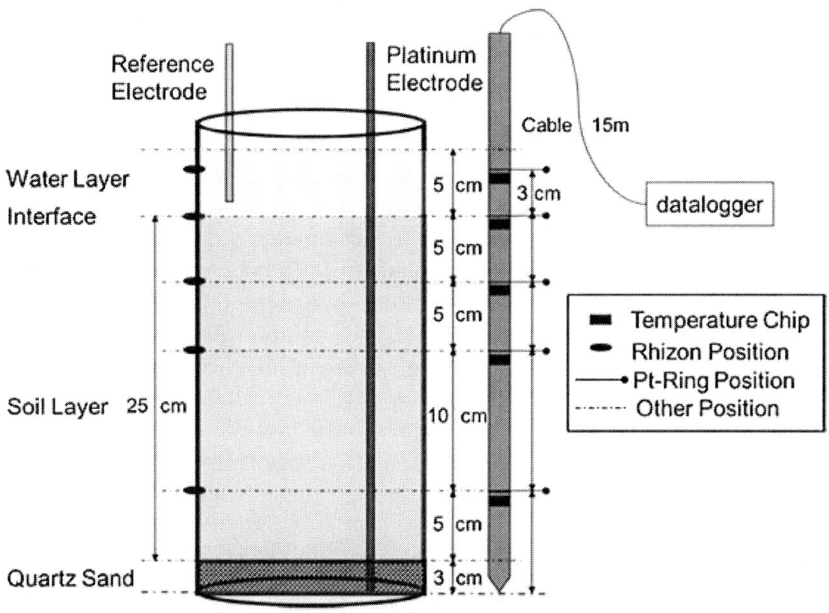

Figure 1. Experimental setup of the pots used to monitor the Eh and chemical composition of the soil solution in the pot experiment with the two rice plant cultivars

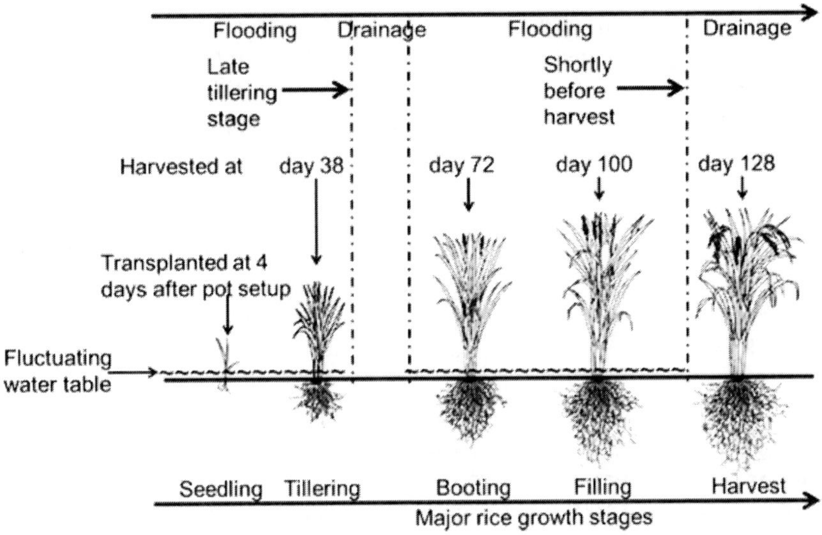

Figure 2. Harvesting times of the rice plants during the major rice growth stages and alternating flooding and drainage periods in the pot experiment. The first drainage period of 8 days was after 58 days of flooding, and the second drainage period of 16 days was after 46 days of flooding

Management and harvesting of the rice plants

Two Indica varieties were used in our pot experiment: Zhongxiang 1 (A16) and Indonesia (A159). These cultivars are commonly grown in Zhejiang Province, P.R. China. Under conventional water management conditions, cultivar A16 has a lower potential to accumulate Cd in brown rice than cultivar A159 (Hu et al. 2013). The seeds of both cultivars were provided by the China National Rice Research Institute (Hangzhou City, Zhejiang Province, P.R. China) and disinfected with 10 % H_2O_2 solution (v:v) before use. After germination, germinated seeds were grown in an uncontaminated soil for 30 days before transplanting the seedlings to the pots. Four days after setting up the pots and flooding the potted soil for the first time, three clusters, with each cluster containing 6 seedlings, were transplanted to each pot. The clusters were evenly positioned over the soil surface of each pot. During rice growth, an extra amount of 0.35 g N as urea was applied to each pot at the early rice tillering stage (day 10). An extra similar dose of urea was applied at the early rice grain filling stage (day 80). A water layer of 5 cm above the soil surface of each pot was maintained by adding deionized water three times each day at 7 a.m., 2 p.m., and 7 p.m. Three pots for each cultivar were destructively harvested at the rice tillering stage (day 38), booting stage (day 72), filling stage (day 100), and ripening stage (day 128) (Fig. 2). The harvested rice plants were first thoroughly washed with tap water and then with deionized water to remove dust and adhering soil particles. Subsequently, the rice plants were separated into different plant tissues including roots, stem, leaves, and grains. These tissues were wrapped in a clean paper envelope and oven-dried at 105 °C for 30 min and further dried at 80 °C to a constant weight. The dried grains were then further separated into the husks and brown rice. Dry weight was determined for all rice tissues. Subsequently, all rice tissues were milled by a stainless steel grinding machine to measure the trace metal contents (see "Chemical analysis" section).

Chemical analysis

Soil analyses

The pH was measured in a settling 1:10 (w:v) suspension of soil in 0.01 M $CaCl_2$ (Houba et al.2000); the CEC was measured using 1 M NH_4OAc buffered at pH 7.0 (USDA 1996), SOM by loss-on-ignition (550 °C) and clay by the sieve and pipette method (Houba et al. 1997). Dissolved organic carbon and sulfur were measured in the same $CaCl_2$ extract as used for the measurement of pH after filtration through a 0.45-µm filter (Schleicher & Schuell 602H). Dissolved organic carbon was measured by a TOC analyzer (Multi C/N 3000). Sulfur was measured by inductively coupled plasma atomic emission spectrometry (ICP-AES; IRIS Intrepid II) and ion chromatography (Dionex, ICS-2100) (Dick and Tabatabai 1979). Ion chromatography specifically measures the inorganic SO_4^{2-} form of sulfur. Since the results of both ICP-AES and ion chromatography were very similar (results not shown), we, therefore, interpreted sulfur measured in the $CaCl_2$ extract by ICP-AES as $S–SO_4^{2-}$ (Pan et al. 2014). Short-range ordered Al-, Fe-, and Mn-(hydr)oxides were determined by using the acid ammonium oxalate extraction method (Novozamsky et al. 1986). The Al, Fe, and Mn concentrations in the ammonium oxalate extracts (i.e., Al-ox, Fe-ox, and Mn-ox) were measured by ICP-AES. The size of the reactive trace metal pools was determined by an extraction of soil with 0.43 M HNO_3, whereas the size of the total trace metal pools was determined by soil digestion with Aqua Regia (Houba et al. 1997). The Cd, Cu, Ni, Pb, and Zn concentrations in the 0.43 M HNO_3 extracts and Aqua Regia digests were measured by ICP-AES. Samples with trace metal concentrations below the detection limit of the ICP-AES were re-measured by an inductively coupled plasma mass spectrometer (ICP-MS; Thermo Electron X7). All physico-chemical soil properties were determined in triplicate, and their average ± standard deviation is presented in Table 1.

Table 1. Selected physico-chemical properties of the paddy soil used in the pot experiment

Soil property	Unit	Value	Standard[g]
Clay (<2 μm)[a]	%	19.7 ± 0.1	–
SOM[b]	%	9.8 ± 0.1	–
pH[c]		5.2 ± 0.1	–
DOC[c]	mg C L^{-1}	150 ± 5	
S-SO_4^{2-} [c]	mmol kg^{-1}	4.2 ± 0.1	
CEC	cmol [+] kg^{-1}	11.5 ± 0.3	–
Al-ox[d]	mmol kg^{-1}	38 ± 1	–
Fe-ox[d]	mmol kg^{-1}	95 ± 1	–
Mn-ox[d]	mmol kg^{-1}	1.7 ± 0.1	–
Reactive Cu[e]	mmol kg^{-1}	3.97 ± 0.03	–
Total Cu[f]	mmol kg^{-1}	7.38 ± 0.08	0.79
Reactive Cd[e]	μmol kg^{-1}	50.1 ± 0.2	–
Total Cd[f]	μmol kg^{-1}	62.0 ± 0.5	2.7
Reactive Pb[e]	μmol kg^{-1}	145 ± 5	–
Total Pb[f]	μmol kg^{-1}	222 ± 10	1207
Reactive Zn[e]	mmol kg^{-1}	0.60 ± 0.03	–
Total Zn[f]	mmol kg^{-1}	2.31 ± 0.08	3.06
Reactive Ni[e]	μmol kg^{-1}	164 ± 5	–
Total Ni[f]	μmol kg^{-1}	767 ± 17	682

All soil properties were determined in triplicate and their average ± standard deviation is presented

[a] Clay by the sieve and pipette method (Houba et al. 1997)

[b] Soil organic matter (SOM) by loss-on-ignition (550 °C) (Houba et al. 1997)

[c] 1:10 (w:v) 0.01 M $CaCl_2$ extraction (Houba et al. 2000)

[d] Extraction with acid ammonium oxalate (Novozamsky et al. 1986)

[e] 1:10 (w:v) 0.43 M HNO_3 (Houba et al. 1997)

[f] Aqua Regia (Houba et al. 1997)

[g] Grade II (pH < 6.5) of the Chinese environmental quality standard for soils (GB 15618-1995)

Soil solution analyses

The pH of the soil solution samples was measured directly after sampling with a combined glass electrode, and DOC was measured by a TOC analyzer. In the acidified subsamples, Fe, Mn, and S were measured by ICP-AES and Cd, Cu, Ni, Pb, and Zn by ICP-MS.

Plant analysis

For analyzing the trace metal contents in the different rice plant tissues, 0.5 g milled tissue material was digested by using a mixture of 5 ml HNO_3 and 3 ml $HClO_4$ (Huang and Schulte1985). The Cu, Zn, Pb, Cd, and Ni concentrations in the digests were measured by ICP-MS.

RESULTS AND DISCUSSION

Physico-chemical soil properties

Selected physico-chemical properties of the paddy soil are summarized in Table 1. The total Cd, Cu, Ni, and Zn contents exceed the Chinese environmental quality standard for soils (GB 15618-1995; Grade II for soil pH < 6.5) about 23 (Cd), 9 (Cu), and 1.1 (Ni) times, respectively. Hence, this soil is especially heavily contaminated with Cd and Cu. The soil is acidic and the size of the reactive trace metal pools (0.43 M HNO_3) relative to the total pools (Aqua Regia) increases in the order Ni (21 %) < Zn (26 %) < Cu (54 %) < Pb (65 %) < Cd (81 %). Consequently, Cd is mostly present in highly reactive forms, which can exchange with the soil solution and become available with time for uptake by plants, whereas Zn and Ni are largely contained within chemically inert forms which are rather immobile (Römkens et al. 2009a). These results are very similar to those of Römkens et al. (2009a) for paddy soils with similar physico-chemical properties. The available S–SO_4^{2-} content in our paddy soil (Table 1) was in the upper range of the available S–SO_4^{2-} contents (0.06–6.6 mmol kg^{-1}) reported for other paddy soils in different regions (Lefroy et al. 1992; Fulda et al. 2013).

Effect of alternating flooding and drainage on Eh, pH, and Mn, Fe, S, and DOC concentrations

In Fig. 3, the Eh, pH, and Mn, Fe, S, and DOC concentrations in the overlying water layer and the soil solution of the different soil layers are presented for the pot experiment with rice plant cultivar A159. Since the results for cultivar A16 are very similar to those for cultivar A159, the results for the former cultivar are presented in the Appendix (Fig. 7). The Eh showed a strong contrast between the water layer and the three lower soil layers: Aerobic conditions were maintained in the water layer with an Eh of around 500 mV throughout the pot experiment, whereas anaerobic conditions with an Eh as low as about −300 mV were present at a depth of 5, 10, and 20 cm below the soil–water interface during the first flooding period. The Eh at the interface between the soil surface and the water layer decreased to −200 mV during the first 20 days, then increased to about 550 mV, and remained high during the remainder of the pot experiment. During the first drainage period, the Eh rapidly reached a level of 400 mV in all soil layers, due to re-aeration and oxidation of the potted soil. During the second flooding period, the Eh decreased to a level of about −200 mV in the two lower soil layers. The Eh at a depth of 5 cm was higher than the Eh in the two lower soil layers, but the temporal variability was high. From visual observations during the sampling of rice plant roots, most of the rice plant roots appeared to be present in the upper 5 cm of the potted soil. Therefore, the high temporal variation in Eh at a depth of 5 cm, which had a diurnal pattern, is probably caused by internal aeration of the rice plants during the daily photosynthesis period, leading to downward transport of O_2 from their shoots to the submerged roots and leakage of O_2 to the local surrounding soil environment (Colmer 2003; Mongon et al. 2014). Finally, aerobic conditions prevailed in all soil layers during the second drainage period except for the bottom layer, which remained slightly anaerobic.

According to the measured Eh range for our paddy soil, redox reactions including denitrification and reduction of Mn(III/IV) to Mn(II), Fe(III) to Fe(II), and SO_4^{2-} to S^{2-} can occur (Fig. 3). Hence, reductive dissolution of Mn- and Fe-(hydr)oxides during flooding can explain the clearly elevated Mn and Fe concentrations in soil solution in the three lower soil layers as compared to those in the water layer. The lower Mn and Fe concentrations during drainage and subsequent oxidation of the paddy soil are probably due to precipitation of these metals in the form of Mn (III/IV)- and Fe(III)-(hydr)oxides (Schulz-Zunkel and Krueger 2009; Shaheen et al. 2014). Reduction of SO_4^{2-} to S^{2-} is likely to occur in the three lower soil layers during flooding according to the Eh values measured in these layers. Therefore, the gradual decrease in the sulfur concentration in soil solution over time can be explained by the formation of poorly soluble sulfide minerals (Murase and Kimura 1997; Du Laing et al. 2009; Schulz-Zunkel and Krueger 2009). Since all above-mentioned reduction reactions consume protons, the pH increased from acidic to an almost neutral level in all soil layers during flooding, whereas the pH in the water layer remained relatively constant. During drainage, the pH in all soil layers decreased again. The DOC concentrations in the three lower anaerobic soil layers during flooding were clearly elevated compared to those in the aerobic overlying water layer. This can be attributed to reductive dissolution of Mn- and Fe-(hydr)oxides and the

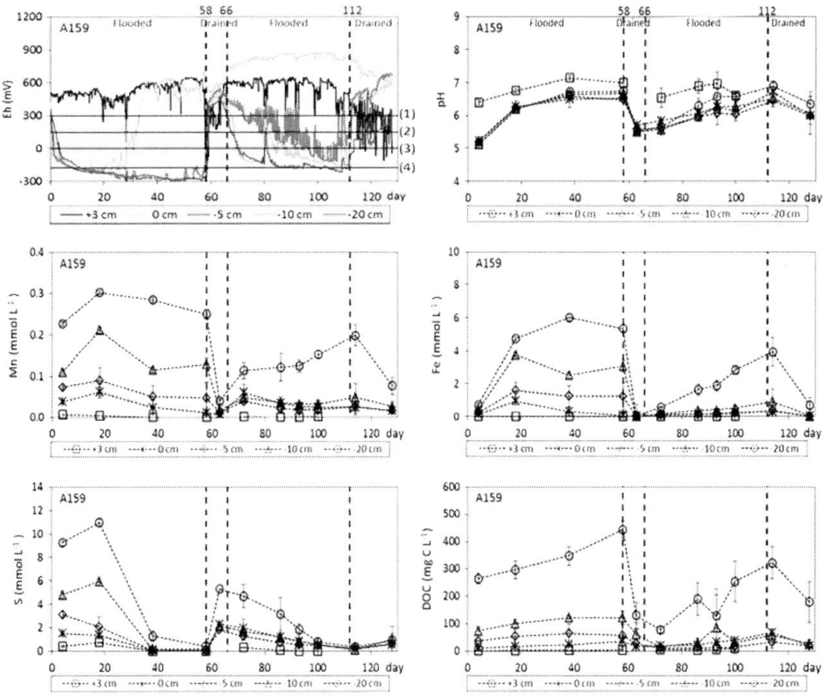

Figure 3. Eh, pH, and Mn, Fe, S, and DOC concentrations in the soil solution for the pot experiment with rice plant cultivar A159 (+3 cm refers to 3 cm above the interface, 0 cm equals the interface between the soil surface and the water layer, and −5, −10 cm, −20 cm refer to 5, 10, and 20 cm below the interface, respectively). *The numbers at the topx*-axis of the *two upper figures* indicate the time at which the drainage and flooding periods started. In figure with Eh data, 1–4 refer to the following reactions: (1) denitrification of NO_3^- to N_2, (2) and (3) reductive dissolution of MnO_2 and $Fe(OH)_3$ to Mn^{2+} and Fe^{2+}, respectively, and (4) reduction of SO_4^{2-} to S^{2-}

elevated pH, both leading to the release of adsorbed DOM (Kaiser et al. 1997; Pan et al. 2014). Also, the release of fermentation products might have contributed to the elevated DOC concentrations, because mineralization of organic matter can be incomplete under anaerobic conditions (Vink et al.2010). The DOC concentrations decreased in all soil layers during drainage.

Effects of alternating flooding and drainage conditions on the solubility of Cu, Cd, Pb, Zn, and Ni

In Fig. 4, the trace metal concentrations in the overlying water layer and the soil solution of the different soil layers are presented for the pot experiment with rice plant cultivar A159. The results for cultivar A16 are very similar to those for cultivar A159 and are, therefore, presented in the Appendix (Fig. 8). At the initial sampling occasion (day 4), the solubility of all trace metals except Pb clearly increased with depth. This gradient is probably due to vertical displacement of trace metals and DOC (Fig. 3) via downward transport during the saturation of the potted soil by adding deionized water at the top of the soil surface. In a previous column experiment with the same paddy soil, as described in Pan et al. (2014, 2016), the opposite trend was found for the initial sampling occasion, with trace metal and DOC concentrations clearly decreasing with depth. However, this could be explained by Pan et al. (2014, 2016) by upward transport because the columns in their experiment were saturated from the bottom. In this study, the solubility of Cu, Cd, Pb, Zn, and Ni in all soil layers decreased markedly during the first flooding period, which might be due to Cu, Cd, Pb, Zn, and Ni precipitation in the form of poorly soluble sulfide minerals after SO_4^{2-} had been reduced to S^{2-} (Weber et al. 2009; de Livera et al. 2011; Fulda et al. 2013). With x ranging between 1.22 and 1.29 for Cu_xS mineral stoichiometry (Pattrick et al. 1997; Pan et al. 2016) and a stoichiometry of 1:1 for Cd:S, Pb:S, Zn:S, and Ni:S minerals (Allison et al. 1991), there would be enough sulfur in this paddy soil to fully precipitate the reactive forms of all five trace metals as sulfide minerals when all available $S{-}SO_4^{2-}$ would be reduced to S^{2-} (Table 1). Additionally, the increase in pH during flooding (Fig. 3) results in an increased trace metal binding to SOM, contributing to the observed decrease in the trace metal solubility. This might especially be relevant for the water layer and the interface between the soil surface and the water layer where no sulfate-reducing conditions were observed (Fig. 3).

During the first drainage period of our paddy soil, the trace metal concentrations increased in all soil layers, now probably due to the oxidation of sulfide minerals after re-aeration of the soil and the development of aerobic conditions and the decrease in pH (Fig. 3). During the second flooding period, the trace metal concentrations in all soil layers decreased again. For the interface between the soil surface and the water layer and the soil layer at a depth of 5 cm, this is the result of an increase in pH, because the Eh values were too high to expect sulfate-reducing conditions (Fig. 3) and consequent precipitation of trace metal sulfide minerals. For the two lower soil layers, however, precipitation of trace metal sulfide minerals might just

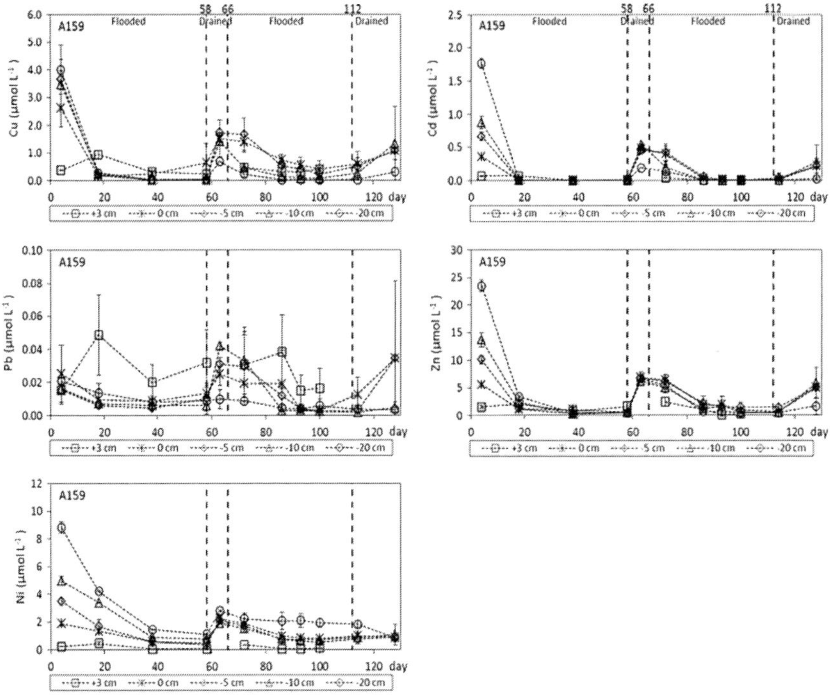

Figure 4. Total dissolved concentrations of Cu, Cd, Pb, Zn, and Ni in the soil solution for the pot experiment with rice plant cultivar A159 (+3 cm refers to 3 cm above the interface, 0 cm equals the interface between the soil surface and the water layer, and −5, −10 cm, −20 cm refer to 5, 10, and 20 cm below the interface, respectively). *The numbers at the topx*-axis of the *two upper figures* indicate the time at which the drainage and flooding periods started

be possible according to the measured Eh values in these layers (Fig. 3). The Ni concentrations in the bottom soil layer are remarkably high. Since Ni is the trace metal which is the least likely to form sulfide minerals among the five trace metals considered here (Weber et al. 2009; Pan et al.2016), Ni solubility in this paddy soil might not have been controlled by Ni sulfide minerals but rather by binding of Ni to organic matter in the form of DOM and SOM. The elevated DOC concentrations in the bottom soil layer (Fig. 3) are probably the cause of the higher Ni concentrations (Pan et al. 2016). During the second drainage period, the solubility of the trace metals increased again in all soil layers, probably because of re-aeration of the potted soil and the decrease in pH (Fig. 3).

Uptake of trace metals by different rice plant tissues during the major growth stages

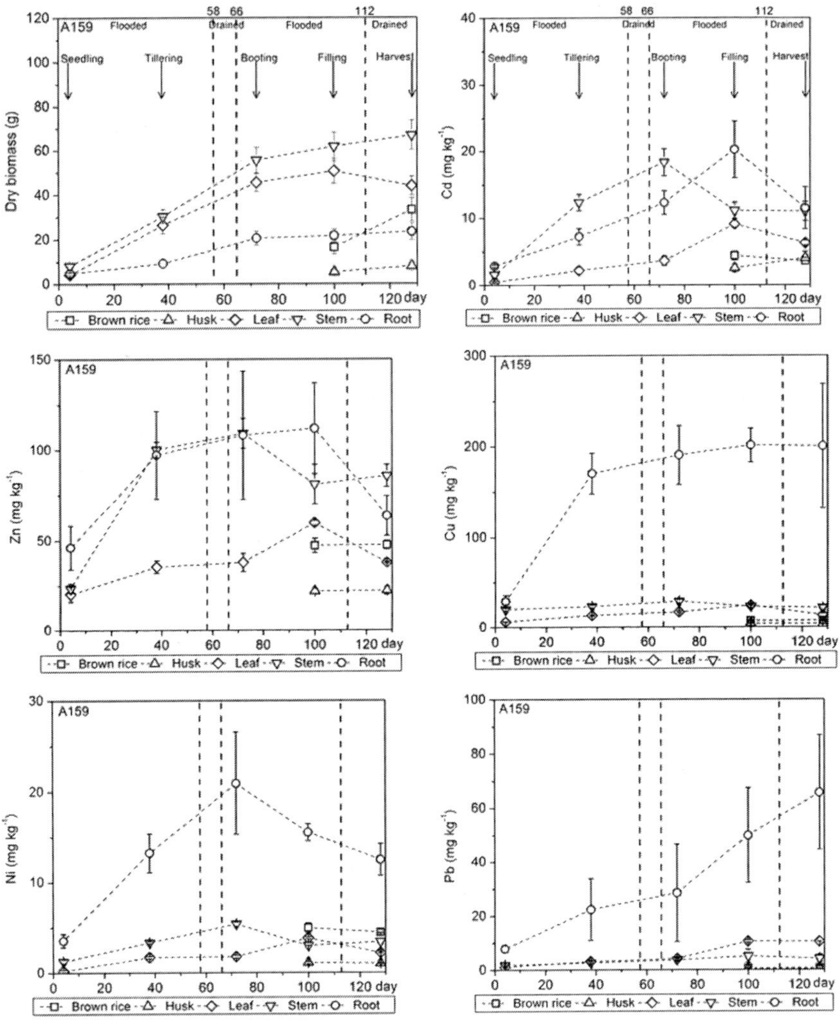

Figure 5. Biomass of the roots and the above-ground plant tissues including brown rice, husks, leaves, and stem and trace metal contents in these plant tissues for rice cultivar A159 during the major rice growth stages. Husks and brown rice were only harvested at the filling stage and final harvest. *The numbers at the top* of the figure indicate the time at which the drainage and flooding periods started

In Fig. 5, the biomass of the different rice plant tissues (roots, stem, leaves, husks, and brown rice) and the trace metal contents in these tissues are presented for the pot experiment with rice plant cultivar A159. The same data for cultivar A16 are presented in the Appendix (Fig. 9). The biomass of the roots and the above-ground plant tissues (stem and leaves) increased at a constant rate but the rate of increase leveled off (root and stem) or the biomass even decreased (leaves) after the booting stage in the second flooding period. For the filling stage and final harvest, the husks and brown rice were included in the sampling of the above-ground plant tissues and these tissues contributed about 50 % to the total above-ground biomass.

The Cd content in the different plant tissues increased at a constant rate but decreased after reaching a maximum at the booting or filling stage in the second flooding period. For the Zn content in the leaves, a similar trend can be observed. The increase in the Zn content in the stem and roots, however, started to lessen after the tillering stage in the first flooding period which was followed by a decrease after the booting or filling stage. The Cd and Zn contents in the husks and brown rice remained constant during the filling stage and final harvest. The increase in the Cu, Ni, and Pb contents in most plant tissues was constant during the first flooding and drainage cycle. For the roots, however, the increase in the Cu content started to level off after the tillering stage, whereas the increase in the Pb content in the roots was constant throughout the pot experiment. The Cu and Ni contents in most plant tissues decreased after the booting or filling stage. Interestingly, the Cu, Ni, and Pb contents in the roots increased at a much higher rate than those in the stem and leaves during the first flooding and drainage cycle. For Cd and Zn, this difference between trace uptake by the roots on the one hand and trace metal accumulation in the stem and leaves on the other was either much less pronounced or even absent. Indeed, large differences are found when calculating the ratios of trace metals in the stem, leaves, husks, and brown rice to trace metals in the roots for Cd, Zn, Cu, Ni, and Pb. For the final harvest, the ratios of Cd and Zn are much higher than the ratios of Cu and Pb for both cultivars (Tables 2, 3). These results reflect a higher translocation of Cd and Zn from the roots into the above-ground plant tissues than for Cu and Pb. In most cases, the ratios for Ni are higher than for Cu and Pb but lower than for Cd and Zn, meaning that Ni has an intermediate translocation behavior. Hence, root uptake and translocation of Cu and Pb into the above-ground plant tissues were strongly regulated, whereas Cd and Zn translocation was more efficient. However, we measured total trace metal contents in the roots in our study and did not distinguish between trace metals adsorbed to the outer root surface and trace metals internalized within the roots, as done before by Kalis et al. (2007) and Duffner et al. (2013) for grass and wheat, respectively. Consequently, it is not possible to be conclusive on whether the exclusion of Cu and Pb from the above-ground plant biomass was due to a mechanism regulating root uptake or an internal mechanism controlling the transfer of Cu and Pb internalized within roots into the above-ground tissues. The Cu, Ni, and Pb contents in the husks and brown rice were constant during the filling stage and harvest. Total Cd uptake by cultivar A159 at the final harvest represents 4.6 % of the total Cd pool in our paddy soil. This is higher than for the other trace metals: 1.6 % for Zn, 0.3 % for Cu, 0.4 % for Ni, and 1.1 % for Pb. Similar results were obtained for cultivar A16. Since the potential of this cultivar to accumulate Cd is lower than for cultivar A159 (Hu et al. 2013), the difference between the relative Cd uptake and the uptake of the other trace metals was less prominent (results not shown) than for cultivar A159. These results agree with the size of the reactive trace metal pools relative to the total trace metal pools in our paddy soil. In this soil, Cd was largely present in reactive forms, whereas the relative size of the reactive Zn, Cu, Ni, and Pb pools was much lower (see "Physico-chemical soil properties" section).

Table 2. Ratios of the trace metal contents in the above-ground rice plant tissues versus those in the roots for rice plant cultivar A159 at the final harvest (day 128)

Ratios	Cd	Zn	Cu	Ni	Pb
Stem:root	0.96 ± 0.15	1.34 ± 0.15	0.11 ± 0.02	0.27 ± 0.04	0.06 ± 0.02
Leaf:root	1.09 ± 0.09	1.18 ± 0.08	0.13 ± 0.02	0.35 ± 0.02	0.32 ± 0.04
Husk:root	0.35 ± 0.07	0.35 ± 0.07	0.02 ± 0.004	0.08 ± 0.02	0.005 ± 0.001
Grain:root	0.31 ± 0.05	0.75 ± 0.06	0.04 ± 0.001	0.36 ± 0.007	0.01 ± 0.002

Table 3. Ratios of the trace metal contents in the above-ground rice plant tissues versus those in the roots for rice plant cultivar A16 at the final harvest (day 128)

Ratios	Cd	Zn	Cu	Ni	Pb
Stem:root	0.99 ± 0.26	1.28 ± 0.15	0.12 ± 0.04	0.16 ± 0.04	0.03 ± 0.01
Leaf:root	0.72 ± 0.10	1.17 ± 0.12	0.17 ± 0.02	0.46 ± 0.11	0.19 ± 0.08
Husk:root	0.20 ± 0.04	0.33 ± 0.07	0.03 ± 0.01	0.08 ± 0.02	0.01 ± 0.002
Grain:root	0.27 ± 0.06	0.52 ± 0.09	0.05 ± 0.01	0.24 ± 0.08	0.003 ± 0.001

Cadmium accumulation in rice grains and food quality standards

The Cd content in the rice grains of rice plant cultivar A159 at the final harvest was 3.6 mg $kg^{-1}$and exceeds about 19 and nine times the food quality standards (FQS) set by the P.R. China and World Health Organization (WHO) (i.e., 0.2 mg Cd kg^{-1}) and Japan and Taiwan (i.e., 0.4 mg Cd kg^{-1}), respectively (Fig. 5). Although cultivar A16 is known to have a lower potential to accumulate Cd in rice grains than cultivar A159 (Hu et al. 2013), the Cd content reached 2.5 mg kg^{-1} and still exceeds about 13 times the FQS from the P.R. China and WHO and six times the FQS from the Japan and Taiwan (Fig. 9). In our study, the rice grains were not polished before chemical analysis and as a consequence, the Cd content was determined here in brown rice, whereas most people eat polished white rice. However, this is not expected to affect our results, because Masironi et al. (1977) and Kokot and Phuong (1999) found very similar Cd contents in unpolished and polished rice grains. So despite the low solubility of Cd in soil solution during both flooding periods (Fig. 4), still significant uptake and accumulation of Cd in the rice grains of both cultivars occurred to levels far above the FQS. Therefore, human exposure to Cd resulting from the consumption of rice cropped on this contaminated paddy soil might be unavoidable, even when using cultivar A16 which exhibits a lower potential to accumulate Cd than cultivar A159 (Hu et al. 2013). Both cultivars tested here are Indica varieties which are commonly grown in Zhejiang Province, P.R. China (Hu et al. 2013). However, Indica varieties are known to have a substantially higher potential to accumulate Cd in rice grains than Japonica varieties (Römkens et al. 2009b). Hence, it remains unclear whether human exposure to Cd might be reduced when cropping Japonica varieties on our paddy soil.

Water management effects on trace metal uptake by rice plants

We expected to find a clear impact of water management of the paddy soil in the pot experiment on the uptake of trace metals by the rice plants over time, with trace metal contents in the various rice plant tissues increasing during both drainage periods when the solubility of trace metals showed a large increase (Fig. 4), as has been found in previous studies (Kashem and Singh 2001; Simmons et al. 2008). Instead, trace metal contents in the roots and above-ground plant tissues seemed to increase at a reasonably constant rate during the first flooding and drainage cycle and decreased after reaching a maximum value at later growth stages in many cases. As such, our experimental results did not show a relation between the high temporal variability in the solubility of trace metals and their uptake by rice plants over time. In our pot experiment, total dissolved trace metal concentrations were measured in soil solution sampled by SMS, which were installed within the bulk soil, whereas plant uptake of trace metals is known to occur predominantly from the rhizosphere (Shuman and Wang 1997). During the second flooding period, the Eh at a depth of 5 cm below the soil surface had a high temporal variability (Fig. 3), which can be explained by leakage of O_2 from the rice plant roots. This O_2leakage apparently maintained aerobic conditions in the rhizosphere, because precipitation of Fe^{3+} in the form of Fe-(hydr)oxides was observed as a thin red layer on the root surface of both cultivars harvested from the pot experiment. This phenomenon is called iron plaque formation and is ubiquitous for wetland plants such as rice (Weis and

Weis 2004; Liu et al. 2006). Iron plaque formation was not limited to the second flooding period when a diurnal pattern in Eh at a depth of 5 cm below the soil surface was measured in the bulk soil but occurred during the first flooding period as well when such a pattern was not detected and apparently limited to the environment close to the roots (Fig. 6). The presence of iron plaque extended over the entire root length of about 5–15 cm for both cultivars. Although the role of iron plaque itself on root uptake and translocation of Cd into shoots of rice plants has been suggested to be insignificant (Liu et al. 2007), the presence of iron plaque on the root surface of both rice plant cultivars in our pot experiments can only be explained from aerobic conditions prevailing in the proximity of the roots. The solubility of trace metals in this local aerobic soil environment surrounding the rice plant roots is likely to be higher than in the anaerobic bulk soil under flooded conditions when trace metal solubility was controlled by sulfide minerals. The aerobic conditions maintained near the roots might have facilitated a continuous supply of trace metals to the root surface, causing the uptake of trace metals by rice plants to be constant in time. Consequently, trace metal solubility in the rhizosphere should be taken into account to further the understanding of water management effects of paddy fields on the uptake of trace metals by rice plants.

Figure 6. Photographs taken from the rice plants harvested at the rice tillering stage (day 38) showing iron plaque formation on the root surface of rice plant cultivar A159

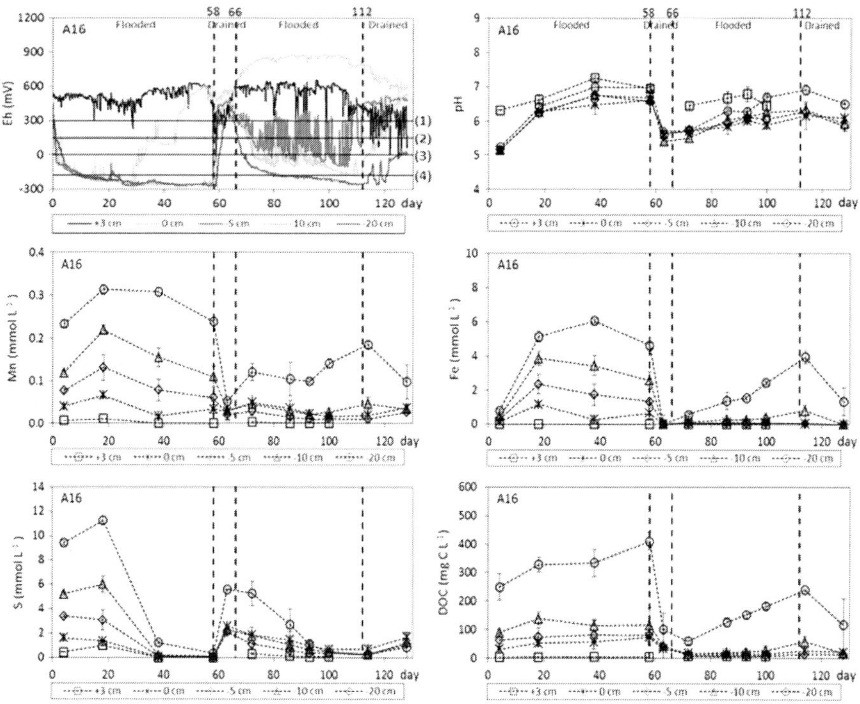

Figure 7 Eh, pH, and Mn, Fe, S, and DOC concentrations in the soil solution for the pot experiment with rice plant cultivar A16 (+3 cm refers to 3 cm above the interface, 0 cm equals the interface between the soil surface and the water layer, and −5, −10 cm, −20 cm refer to 5, 10, and 20 cm below the interface, respectively). *The numbers at the topx*-axis of *the two upper figures* indicate the time at which the drainage and flooding periods started. In figure with Eh data, 1–4 refer to the following reactions: (1) denitrification of NO_3^- to N_2, (2) and (3) reductive dissolution of MnO_2 and $Fe(OH)_3$ to Mn^{2+} and Fe^{2+}, respectively, and (4) reduction of SO_4^{2-} to S^{2-}

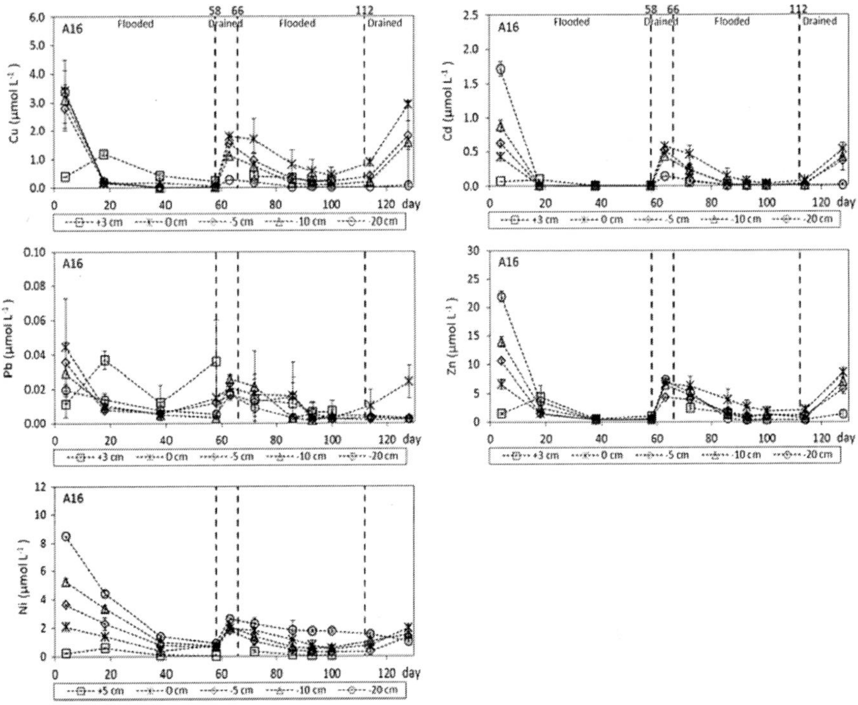

Figure 8. Total dissolved concentrations of Cu, Cd, Pb, Zn, and Ni in the soil solution for the pot experiment with rice plant cultivar A16 (+3 cm refers to 3 cm above the interface, 0 cm equals the interface between the soil surface and the water layer, and −5, −10 cm, −20 cm refer to 5, 10, and 20 cm below the interface, respectively). *The numbers at the top x*-axis of *the two upper figures* indicate the time at which the drainage and flooding periods started

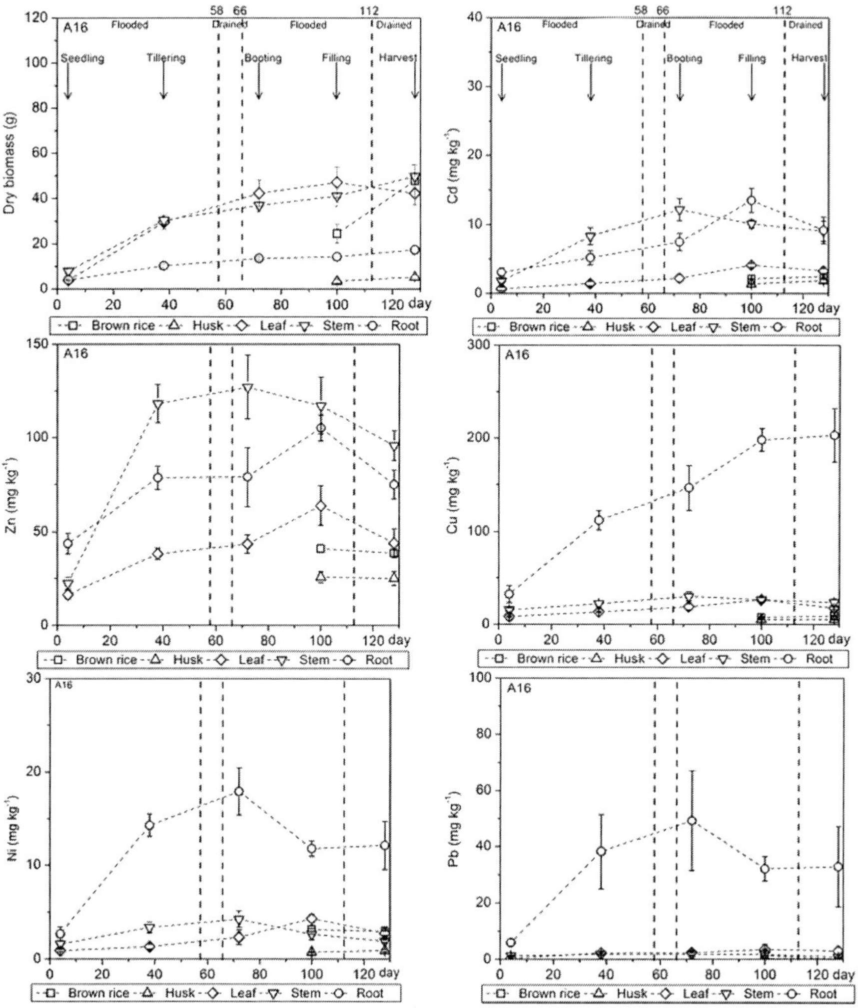

Figure 9. Biomass of the roots and the above-ground plant tissues including brown rice, husks, leaves, and stem and trace metal contents in these plant tissues for rice cultivar A16 during the major rice growth stages. Husks and brown rice were only harvested at the filling stage and final harvest. *The numbers at the top* of the figure indicate the time at which the drainage and flooding periods started

CONCLUSIONS

Two successive flooding and drainage cycles of a field-contaminated paddy soil greatly affected the solubility of trace metals at different depths along the soil profile. During flooding, the solubility of Cd, Zn, Cu, Ni, and Pb in all soil layers decreased markedly, due to the precipitation of trace metal sulfide minerals under anaerobic conditions. Despite its low solubility during flooding, the Cd content in the rice grains exceeds for both cultivars by far the FQS set by individual countries and the WHO. The increase in the trace metal contents in the roots, stem, and leaves was reasonably constant during the first flooding and drainage cycle, but the trace metal contents decreased after reaching a maximum value during the second cycle in most cases. As such, the high temporal variability in trace metal solubility was not reflected in trace metal uptake by rice plants over time. This is probably caused by the presence of aerobic conditions near the root surface, leading to a higher trace metal solubility in the rhizosphere than in the bulk soil and a continuous supply of trace metals to the root surface over time. Consequently, trace metal solubility in the rhizosphere should be taken into account when quantifying the effects of water management on the uptake of trace metals by rice plants. To enable rice production on trace metal contaminated paddy soils, either selection and breeding of rice cultivars further minimizing Cd accumulation or in situ immobilization of trace metals by using stabilizing amendments should be investigated as a measure to mitigate human health risks from trace metal contamination.

ACKNOWLEDGMENTS

This research was financially supported by the CAS-KNAW Joint Ph.D. Training Programme (No. 5237829), the National Natural Science Foundation of China (No. 40971250), the Key Project of the National Natural Science Foundation of China (No. 41230858), and the Environmental Protection Public Welfare Project funded by the Chinese Ministry of Environmental Protection (No. 201409042-03). This research is also part of the strategic research program KBIV-"Sustainable spatial development of ecosystems, landscapes, seas and regions," which is funded by the Dutch Ministry of Economic Affairs. The authors gratefully acknowledge Thom Kuyper for his comments on a previous version of our paper.

REFERENCES

1. Allison, J. D., Brown, D. S., & Novo-Gradac, K. J. (1991). MINTEQA2/PRODEFA2, A geochemical assessment model for environmental systems: Version 3.0 user's manual. US EPA, Athens, GA. EPA/600/3–91/021.

2. Amery, F., Degryse, F., Degeling, W., Smolders, E., & Merckx, R. (2007). The copper-mobilizing-potential of dissolved organic matter in soils varies 10-fold depending on soil incubation and extraction procedures. *Environmental Science and Technology, 41*(7), 2277–2281.

3. Brus, D. J., Li, Z. B., Song, J., Koopmans, G. F., Temminghoff, E. J. M., Yin, X. B., et al. (2009). Predictions of spatially averaged cadmium contents in rice grains in the Fuyang Valley, P.R. China. *Journal of Environmental Quality, 38*(3), 1126–1136.

4. Colmer, T. D. (2003). Aerenchyma and an inducible barrier to radial oxygen loss facilitate root aeration in upland, paddy and deep-water rice (*Oryza sativa* L.). *Annals of Botany, 91*(2), 301–309.

5. de Livera, J., McLaughlin, M. J., Hettiarachchi, G. M., Kirby, J. K., & Beak, D. G. (2011). Cadmium solubility in paddy soils: Effects of soil oxidation, metal sulfides and competitive ions.*Science of the Total Environment, 409*(8), 1489–1497.

6. Dick, W. A., & Tabatabai, M. A. (1979). Ion chromatographic determination of sulfate and nitrate in soils. *Soil Science Society of America Journal, 43*(5), 899–904.

7. Du Laing, G., Rinklebe, J., Vandecasteele, B., Meers, E., & Tack, F. M. G. (2009). Trace metal behaviour in estuarine and riverine floodplain soils and sediments: A review. *Science of the Total Environment, 407*(13), 3972–3985.

8. Duffner, A., Hoffland, E., Weng, L. P., & van der Zee, S. E. A. T. M. (2013). Predicting zinc bioavailability to wheat improves by integrating pH dependent nonlinear root surface adsorption. *Plant and Soil, 373*(1–2), 919–930.

9. Fulda, B., Voegelin, A., & Kretzschmar, R. (2013). Redox-controlled changes in cadmium solubility and solid-phase speciation in a paddy soil as affected by reducible sulfate and copper.*Environmental Science and Technology, 47*(22), 12775–12783.

10. Grybos, M., Davranche, M., Gruau, G., & Petitjean, P. (2007). Is trace metal release in wetland soils controlled by organic matter mobility or

Fe-oxyhydroxides reduction? *Journal of Colloid and Interface Science, 314*(2), 490–501.

11. Houba, V. J. G., Temminghoff, E. J. M., Gaikhorst, G. A., & van Vark, W. (2000). Soil analysis procedures using 0.01 M calcium chloride as extraction reagent. *Communications in Soil Science and Plant Analysis, 31*(9–10), 1299–1396.

12. Houba, V. J. G., van der Lee, J. J. G., & Novozamsky, I. (1997). *Soil and plant analysis. Part 1: Soil analysis procedures*. Wageningen: Wageningen University.

13. Hu, P. J., Li, Z., Yuan, C., Ouyang, Y. N., Zhou, L. Q., Huang, J. X., et al. (2013). Effect of water management on cadmium and arsenic accumulation by rice (*Oryza sativa* L.) with different metal accumulation capacities. *Journal of Soils and Sediments, 13*(5), 916–924.

14. Huang, C. Y. L., & Schulte, E. E. (1985). Digestion of plant-tissue for analysis by ICP Emission-Spectroscopy. *Communications in Soil Science and Plant Analysis, 16*(9), 943–958.

15. Kaiser, K., Guggenberger, G., Haumaier, L., & Zech, W. (1997). Dissolved organic matter sorption on subsoils and minerals studied by C-13-NMR and DRIFT spectroscopy. *European Journal of Soil Science, 48*(2), 301–310.

16. Kalis, E. J. J., Temminghoff, E. J. M., Visser, A., & van Riemsdijk, W. H. (2007). Metal uptake by *Lolium perenne* in contaminated soils using a four-step approach. *Environmental Toxicology and Chemistry, 26*(2), 335–345.

17. Kashem, M. A., & Singh, B. R. (2001). Metal availability in contaminated soils: II. Uptake of Cd, Ni and Zn in rice plants grown under flooded culture with organic matter addition. *Nutrient Cycling in Agroecosystems, 61*(3), 257–266.

18. Kögel-Knabner, I., Amelung, W., Cao, Z. H., Fiedler, S., Frenzel, P., Jahn, R., et al. (2010). Biogeochemistry of paddy soils. *Geoderma, 157*(1–2), 1–14.

19. Kokot, S., & Phuong, T. D. (1999). Elemental content of Vietnamese rice—Part 2. Multivariate data analysis. *Analyst, 124*(4), 561–569.

20. Koopmans, G. F., & Groenenberg, J. E. (2011). Effects of soil oven-drying on concentrations and speciation of trace metals and dissolved

organic matter in soil solution extracts of sandy soils.*Geoderma, 161*(3–4), 147–158.

21. Kuo, S., Lai, M. S., & Lin, C. W. (2006). Influence of solution acidity and $CaCl_2$ concentration on the removal of heavy metals from metal-contaminated rice soils. *Environmental Pollution,144*(3), 918–925.

22. Lefroy, R. D. B., Mamaril, C. P., Blair, G. J., & Gonzales, P. J. (1992). Sulphur cycling in rice wetlands. *Sulphur cycling on the continents: Wetlands, terrestrial ecosystems and associated water bodies* (pp. 279–300). Chichester: Wiley.

23. Liu, J. G., Qian, M., Cai, G. L., Yang, J. C., & Zhu, Q. S. (2007). Uptake and translocation of Cd in different rice cultivars and the relation with Cd accumulation in rice grain. *Journal of Hazardous Materials, 143*(1–2), 443–447.

24. Liu, W. J., Zhu, Y. G., Hu, Y., Williams, P. N., Gault, A. G., Meharg, A. A., et al. (2006). Arsenic sequestration in iron plaque, its accumulation and speciation in mature rice plants (*Oryza sativa* L.). *Environmental Science and Technology, 40*(18), 5730–5736.

25. Lofts, S., Spurgeon, D. J., Svendsen, C., & Tipping, E. (2004). Deriving soil critical limits for Cu, Zn, Cd, and pH: A method based on free ion concentrations. *Environmental Science and Technology, 38*(13), 3623–3631.

26. Masironi, R., Koirtyohann, S. R., & Pierce, J. O. (1977). Zinc, copper, cadmium and chromium in polished and unpolished Rice. *Science of the Total Environment, 7*(1), 27–43.

27. Meharg, A. A., Norton, G., Deacon, C., Williams, P., Adomako, E. E., Price, A., et al. (2013). Variation in rice cadmium related to human exposure. *Environmental Science and Technology,47*(11), 5613–5618.

28. Mongon, J., Konnerup, D., Colmer, T. D., & Rerkasem, B. (2014). Responses of rice to Fe^{2+} in aerated and stagnant conditions: Growth, root porosity and radial oxygen loss barrier.*Functional Plant Biology, 41*(9), 922–929.

29. Murase, J., & Kimura, M. (1997). Anaerobic reoxidation of Mn^{2+}, Fe^{2+}, S^0 and S^{2-} in submerged paddy soils. *Biology and Fertility of Soils, 25*(3), 302–306.

30. Novozamsky, I., van Eck, R., Houba, V. J. G., & van der Lee, J. J. (1986). Use of ICP-AES for determination of iron, aluminium and phosphorus

in Tamm's soil extracts. *Netherlands Journal of Agricultural Science, 34*, 185–191.

31. Pan, Y. Y., Bonten, L. T. C., Koopmans, G. F., Song, J., Luo, Y. M., Temminghoff, E. J. M., et al. (2016). Solubility of trace metals in two contaminated paddy soils exposed to alternating flooding and drainage. *Geoderma, 261*, 59–69.

32. Pan, Y. Y., Koopmans, G. F., Bonten, L. T. C., Song, J., Luo, Y. M., Temminghoff, E. J. M., et al. (2014). Influence of pH on the redox chemistry of metal (hydr)oxides and organic matter in paddy soils. *Journal of Soils and Sediments, 14*(10), 1713–1726.

33. Pattrick, R. A. D., Mosselmans, J. F. W., Charnock, J. M., England, K. E. R., Helz, G. R., Garner, C. D., et al. (1997). The structure of amorphous copper sulfide precipitates: An X-ray absorption study. *Geochimica et Cosmochimica Acta, 61*(10), 2023–2036.

34. Peijnenburg, W. J. G. M., Zablotskaja, M., & Vijver, M. G. (2007). Monitoring metals in terrestrial environments within a bioavailability framework and a focus on soil extraction.*Ecotoxicology and Environmental Safety, 67*(2), 163–179.

35. Reynolds, B., Stevens, P. A., Hughes, S., & Brittain, S. A. (2004). Comparison of field techniques for sampling soil solution in an upland peatland. *Soil Use and Management, 20*(4), 454–456.

36. Römkens, P. F., Guo, H. Y., Chu, C. L., Liu, T. S., Chiang, C. F., & Koopmans, G. F. (2009a). Characterization of soil heavy metal pools in paddy fields in Taiwan: Chemical extraction and solid-solution partitioning. *Journal of Soils and Sediments, 9*(3), 216–228.

37. Römkens, P. F. A. M., Guo, H. Y., Chu, C. L., Liu, T. S., Chiang, C. F., & Koopmans, G. F. (2009b). Prediction of Cadmium uptake by brown rice and derivation of soil–plant transfer models to improve soil protection guidelines. *Environmental Pollution, 157*(8–9), 2435–2444.

38. Schulz-Zunkel, C., & Krueger, F. (2009). Trace metal dynamics in floodplain soils of the River Elbe: A review. *Journal of Environmental Quality, 38*(4), 1349–1362.

39. Shaheen, S. M., Rinklebe, J., Rupp, H., & Meissner, R. (2014). Lysimeter trials to assess the impact of different flood–dry-cycles on the dynamics of pore water concentrations of As, Cr, Mo and V in a contaminated floodplain soil. *Geoderma, 228*, 5–13.

40. Shen, J. B., & Hoffland, E. (2007). In situ sampling of small volumes of soil solution using modified micro-suction cups. *Plant and Soil, 292*(1–2), 161–169.

41. Shuman, L. M., & Wang, J. (1997). Effect of rice variety on zinc, cadmium, iron, and manganese content in rhizosphere and non-rhizosphere soil fractions. *Communications in Soil Science and Plant Analysis, 28*(1–2), 23–36.

42. Simmons, R. W., Noble, A. D., Pongsakul, P., Sukreeyapongse, O., & Chinabut, N. (2008). Analysis of field-moist Cd contaminated paddy soils during rice grain fill allows reliable prediction of grain Cd levels. *Plant and Soil, 302*(1–2), 125–137.

43. Sun, M. M., Luo, Y. M., Teng, Y., Jia, Z. J., Li, Z. G., & Deng, S. P. (2013). Remediation of polycyclic aromatic hydrocarbon and metal-contaminated soil by successive methyl-beta-cyclodextrin-enhanced soil washing-microbial augmentation: A laboratory evaluation.*Environmental Science and Pollution Research, 20*(2), 976–986.

44. Tipping, E., Rieuwerts, J., Pan, G., Ashmore, M. R., Lofts, S., Hill, M. T. R., et al. (2003). The solid-solution partitioning of heavy metals (Cu, Zn, Cd, Pb) in upland soils of England and Wales. *Environmental Pollution, 125*(2), 213–225.

45. Tsukahara, T., Ezaki, T., Moriguchi, J., Furuki, K., Shimbo, S., Matsuda-Inoguchi, N., et al. (2003). Rice as the most influential source of cadmium intake among general Japanese population. *Science of the Total Environment, 305*(1–3), 41–51.

46. USDA. (1996). Soil survey laboratory methods manual. Soil survey investigations, Report No. 42. USDA, Washington, DC.

47. Vink, J. P. M., Harmsen, J., & Rijnaarts, H. (2010). Delayed immobilization of heavy metals in soils and sediments under reducing and anaerobic conditions; consequences for flooding and storage. *Journal of Soils and Sediments, 10*(8), 1633–1645.

48. Weber, F. A., Voegelin, A., & Kretzschmar, R. (2009). Multi-metal contaminant dynamics in temporarily flooded soil under sulfate limitation. *Geochimica et Cosmochimica Acta, 73*(19), 5513–5527.

49. Weis, J. S., & Weis, P. (2004). Metal uptake, transport and release by wetland plants: Implications for phytoremediation and restoration. *Environment International, 30*(5), 685–700.

50. Weng, L. P., Temminghoff, E. J. M., & van Riemsdijk, W. H. (2001). Contribution of individual sorbents to the control of heavy metal activity in sandy soil. *Environmental Science and Technology, 35*(22), 4436–4443.

51. Yu, H., Wang, J. L., Fang, W., Yuan, J. G., & Yang, Z. Y. (2006). Cadmium accumulation in different rice cultivars and screening for pollution-safe cultivars of rice. *Science of the Total Environment, 370*(2–3), 302–309.

52. Zeng, F. R., Ali, S., Zhang, H. T., Ouyang, Y. B., Qiu, B. Y., Wu, F. B., et al. (2011). The influence of pH and organic matter content in paddy soil on heavy metal availability and their uptake by rice plants. *Environmental Pollution, 159*(1), 84–91.

Chapter 11

Distribution, Enrichment and Ecological Risk Assessment of Six Elements in Bed Sediments of a Tropical River, Chottanagpur Plateau: A Spatial and Temporal Appraisal

Kumar Manoj, Pratap Kumar Padhy*

Department of Environmental Studies, Institute of Science, Visva-Bharati University, Santiniketan, India

ABSTRACT

Distribution and enrichment of six elements (iron, zinc, copper, lead, cadmium and manganese) in surface bed sediments, collected from seventeen selected locations during pre-monsoon and post- monsoon periods, of the tropical Chottanagpur plateau river Subarnarekha along with the ecologi- cal risks involved were investigated. Owing to the rich occurrence of mineral resources, the Sub- arnarekha river basin has a large scale presence of industrial and mining units especially in the Indian State of Jharkhand. An assessment, which involved examining distribution pattern of ele- ments, comparative studies with sediment quality guidelines (SQGs) and geochemical background values and a sequential and integrated index analyses approach (containing contamination factor (CF), pollution load index (PLI), contamination degree (CD), enrichment factor (EF), geo-accumu- lation index (Igeo) and potential ecological risk index (PERI)), was followed to estimate enrich- ment and risks of elements in the bed sediments. Sediments collected from areas having abun- dance of population, industrial conglomerates and mining units recorded elevated element con- centrations, which exceeded SQGs, and significantly higher values of CF, CD, PLI, EF, Igeo and PERI. Cadmium demonstrated surprising regularity in its enrichment; contributed most to the ecologi- cal risks; and high toxicity risks due to cadmium exceeded 64% of the sites. Moreover, chronic exposures of other elements would also lead to similar ecological risks. In addition to revealing potential ecological risks due to cadmium and other elements our investigation markedly high- lighted anthropogenic control over sediment quality deterioration and some immediate sedi- ment quality management strategies are needed to remediate and control river bed contamina- tion.

Keywords: Anthropogenic Impacts, Ecological Risk Index, Geo-Accumulation Index, Trace Elements, Pollution Load Index, Sediment Contamination

1. INTRODUCTION

Developing countries in the last few decades have undergone rapid industrialization, urbanization, population growth and change in land use pattern. These factors have directly influenced the aquatic systems and contri- buted to their deterioration in quality. In India, most of the industrial conglomerates and urban areas, being lo- cated along the river basins, discharge their wastes into the river streams. In addition, application of pesticides and chemical fertilizers in agricultural fields, urban runoffs and atmospheric depositions also contribute to the pollution of water bodies. Sediments are regarded as sensitive indicators for monitoring toxic and hazardous contaminants entering aquatic environments [1] . Though, quality assessment and monitoring of natural streams have become regular part of the national river conservation plans, investigations on sediment quality also require equal attention as they are potential sources for mobility and bioavailability of contaminants like trace elements in the aquatic systems [2] - [4] . A thorough analysis of nature of sediments is required to identify the natural li- thogenic and anthropogenic sources of contaminants [5] [6] . Investigation of elements distribution in sediments is also important to understand their behaviour and transport in the fluvial or aquatic environment [7] . Sediments work as adsorbent pads for elements which can remobilize in due course under changed physical and chemical conditions of the water bodies [8] . However, sediments also act as habitat and major source of nutrients for aq- uatic organisms [9] . Recent examinations of river beds put more emphasis on the presence of trace elements in sediments because of their persistent, non-degradable nature and potential to enter the food chain. Elements en- tering the river system become bound to particulate matter, which after settling down become part of the sedi- ments [10] . Though, elements such as Fe, Zn and Cu are required for the normal functions of the body, at con- centrations higher than the required value they can be detrimental to the physiology and health of the living or- ganisms. On the other hand, elements like Cd and Pb are toxic even at minute concentrations [5] [11] - [14] . Reg- ular environmental monitoring and assessment of aquatic systems for the presence of elements is indispensable to examine possible impacts of developmental projects on their sediment environment. Since sediment environ- ment directly influences the physicochemical and ecological dynamics of aquatic resources, a deteriorated sedi- ment quality may severely impact the supports that sustain aquatic life and the surrounding ecosystem, which also includes human beings, dependent on it [15] .

The Subarnarekha river basin, having catchment area of more than 19,200 km^2 and providing about 0.4% of the country's total surface water resources [16] , is one of the most important inter-state river basins of India. The river rises on the eastern slopes of the Chottanagpur plateau in Ranchi district (Jharkhand state) and flows through three states namely Jharkhand, West Bengal and Odisha before draining into the Bay of Bengal towards the eastern coast. The river predominantly flows through tribal belt and is extensively used for domestic and industrial purposes. It is the main receiving body of wastes from populated and industrialized regions like Ranchi, Adityapur, Jamshedpur and Ghatsila. Effluents of heavy and small scale industries located along the basin and the domestic and municipal wastes of the rural and urban areas are discharged into the river. Conventional water treatment plants are limited and inadequate which put additional stress on the ecology of the river. Owing to the rich occurrence of major mineral resources the basin has significant presence of mining and mineral processing units. Based on above mentioned information and the relatively new occurrence of two activities, that is, large scale quarrying of the river bed for construction materials (minor minerals) and encroachment, examination of the river basin becomes essential for ecological reasons. In this study geochemical distribution, enrichment and ecological risk of six elements were investigated. Eco-toxicological potential of Cd and other elements were de- scribed in terms of both threshold effect concentration (TEC) and probable effect concentration (PEC) of the se- diment quality guidelines (SQGs) as well as ecological risk indices computed. The objectives of our investiga- tion were: to determine spatial and temporal distribution patterns of six element concentrations in surface sedi- ments of the Subarnarekha river in the Indian state of Jharkhand, where the river shows maximum presence; to assess control of anthropogenic activities on sediment contamination of elements; and to find out potential ecological risks to the benthic biota and other freshwater organisms associated with element concentrations. Addi- tionally, since the National River Water Quality Monitoring Network in India mainly involves studying water quality, this study was a step in the direction to acquaint the public and the policy makers about sediment quality of the river to develop management strategies needed to remediate and control river bed contamination.

2. MATERIALS AND METHODS

2.1. Study Area

The study area, located between latitudes 22˚13'14"N and 23˚27'00"N; longitudes 85˚11'00"E and 86˚42'50"E, consisted of three districts namely Ranchi, Saraikela Kharswan and Purba Singhbhum of the Jharkhand state, India. Subarnarekha river flows through these three districts before entering the Indian state of West Bengal near Gopiballabhpur in the West Medinipore district (Figure 1). The river during its course passes Ranchi uplands, Bhagmundi hills (Ayodhya pahar), Dalma and Dhanjori ranges. Its bed belongs to the hard rock of the Pre- Cambrian or Archaean formations comprising mainly quartzites, granites, gneiss, dolomites, mica-schists, phyl- lites [17] and Dhanjori groups of volcanic rocks. Kharkai river, most important tributary of the river Subarna- rekha, passes through one of the most important industrialized belts of Asia (Adityapur industrial area) and joins it near Sonari towards north-western periphery of the steel city Jamshedpur. In addition, Kanchi, Karkari, Garra, Gurma and Sankh are other significant tributaries that make the river basin. Downstream of the Jamshedpur city Garra, Gurma and Sankh rivers pass through key mining belts and join the river Subarnarekha at different places. Many heavy mining and processing belts, industrial conglomerates and populated areas are located along the ba- sin which make river Subarnarekha as the main receiving body of industrial discharges, mine drainages, partially treated or untreated domestic wastes etc.

2.2. Sediment Sampling

Sediment samples were collected from seventeen selected locations along the Subarnarekha river (Figure 1) during May, 2011 (pre-monsoon) and November, 2011 (post-monsoon) periods. Sampling sites were selected based on surveys, that is, areas having high anthropogenic activities and areas relatively free of anthropogenic influences, and accessibility. Surface sediment samples (3 grabs per sample) were taken at a depth of about 5 cm and immediately transferred into pre-cleaned polythene bags [18] . The collected samples were oven dried at 40˚C for 48 hours; passed through a 1 mm plastic sieve to remove plant parts, debris and gravel-sized materials; and then sieved through a nylon sieve (aperture 125 μm). Use of fractions <125 μm for characterization of the bed sediments has also been reported earlier [3] [15] . The lower particle size sediment fractions were ground using mortar and pestle for homogenization and prepared for further analysis of physicochemical properties and concentration of six elements [19] . All necessary precautions and care were taken during drying, sieving, grind- ing and storage of sediment samples to avoid any kind of contamination.

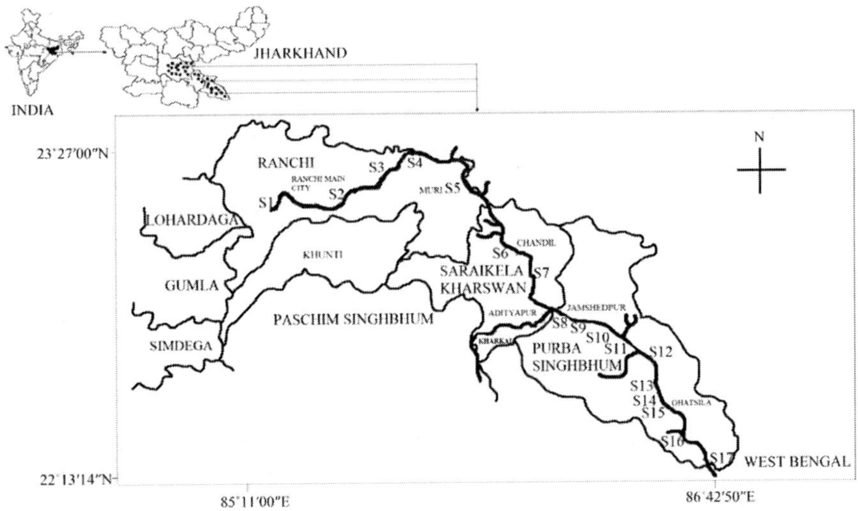

Figure 1. Study area and the sampling locations; black spots represent study area; capital "S" letter indicates sampling site.

2.3. Physicochemical and Statistical Analyses

The pH of the sediment samples was measured using a digital pH meter (pHTestr 30, Eutech instruments, Oak- lon) in 1:2.5 sediment to water ratio. The suspension was allowed to stand overnight prior to pH measurement [3] . Electrical conductivity (EC) was determined using Systronics Conductivity Meter (Model 361) in sediment saturation extracts as described by [20] . Organic Carbon (OC) was measured by partial oxidation method as out- lined by [21] . Widely used HNO_3 + $HClO_4$ acid digestion procedure for total element analysis in sediment sam- ples and as described by [22] was carried out. Briefly 1 gm of sample was digested on hot plate with acid mix- ture containing 10 ml of concentrated HNO_3 and 5 ml 70% $HClO_4$. After cooling, the mixture was boiled with 20 ml distilled water to release any fumes. The digested samples were filtered through Whatman No. 42 filters and made up to 25 ml by adding distilled water in a volumetric flask. Concentrations of six elements, namely, Fe, Zn, Cu, Pb, Cd and Mn were determined with Atomic Absorption

Spectrophotometer (AAS: Varian AA50) at wavelengths (nm)—Fe = 248.3, Zn = 213.9, Cu = 324.8, Pb = 217, Cd = 228.8, Mn = 279.5. Before running sample analysis, the AAS was subjected to calibration and standardization using appropriate standards. The ba- sis of element quantification was calibration curves of standard solution of elements (Fe: HC001722, Zn: HC117124, Cu: HC114735, Pb: HC125468, Cd: HC123670, Mn: HC241921) purchased from Merck (KGaA), Germany. To check the accuracy of the data, calibration was verified by analyzing a midpoint calibration stan- dard and a calibration blank with each batch of samples run. Reagent blank, prepared to assess presence of ana- lytes or interference, was found to be devoid of any analytes. The analytical precision was also verified by run- ning one replicate and one blank spike sample after each batch of sample [7] . The accuracy of the procedure was also checked running five replicates of the standard solution of each element (expressed as mean ± standard deviation)—Fe: 0.998 ± 0.053, Zn: 0.55 ± 0.029, Cu: 0.126 ± 0.014, Pb: 0.22 ± 0.021, Cd: 0.049 ± 0.0045, Mn: 0.254 ± 0.027. The glassware were washed with 14% HNO_3 and rinsed thoroughly with Milli-Q quality water prior to use. The Milli-Q quality water was used throughout the experiments. All the chemicals and reagents used were of analytical grade purchased from Sigma-Aldrich (USA) and Merck (Germany and India). The raw data were subjected to basic statistical analysis to identify some descriptive statistical factors. Interrelationships between variables were determined by correlation analysis applying Pearson correlation coefficient.

2.4. Calculation of Enrichment Factors and SQGs

In the present study six indices, namely, contamination factor (CF), contamination degree (CD), pollution load index (PLI), enrichment factor (EF), geo-accumulation index (Igeo) and potential ecological risk index (PERI) were prepared and evaluated stepwise to determine degree of anthropogenic influence on the sediment quality. An integrated index is a collection of single indices determined separately and then combined to designate the overall quality of the sediment. Three integrated indices, namely, PLI, CD and PERI were used in this study. Methods of calculation and classification criteria of studied indices with references are presented in Table 1. CF, PLI and CD calculations involve comparing concentration of elements with their geochemical background ref- erence value and then computing the indices. While CF is calculated for individual elements, CD and PLI pro- vide overall information about the sediment contamination of study sites. Similar to CF, Igeo [27] is also com- puted for individual elements which supply valuable knowledge about the degree and extent of element load in the sediments. EF, as introduced by [25] , is used to assess the degree of anthropogenic influence on element load in the sediments, and also to differentiate between elements originating from the natural or anthropogenic activi- ties [9] . PERI, originally developed by [23] , provides a mean to quantitatively evaluate ecological risk posed by multiple element pollutions in the bed sediments of aquatic systems [30] . It takes into account concentration of elements, their potential ecological risk factors (PERF), and the sedimentological toxic response factors (TRF). TRF assigned to elements used in the calculation of PERI are: Cu = Pb = 5; Cd = 30; Zn = 1 [29] . Due to ab-

Table 1. Sediment quality indices and their classification systems.

Index	Equation	Classification	References
CF	$CF = C_n / C_b$	CF < 1 (low CF) 1 ≤ CF < 3 (moderate CF) 3 ≤ CF < 6 (considerable CF) CF ≥ 6 (very high CF)	[18] [23]
PLI	$PLI = (CF_1 \times CF_2 \times \cdots \times CF_n)^{1/n}$	PLI = 0 (perfection) PLI = 1 (baseline levels of pollutants present) PLI > 1 (progressive deterioration of site)	[24]
CD	$CD = \sum_{i=1}^{n} CF$	CD < 6 (low CD) 6 ≤ CD < 12 (moderate CD) 12 ≤ CD < 24 (considerable CD) CD ≥ 24 (very high CD) indicating alarming anthropogenic contamination	[23] [18]
EF	$EF = (M/R)_{Sample} / (M/R)_{Background}$	EF: 2 - 5 (moderate enrichment) EF: 5 - 20 (significant enrichment) EF: 20 - 40 (very high enrichment) EF > 40 (extremely high enrichment)	[25] [26]
Igeo	$Igeo = \log_2 C_n / 1.5 \times B_n$	Igeo < 0 Class 0 (uncontaminated) 0 < Igeo < 1 Class 1 (uncontaminated to moderately contaminated) 1 < Igeo < 2 Class 2 (moderately contaminated) 2 < Igeo < 3 Class 3 (moderately to heavily contaminated) 3 < Igeo < 4 Class 4 (heavily contaminated) 4 < Igeo < 5 Class 5 (heavily to extremely contaminated) Igeo ≥ 5 Class 6 (extremely contaminated)	[3] [27] [28]
PERI	$PERI = \sum_{i=1}^{n} E_r^i$ $E_r^i = T_r^i \times CF_i$	E_r^i < 30 (low risk) E_r^i : 30 - 60 (moderate risk) E_r^i : 60 - 120 (considerable risk) E_r^i : 120 - 240 (high risk) E_r^i > 240 (significantly high risk) PERI < 110 (low risk) PERI: 110 - 220 (moderate risk) PERI: 220 - 440 (high risk) PERI > 440 (significantly high risk)	[29] [23]

CF = Contamination Factor; PLI = Pollution Load Index; CD = Contamination Degree; EF = Enrichment Factor; Igeo = Geo-Accumulation Index; PERI = Potential Ecological Risk Index; Co = Observed Element Concentration; C_b = Background Concentration of the Element; M = Element; R = Reference Element; B_n = Background Concentration; E_r^i = Monomial Ecological Risk Index; T_r^i = Toxic Response Factor

sence of any background concentration of elements from the studied area, average shale standards of trace ele- ments described by [31] were taken as geochemical reference values in the present study. Fe was employed as normalization element to calculate EF values due to its abundance and controlling influence on the distribution of other trace elements [32] [33] . Based on positive correlations between sediment chemistry and sediment toxicity, [34] explained consensus-based sediment quality guidelines (SQGs) for classifying sediments as toxic and not-toxic. According to authors, SQGs are important tools for identifying contaminated sediment hotspots and also for assessing possible effects of contaminated sediments on benthic organisms. Threshold effect concentra- tion (TEC) and probable effect concentration (PEC) are commonly employed for sediment contamination inves- tigations with respect to elements [1] [35] as these guidelines provide predictive ability of adverse biological ef- fects to the sediment dwelling fauna and other aquatic organisms [35] . TEC is defined as the concentration be- low which adverse effects are unlikely to be observed whereas PEC denotes concentration above which adverse effects are frequently expected to occur [34] . Another tool developed to study sediment toxicity is the mean se- diment quality guideline quotients (mSQGQs). In the present study mSQGQs were derived following [36] . The quotients were, however, adjusted and derived from four analytes: Zn, Cu, Pb and Cd. The methodology em- ployed to investigate Subarnarekha river bed sediments is schematically represented in Figure 2.

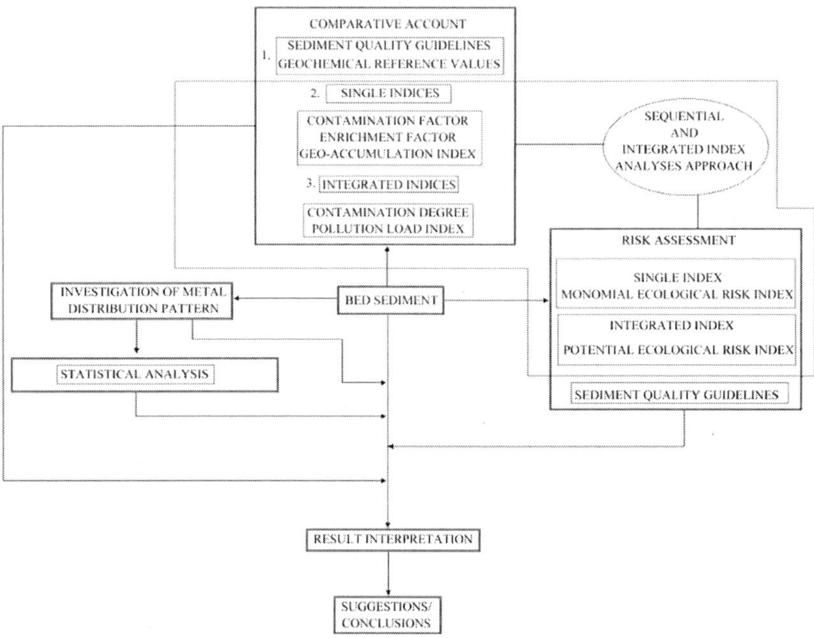

Figure 2. A schematic representation of methodology employed to characterize Subarnarekha bed sediments.

3. RESULTS AND DISCUSSION

3.1. Determination of pH, EC and OC

The pH, EC and OC (%) values obtained in the sediments of the river Subarnarekha during pre-monsoon study are displayed in Table 2; a summary of post-monsoon data of these measured parameters is also given in Table 3. The pH values ranged from 5.76 to 6.83 and 5.91 to 6.81 during pre-monsoon and post-monsoon periods re- spectively indicating slightly acidic nature of the sediments. This might have resulted from the formations of humic acid substances as well as discharge of acidic industrial, mining and domestic wastes into the river. EC of the sediment implies ability of the sediment to conduct electric current [37] . Lowest EC value was obtained at S1, which is a comparatively non-polluted site, while highest EC was recorded at S14 during both periods of study. Post-monsoon sediments had higher EC level (0.153 - 1.147 ms/cm) than pre-monsoon sediments. OC values ranged from 2.97% to 5.34% during pre-monsoon and showed almost similar pattern in the second study period. Significantly higher values of OC were observed at S9, S14 and S15. In addition to adsorption and other related processes, elements are also incorporated into the sediment by organic complexation. Organic materials from river water and decomposed matter get enriched into the sediment [38] . However, smaller fractions of se- diments display more efficient relationship with OC as compared to larger fractions.

3.2. Element Concentrations and SQGs

Concentrations of six studied elements recorded in collected surface sediments (of pre-monsoon season) from the Subarnarekha river are displayed in Table 2; a summary of post-monsoon data of the elements is also given in Table 3. Elements displayed wide variations in their distribution suggesting control of anthropogenic activi ties on sediment chemistry. Although similar spatial variability of elements was recorded for the Subarnarekha river sediments in both pre-monsoon and post-monsoon periods, the latter displayed higher concentration load than the former. This could be due to more runoffs from industrial, mining, urban and agricultural areas during rainy season. Elements released into aquatic environments become bound to the particulate matters which in due course settle down and become part of the sediments [10] . Overall concentration of trace elements followed the order: Fe > Mn > Cu > Zn > Pb > Cd. The sampling site S14 recorded highest concentration of Fe, Zn, Cu, Pb and Cd while S9 showed maximum Mn value. Four sampling stations S8, S9, S14 and S15 regularly displayed more element concentrations compared to others. S8 and S9 are located in the Jamshedpur city, a highly popu- lated and

Table 2. Recorded values of pH, EC and OC and distribution of six measured elements in sediments of the Subarnarekha river, India.

Sampling locations	Physicochemical parameters and elements								
	pH	EC (mS/cm)	OC (%)	Fe (mg/kg)	Zn (mg/kg)	Cu (mg/kg)	Pb (mg/kg)	Cd (mg/kg)	Mn (mg/kg)
S1	6.83	0.142	3.32	12631.00	30.11	11.71	11.70	0.60	450.35
S2	6.41	0.407	4.76	30319.00	30.56	19.30	21.41	1.30	312.78
S3	6.76	0.315	3.85	27327.00	13.10	24.71	13.02	0.57	500.83
S4	6.83	0.378	2.97	28770.00	17.66	8.34	5.70	0.50	252.30
S5	6.20	0.432	4.69	30286.00	31.78	13.72	18.41	1.60	665.60
S6	6.43	0.526	3.85	36077.00	45.00	24.90	21.23	0.70	712.90
S7	6.47	0.683	3.46	30319.00	58.61	91.80	18.07	1.26	960.21
S8	6.30	0.812	4.39	40889.00	36.34	28.90	18.42	1.61	1945.58
S9	6.18	0.875	5.08	56736.00	60.74	40.75	22.20	1.68	2449.30
S10	6.75	0.831	3.27	36578.00	37.72	28.76	13.50	1.10	1032.36
S11	6.72	0.782	3.14	21814.00	31.50	28.37	8.01	0.70	574.80
S12	6.14	0.894	4.75	31910.00	37.22	28.20	13.81	1.34	1343.54
S13	6.25	0.647	4.39	20293.00	26.91	16.70	8.13	1.20	434.31
S14	5.76	1.128	5.34	125516.00	288.80	2121.70	40.10	3.08	706.81
S15	5.85	0.984	5.12	36786.00	68.50	94.70	24.68	2.30	586.24
S16	6.12	0.817	3.63	30586.00	28.37	42.23	11.40	1.41	446.30
S17	6.48	0.685	4.07	13237.00	9.20	2.00	4.80	1.24	468.71
Minimum	5.76	0.142	2.97	12631.00	9.20	2.00	4.80	0.50	252.30
Maximum	6.83	1.128	5.34	125516.00	288.80	2121.70	40.10	3.08	2449.30
Mean	6.38	0.667	4.12	35886.71	50.13	154.49	16.14	1.30	814.20
Median	6.41	0.685	4.07	30319.00	31.80	28.00	13.80	1.24	586.20
[a]SD	0.33	0.264	0.76	25311.42	63.54	507.58	8.60	0.66	594.41
Skewness	−0.23	−0.317	0.056	3.04	3.70	4.11	1.23	1.26	1.83
Kurtosis	−0.68	−0.526	−1.331	10.85	14.51	16.90	2.61	2.28	2.99
[b]TEC	—	—	—	—	121.00	31.60	35.80	0.99	—
[b]PEC	—	—	—	—	459.00	149.00	128.00	4.98	—

[a]Standard deviation; [b]Consensus based sediment quality guidelines (CSQGs).

Table 3. A synopsis of analyzed results of the Subarnarekha bed sediments in post-monsoon period.

	Parameters, CD, PLI and PERI		CF		EF		Igeo	
	Min	Max	Min	Max	Min	Max	Min	Max
pH	5.83	6.87	—	—	—	—	—	—
EC	0.153	1.147	—	—	—	—	—	—
OC	2.92	5.47	—	—	—	—	—	—
Fe	12.638	125.603	0.268	2.662	—	—	−2.486	+0.828
Zn	8.73	302.96	0.092	3.189	0.248	1.206	−4.029	+1.088
Cu	2.17	2184.63	0.048	48.550	0.182	19.320	−4.959	+5.017
Pb	4.92	53.91	0.246	2.696	0.509	2.384	−2.608	+0.846
Cd	0.54	4.79	1.8	15.967	3.096	10.757	+0.263	+3.412
Mn	257.83	2478.53	0.385	2.916	0.359	2.665	−2.306	+0.959
CD	3.512	74.02	—	—	—	—	—	—
PLI	0.293	5.069	—	—	—	—	—	—
PERI	60.44	738.43	—	—	—	—	—	—

Values of pH in pH units; EC in mS/cm; OC in %; elements in mg/kg; Min = Minimum; Max = Maximum.

industrialized belt of Jharkhand. Moreover, river Kharkai, which carries industrial and domestic wastes from the Adityapur industrial area, one of the largest in Asia, joins the river Subarnarekha near S8. The river at S14, located nearby a copper factory site and upstream of Ghatsila town, and S15, situated downstream of Ghat- sila, is a major industrial and domestic discharges receiving body. Sampling sites located nearby industrial, mining and urban areas either exceeded TEC with respect to Cd, Pb, Cu and Zn or closely approached the benchmark, while for PEC, concentration of elements at sampling sites S14 and S15 was relatively very high. Analyzed appraisal with SQGs (TECs and PECs) and the results of temporal study indicated adverse eco-toxic effects from long term exposure of elements which was more alarming for Cd. Elements also exceeded their av- erage shale standard concentrations in sampling sites characterized by high anthropogenic activities. More pro-nounced results were obtained at S7, S8, S9, S14 and S15 where elements surpassed their geochemical back- ground values as high as forty seven times indicating high enrichment of sediments with elements. Accumula- tion of elements in bed sediments of the river Subarnarekha as compared to some other river sediments around the world, even though each area has its own geochemical characteristics, indicate higher load for Fe, Cu and Mn and substantial enrichment for Cd (Table 4). The spatial and temporal investigation revealed Cd as the big- gest contaminant and chief cause of concern. Cd concentration was noted above its geochemical background value throughout the studied area in both study periods; however, its concentration was significantly higher at sites characterized by dominant anthropogenic activities. Contamination of sediments of freshwater systems with Cd is increasingly becoming a major problem in developing countries worldwide. Some cases include: Gomti river (India) 0.70 to 7.90 mg/kg [40] ; Achankovil river (India) 3.67 to 11.43 mg/kg [7] ; Hindon river (In- dia) 1.30 to 3.28 mg/kg [10] ; Buriganga (Bangladesh) 3.5 to 7.8 mg/kg and 4.1 to 9.5 mg/kg in summer and winter seasons respectively [43] , Lianshan and Wuli rivers (China) 25.53 to 98.78 mg/kg and 8.04 to 17.75 mg/kg respectively [30] . Not only rivers, Cd contamination are also being recorded for the wetland systems. For examples, Bangalore urban lakes (India) recorded 4.68 to 14.25 mg/kg Cd [44] and Vembanad wetland system in Kerala (India) 0.26 to 0.73 mg/kg [1] . Results of only one season (pre-monsoon) are displayed in tables and figures, unless otherwise mentioned.

Table 4. Comparison of mean element concentrations recorded in present investigation with some other studies.

Rivers	Parameters					
	Fe	Cu	Zn	Pb	Cd	Mn
Yarmouk[a]	25883.22	-	172.61	67.20	8.48	594.50
Gomti[b]	6735.77	10.705	42.83	51.04	2.53	220.33
Achankovil[c]	11858	224	415	72	6	699
Orogodo[d]	24.60	2.50	1.26	4.39	0.49	1.60
Ganga[e]	5816.50	12.82	49.02	2.50	3.08	148.50
Hindon[f]	233.72	59.33	58.29	41.20	2.29	150.88
Dikrong[g]	1.76	190.10	26.40	39.10	-	548.80
Euphrates[h]	661.70	14.14	67.66	0.59	11.22	37.70
Shale standard[i]	47200	45	95	20	0.3	850
Present study	35886.71	154.49	50.13	16.14	1.30	814.20

[a][39]; [b][40]; [c][7]; [d][37]; [e][41]; [f][10]; [g][33]; [h][42]; [i][31]; All values in mg/kg

3.3. Sequential and Integrated Index Analyses Approach

CF is used to illustrate the contamination of particular toxic substance at a given site [1] . The characteristic ter- minologies of CF and its calculated values in the sediment samples from the Subarnarekha river are given in Table 1 and Table 5 respectively. S14 displayed moderate contamination for Fe; considerable contamination for Zn; moderate contamination for Pb; and very high contamination for Cd and Cu in both pre-monsoon (10.27 and 47.15) and post-monsoon periods (15.97 and 48.55) indicating serious anthropogenic contribution to the conta- mination load of sediments at this site. Meanwhile contaminations of sediments, with reference to one or more elements, were also reported from S2, S6, S7, S8, S9, S12 and S15. Of the two study periods, the sediment was more contaminated in post-monsoon period. Trace element Cd revealed as the most severe component causing moderate to very high contamination of the sediments in both periods of study. Similar pattern was noted for contamination degree (CD) (Figure 3(a)), where sampling sites having dominant anthropogenic activities dis- played high CD. This demonstrates regular monitoring of the sediments for the presence of trace elements espe- cially cadmium is required. In geochemical investigations, PLI is used as a resourceful tool to measure and compare sediment contamination. Analyzed sediments at S7 (1.231), S8 (1.158) and S9 (1.524) towards up- stream segment, and S14 (4.337) and S15 (1.407) towards downstream region displayed higher PLI values and progressive deterioration in quality (Figure 3(a)). S14 recorded highest PLI closely followed by S15. Compara- tively higher PLIs (0.293 - 5.069) were noticed in post-monsoon study (Table 3), which indicated more enrich- ment of metals. Higher PLI values in sediments demonstrated substantial anthropogenic impacts on the sediment quality whereas lower PLI values pointed to no considerable anthropogenic activities. This investigation strengthened employment of CD and PLI as effective instruments for assessing the environmental geochemistry of se- diments and could be used individually or in combination as they closely complemented each other. Moreover, they easily convey information to the public and policy makers to ascertain the contamination load of the sedi- ments to take necessary remedial measures. To measure degree and extent of element contamination in freshwa- ter resources EF and Igeo can also be used as reference systems [10] . Geo-accumulation index consists of a seven grade sediment quality classification scheme varying from uncontaminated to extremely contaminated

Table 5. Contamination factor of six studied elements in bed sediments.

Sampling locations	Contamination factor					
	Fe	Zn	Cu	Pb	Cd	Mn
S1	0.268	0.317	0.260	0.585	2.0	0.506
S2	0.642	0.322	0.429	1.071	4.333	0.368
S3	0.579	0.138	0.549	0.651	1.9	0.589
S4	0.610	0.186	0.185	0.285	1.667	0.297
S5	0.642	0.335	0.305	0.921	5.333	0.783
S6	0.764	0.474	0.533	1.062	2.333	0.839
S7	0.642	0.617	2.04	0.904	4.2	1.130
S8	0.866	0.383	0.642	0.921	5.637	2.289
S9	1.20	0.639	0.906	1.11	5.6	2.882
S10	0.775	0.397	0.639	0.675	3.667	1.215
S11	0.462	0.332	0.630	0.401	2.333	0.676
S12	0.676	0.392	0.627	0.691	4.467	1.581
S13	0.430	0.283	0.371	0.407	4.0	0.511
S14	2.659	3.04	47.147	2.008	10.267	0.832
S15	0.779	0.721	2.104	1.234	7.667	0.690
S16	0.648	0.299	0.938	0.570	4.70	0.525
S17	0.280	0.097	0.045	0.240	4.133	0.551

categories. According to [40] Class 6 indicates 64 fold enrichment above the geochemical background value. The EF and Igeo values calculated for studied elements in all sediment samples and variations obtained (pre- monsoon data) are graphically presented in Figure 3(b) and Figure 3(c) respectively. A brief post-monsoon summary of these indices is also given in Table 3. The EF and Igeo exhibited similar spatial variations in both study periods; however, temporally post-monsoon period was more contaminated. Analysis of results

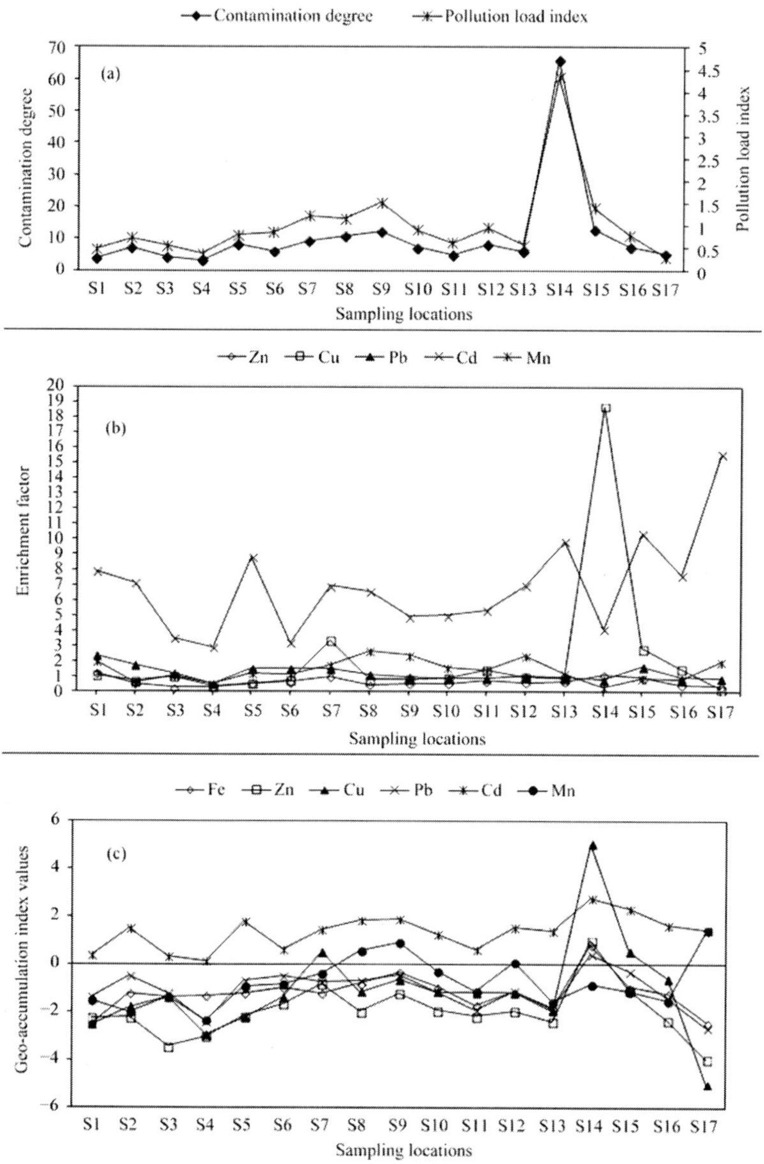

Figure 3. Contamination degrees and pollution load indices of sediments at different sites (a), enrichment factor values (b) and sediment geo-accumulation indices (c) of trace elements in Subarnarekha river bed sediments.

displayed sediment samples collected from urban-industrial areas enriched with Cu and Mn. Sites S7 and S15 showed moderate enrichment for Cu while S14 demonstrated significantly high enrichment for the element. Moderate enrichment for Mn was noted at S8, S9 and S12. Sediments of Subarnarekha were found to be uniformly en- riched with Cd which varied from moderate to significant enrichment. However, with some exceptions, sam- pling sites dominated by industrial, mining and other anthropogenic activities presented elevated Cd enrichment. Analyzed results showed sediment at S14 was moderately contaminated with Fe and Zn. The Igeo of Cu varied from uncontaminated to extremely contaminated, and the latter was recorded at S14 (+5.045). Sediments of S7 and S15 suggested uncontaminated to moderate contamination of Cu. For Pb only S14 exhibited some degree of contamination. Presence of Cd again demonstrated severe contamination of the Subarnarekha sediments where its Igeo ranged from +0.152 (S4) to +2.775 (S14).

The indices mentioned above deal with anthropogenic input and estimate contamination not toxicity. While contamination is related with anthropogenic input, pollution describes toxicity. As lucidly explained by [45] , the contamination is said to happen when a substance is present where it should not occur in general or at concen- trations above the background level. On the other hand, pollution is that contamination that can cause adverse biological effects to the natural biotic communities. Thus, in addition to quantify accumulation of trace elements in sediments and apportion anthropogenic impacts, a tool is needed to express potential effects of presence of multiple elements in the sediments, that is contamination situation, on ecological risks or harm to the local eco- system [29] [30] . Potential ecological risk index (PERI) and also SQGs can be applied in ecological risk as- sessments of trace element pollutants in soils and sediments. However, SQGs (as described earlier) are element- specific and do not take into account multiple element mixtures which are normally expected in the environ- mental studies. These are used for initial determination to assess whether or not contaminants of potential con- cern could cause any harm to the receptors of potential concern [45] . To overcome this gap mSQGQs have been developed; which involve chemical mixtures normalized to their appropriate SQGs to determine biological ef- fects. [36] expressed a broad range of toxicity incidence, between low (<0.1) and high (>2.3), based on their mSQGQ method. We recalibrated their toxicity quotients based on average (%) survival of amphipods and de- veloped a new modified terminology on mSQGQ classification system, which can be expressed as: <0.1 (very low incidence of toxicity), 0.1 - 0.25 (low incidence of toxicity), 0.25 - 1.0 (moderate incidence of toxicity), 1.0 - 2.3 (high incidence of toxicity) and >2.3 (extremely high incidence of toxicity). Sampling site S14 displayed adjusted mSQGQ very close to high incidence of toxicity (0.813) during post-monsoon season which was almost 2.3 times than the pre-monsoon value (0.359). According to [35] SQGs should be applied either as a screening approach or a weight-of-evidence procedure for the risks associated with the contaminated sediments. An integrated index system like PERI provides more valuable insights into eco-toxicological effects of elements polluting bed sediments. Except five non-polluted sites, the PERI of studied sediments varied from moderate risk to significantly high risk (116.98 - 556.83) during pre-monsoon period, indicating that sediments in the Subarnarekha river were exceedingly polluted by elements (Figure 4), especially present in discharges of indus- trial and urban wastes. The river showed similar spatial distribution of ecological risk in post-monsoon period; however, corresponding PERI recorded was higher which varied from 128.76 - 738.43. Cd displayed highest monomial potential ecological risk, which varied from 50.01 to 308.01 (pre-monsoon) and 54.00 to 479.01 (post-monsoon) followed by Cu (235.74 - 242.75) at S14. Analyzed results always showed pronounced Cd con-

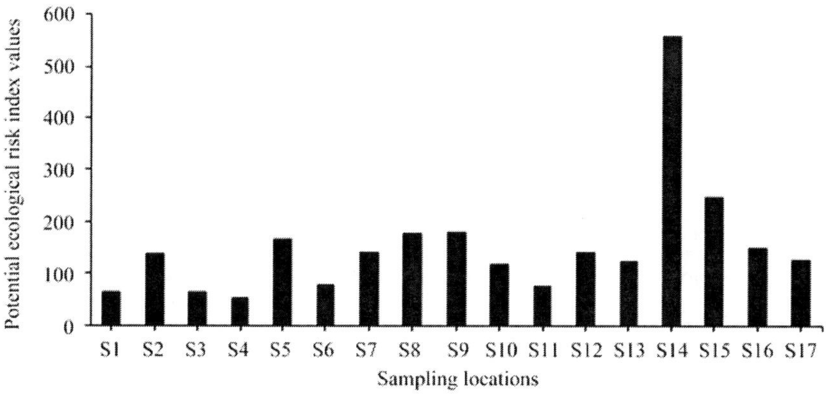

Figure 4. Potential ecological risk indices determined in Subarnarekha river sediments.

tribution in PERI of bed sediments because of its higher toxicity response factor as compared to other elements and excessive enrichment. Authors like [29] [30] [46] in their studies on soils and sediments have also demon- strated enhanced role of Cd in causing ecological risks. Acute toxicity due to Cd in the Subarnarekha bed sedi- ments would be expected at 64% of the sites. Assessment of distribution and degree of accumulation of Cd and other elements in bed sediments not only provide information on anthropogenic enrichment, but also potential ecological risks associated with their exposure. Enrichment of Cd and other elements in aquatic beds can affect entire range of biotic spectrum ranging from benthic biota to the organisms higher up the food chains due to their persistence, bio-accumulative and injurious properties. In aquatic ecosystems both abiotic and biotic com- ponents continuously interact with each other and presence of toxic elements beyond the restoring capacity can severely impact the ecological functioning of the aquatic systems in long term, which in short term may not be visible to us. Here, it should also be noted that, although, both PLI and PERI are based on CF values, the former does not take into account any sedimentological toxicity factor whereas the latter uses a toxicity factor for its computation. PLI portrays contamination situation and not ecological risk, and therefore, we suggest, the term PLI should be replaced with CLI (contamination load index). The clear-cut definitions of contamination and pollution also support this observation.

Correlation analyses are commonly employed to determine relationships between physicochemical compo- nents as well as identification of factors and sources of variables. Results of Pearson correlation analysis applied in the present investigation and the corresponding matrix generated are displayed in Table 6. Most of the ele- ments showed strong correlation with OC at $P < 0.05$ and $P < 0.01$ suggesting presence of organic matter influ- enced elements' accumulation in the sediments. Impact of organic pollution load in causing enhanced element enrichment in bed sediments was also reported for the Hindon river in India [10] . Elements showing signifi- cantly high relationship in correlation analysis may indicate same origin and controlling factors [47] . Very strong correlation was noticed between Fe and most of the elements (Zn, Cu, Pb and Cd) at $P < 0.01$ suggesting some common sources of elements in the Subarnarekha. In addition to industrial effluents and urban wastewater, possibly urban runoffs and atmospheric deposition were also contributing to the element pollution of the river. Moreover, as the river basin has rich occurrence of ores containing Fe, Zn, Cu and Pb, their chemical weathering and leaching might also be contributing elements in the sediments. However, elevated concentrations and do- minance of these elements at industrial hubs and predominantly urban regions like S7, S8, S9, S14 and S15 re- vealed that industrial and other anthropogenic activities were overwhelmingly responsible for the presence of these elements in the bed sediments. Similarly strong correlation was noted between Zn, Cu, Pb and Cd at $P < 0.01$ demonstrating elements had a common source. Meanwhile, positive correlation was also noticed between Cu, Pb and Cd ($P < 0.01$). The common presence of Cd in correlation analysis indicated that the element had chiefly anthropogenic origin. Towards upper reaches application of mineral fertilizers and consequent agricul- tural runoff were principally responsible for its presence in river. [48] observed higher concentrations of Cd in upper reaches of Siahroud river from agricultural use of phosphorus, zinc and iron fertilizers. The element is used in electroplating, batteries, paints, pigments and in alloys with a range of other elements and also occurs in

Table 6. Pearson correlation analysis comprising physicochemical parameters and elements.

	pH	EC	OC	Fe	Zn	Cu	Pb	Cd	Mn
pH	1	−0.688**	−0.845**	−0.602*	−0.588*	−0.507*	−0.676**	−0.893**	−0.247
EC		1	0.453	0.528*	0.482*	0.395	0.405	0.709**	0.462
OC			1	0.549*	0.494*	0.424	0.693**	0.791**	0.354
Fe				1	0.950**	0.919**	0.855**	0.778**	0.270
Zn					1	0.977**	0.831**	0.783**	0.081
Cu						1	0.737**	0.716**	−0.037
Pb							1	0.785**	0.270
Cd								1	0.268
Mn									1

*Correlation is significant at the 0.05 level (two tailed); **Correlation is significant at the 0.01 level (2-tailed).

sulphide minerals containing Zn, Pb and Cu [49] . Also overall concentration of Cd was reported to be four times its geochemical background. However, no significant relationship was observed for Mn with any other element, possibly because unlike other elements Mn had different sediment deposition features [10] . The Subarnarekha river from its origin to the border of Jharkhand (with West Bengal) receives effluents and discharges of big and small industrial conglomerates, populated urban areas and extensively mined regions. Additionally, open dump- ing of garbage beside river banks and washing of automobiles and vehicles in river water are making sediments of Subarnarekha increasingly contaminated.

Degradation of river bed sediment quality portrays river degradation as a whole because it is the outcome of river water pollution. The problem of river basin degradation is more pronounced in developing countries be- cause expanding economic growth puts terrific pressure on the limited available natural resources. As per the assessment of the renowned economic consultancy the Centre for Economics and Business Research (CEBR) in World Economic League Table Report, the Indian economy, which presently holds 11th position in the world, is growing very fast, and by 2028, it is expected to become the 3rd largest economy after China and United States [50] overtaking Japan. Meanwhile, the population of India is also expected to surpass 1.45 billion. Thus, to sa- tisfy the demand of the growing population along with the economic growth, massive industrialization, urbani- zation, changes in land use pattern and intensive agriculture are inevitable. All these activities, in turn, will af- fect river basins if complete attention is not given. The Subarnarekha basin will have more alarming situation because, according to the International Water Management Institute (IWMI, Colombo, Sri Lanka) research re- port, currently its population density (347 persons per km^2) is already much ahead of the national average of all river basins (282 persons per km^2), and in terms of some major river basins it lies only behind Sabarmati (521), Ganges (449) and Cauvery (389) [51] . The information obtained through this research communication, based on systematic investigation of the sediment quality, provides valuable insights for result-oriented environmental management of the industrial, urban, mining and mineral processing regions. Some immediate requirements on effective management policies and conservation strategies are needed to prevent the current contamination situation developing into pollution problems all through the river basin.

4. CONCLUSION

The bed sediments collected nearby industrial, urban and mining sites exceedingly displayed elevated concen- tration of elements, at times much above their background values, which markedly highlighted anthropogenic control on enrichment and geochemical distribution of trace elements in the Subarnarekha river. Although simi- lar spatial variations in element contents were recorded in pre-monsoon and post-monsoon periods, the latter showed higher concentration load than the former. River sediment at S14 (Moubhandar) emerged as the most contaminated where highest EF for Zn and Cu, maximum CD and PLI (CLI) values, and highest Igeo values for Fe, Zn, Cu, Pb and Cd were recorded. The site was also most polluted as evident from mSQGQ and PERI meas- ures. Based on ecological risk assessment methods (SQGs and PERI), Cd and to some extent Cu revealed to be the most severe polluting agents responsible for degrading the river sediment environment. When dealing with environmental studies, it is essential to study both contamination and pollution; and contaminants and pollutants. To effectively disseminate information about contamination and pollution, PLI (CLI) and PERI respectively are suitable indicators. Cadmium exhibited continuous regularity in its enrichment at all sampling sites; contributed most to the ecological risks and would be a potential toxic to the sediment dwelling organisms at 64% of the sites which demonstrated high toxicity risks due to its exposure. Some immediate sediment quality management strategies are needed to remediate and control river bed contamination. This study also demonstrated that rapid developmental activities can severely affect the ecology of the river. Presence of toxic elements at elevated le- vels in bed sediments indicates higher exposure risk to the benthic biota and other aquatic organisms of the river. The present investigation, along with some other reported literature [40] , reinforces suggestion that contami- nated sediments are significantly responsible for the deterioration in ecological conditions of the rivers world-wide. Sediments act as sink of the trace elements, which under altered physical and chemical conditions of the river can alternately turn into source of the same elements. These elements in dissolved form can become avail- able to the free floating organisms of the river like fishes, and can ultimately enter into the human beings via food chain. The bioavailability of Cd and other elements which is differentially and sequentially determined is the next phase of our investigations along with identifying possible effects of polluted sediments on the bio- chemical aspects of the higher organisms. This study provided some valuable insights employing a combination of sediment assessment approach which can be constructive in designing and planning strategic sediment management programmes and conservation guidelines of the river basins.

ACKNOWLEDGEMENTS

Kumar Manoj, Pratap Kumar Padhy (CSIR), India, in the form of Ju- nior and Senior Research Fellowships and contingency grant to the Kumar Manoj for carrying carry out the re- search work is deeply acknowledged.

REFERENCES

1. Harikumar, P.S., Nasir, U.P. and Rahman, M.P.M. (2009) Distribution of Heavy Metals in the Core Sediments of a Tropical Wetland System. International Journal of Environmental Science and Technology, 6, 225-232.

2. Sekabira, K., Oryem, O.H., Basamba, T.A., Mutumba, G. and Kakudidi, E. (2010) Assessment of Heavy Metal Pollu- tion in the Urban Stream Sediments and Its Tributaries. International Journal of Environmental Science and Technol- ogy, 7, 435-446.

3. Mohiuddin, K.M., Zakir, H.M., Otomo, K., Sharmin, S. and Shikazono, N. (2010) Geochemical Distribution of Trace Metal Pollutants in Water and Sediments of Downstream of an Urban River. International Journal of Environmental Science and Technology, 7, 17-28.

4. Sun, W., Sang, L. and Jiang, B. (2012) Trace Metals in Sediments and Aquatic Plants from the Xiangjiang River, China. Journal of Soils and Sediments, 12, 1649-1657.

5. Milenkovic, N., Damjanovic, M. and Ristic, M. (2005) Study of Heavy Metal Pollution in Sediments from the Iron Gate (Danube River), Serbia and Montenegro. Polish Journal of Environmental Studies, 14, 781-787.

6. Begum, A., Ramaiah, M., Harikrishna, Khan, I. and Veena, K. (2009) Heavy Metal Pollution and Chemical Profile of Cauvery River Water. E-Journal of Chemistry, 6, 47-52.

7. Prasad, M.B.K., Ramanathan, A.L., Shrivastav, S.K., Anshumali and Saxena, R. (2006) Metal Fractionation Studies in Surfacial and Core Sediments in the Achankovil River Basin in India. Environmental Monitoring and Assessment, 121, 77-102.

8. Palumbo-Roe, P., Wragg, J. and Banks, V.J. (2012) Lead Mobilisation in the Hyporheic Zone and River Bank Sedi- ments of a Contaminated Stream: Contribution to Diffuse Pollution. Journal of Soils and Sediments, 12, 1633-1640

9. Olubunmi, F.E. and Olorunsola, O.E. (2010) Evaluation of the Status of Heavy Metal Pollution of Sediment of Agbabu Bitumen Deposit Area, Nigeria. European Journal of Scientific Research, 41, 373-382.

10. Suthar, S., Nema, A.K., Chabukdhara, M. and Gupta, S.K. (2009) Assessment of Metals in Water and Sediments of Hindon River, India: Impact of Industrial and Urban Discharges. Journal of Hazardous Materials, 171, 1088-1095.

11. Nicolau, R., Galera-Cunha, A. and Lucas, Y. (2006) Transfer of Nutrients and Labile Metals from the Continent to the Sea by a Small Mediterranean River. Chemosphere, 63, 469-476.

12. Dan'Azumi, S. and Bichi, M.H. (2010) Industrial Pollution and Heavy Metals Profile of Challawa River in Kanao, Ni- geria. Journal of Applied Sciences in Environmental Sanitation, 5, 23-29.

13. Kar, D., Sur, P., Mandal, S.K., Saha, T. and Kole, R.K. (2008) Assessment of Heavy Metal Pollution in Surface Water. International Journal of Environmental Science & Technology, 5, 119-124.

14. Nair, I.V., Singh, K., Arumugam, M., Gangadhar, K. and Clarson, D. (2010) Trace Metal Quality of Meenachil River at Kottayam, Kerala (India) by Principal Component Analysis. World Applied Sciences Journal, 9, 1100-1107.

15. Harikumar, P.S. and Jisha, T.S. (2010) Distribution Pattern of Trace Metal Pollutants in the Sediments of an Urban Wetland in the Southwest Coast of India. International Journal of Engineering Science and Technology, 2, 840-850.

16. Roy, D., Begam, S., Ghosh, S. and Jana, S. (2013) Calibration and Validation of HEC-HMS Model for a River Basin in Eastern India. ARPN Journal of Engineering and Applied Sciences, 8, 40-56.

17. IRBMS (2012) Integrated River Basin Management Society. Newsletter, No. 38, Ganga.

18. Ahdy, H.H.H. and Khaled, A. (2009) Heavy Metals Contamination in Sediments of the Western Part of the Egyptian Mediterranean Sea. Australian Journal of Basic and Applied Sciences, 3, 3330-3336.

19. Han, Y.M., Du, P.X., Cao, J.J. and Posmentier, E.S. (2006) Multivariate Analysis of Heavy Metal Contamination in Urban Dusts of Xi'an, Central China. Science of the Total Environment, 355, 176-186.

20. Behera, P.K. (2006) Soil and Solid Waste Analysis: A Laboratory Manual. Dominant Publishers and Distributors, New Delhi.

21. Walkley, A. and Black, I.A. (1934) An Examination of the Degtjareff Method for Dtermining Soil Organic Matter and a Proposed Modification of the Chromic Acid Titration Method. Soil Science, 37, 29-38.

22. Hseu, Z.Y. (2004) Evaluating Heavy Metal Contents in Nine Composts Using Four Digestion Methods. Bioresource Technology, 95, 53-59.

23. Hakanson, L. (1980) An Ecological Risk Index for Aquatic Pollution Control: A Sedimentological Approach. Water Research, 14, 975-1001.

24. Tomlinson, D.C., Wilson, J.G., Harris, C.R. and Jeffrey, D.W. (1980) Problems in Assessment of Heavy Metals in the Estuaries and the Formation of Pollution Index. Helgoland Marine Research, 33, 566-575.

25. Sinex, S.A. and Helz, G.R. (1981) Regional Geochemistry of Trace Elements in Checapeake Bay Sediments. Environmental Geology, 3, 315-323.

26. Sutherland, R.A. (2000) Bed Sediment-Associated Trace Metals in an Urban Stream, Oahu, Hawaii. Environmental Geology, 39, 611-627.

27. Müller, G. (1979) Heavy Metals in the Sediment of the Rhine-Changes Seity. Umschau in Wissenschaft und Technik, 79, 778-783.

28. Müller, G. (1981) The Heavy Metal Pollution of the Sediments of Neckars and its Tributary: A Stocktaking. Chemiker Zeitung, 105, 157-164.

29. Wei, B.G., Jiang, F.Q., Li, X.M. and Mu, S.Y. (2010) Heavy Metal Induced Ecological Risk in the City of Urumqi, NW China. Environmental Monitoring and Assessment, 160, 33-45.

30. Li, X., Liu, L., Wang, Y., Luo, G., Chen, X., Yang, X., Gao, B. and He, X. (2012) Integrated Assessment of Heavy Metal Contamination in Sediments from a Coastal Industrial Basin, NE China. PLoS ONE, 7, e39690.

31. Turekian, K.K. and Wedepohl, K.H. (1961) Distribution of the Elements in Some Major Units of the Earth's Crust. Geological Society of America Bulletin, 72, 175-192.

32. Rath, P., Panda, U.C., Bhatta, D. and Sahoo, B.N. (2005) Environmental Quantification of Heavy Metals in the Sediments of the Brahmani and Nandira Rivers, Orissa. Journal of the Geological Society of India, 65, 487-492.

33. Chakravarty, M. and Patgiri, A.D. (2009) Metal Pollution Assessment in Sediments of the Dikrong River, N.E. India. Journal of Human Ecology, 27, 63-67.

34. MacDonald, D.D., Ingersoll, C.G. and Berger, T.A. (2000) Development and Evaluation of Consensus-Based Sediment Quality Guidelines for Freshwater Ecosystems. Archives of Environmental Contamination and Toxicology, 39, 20-31.

35. Luo, W., Lu, Y., Wang, T., Hu, W., Jiao, W., Naile, J.E., Khim, J.S. and Giesy, J.P. (2010) Ecological Risk Assessment of Arsenic and Elements in Sediments of Coastal Areas of Northern Bohai and Yellow Seas, China. Ambio, 39, 367-375.

36. Fairey, R., Long, E.R., Roberts, C.A., Anderson, B.S., Phillips, B.M., Hunt, J.W., Puckett, H.R. and Wilson, C.J. (2001) An Evaluation of Methods for Calculating Mean Sediment Quality Guideline Quotients as Indicators of Contamination and Acute Toxicity to Amphipods by Chemical Mixtures. Environmental Toxicology and Chemistry, 20, 2276-2286.

37. Puyate, Y.T., Rim-Rukeh, A. and Awatefe, J.K. (2007) Metal Pollution Assessment and Particle Size Distribution of Bottom Sediment of Orogodo River, Agbor, Delta State, Nigeria. Journal of Applied Sciences Research, 3, 2056-2061.

38. Kumar, S.P. and Edward, J.K.P. (2009) Assessment of Metal Concentration in the Sediment Cores of Manakudy Estu- ary, South West Coast of India. Indian Journal of Geo-Marine Sciences, 38, 235-248.

39. Abu-Rukah, Y. and Ghrefat, H.A. (2001) Assessment of the Anthropogenic Influx of Metallic Pollutants in Yarmouk River, Jordan. Environmental Geology, 40, 683-692.

40. Singh, V.K., Singh, K.P. and Mohan, D. (2005) Status of Heavy Metals in Water and Bed Sediments of River Gomti- —A Tributary of the Ganga River, India. Environmental Monitoring and Assessment, 105, 43-67.

41. Beg, K.R. and Ali, S. (2008) Chemical Contaminants and Toxicity of Ganga River Sediments from Up and Downstream Area at Kanpur. American Journal of Environmental Sciences, 4, 362-366.

42. Hassan, F.M., Saleh, M.M. and Salman, J.M. (2010) A Study of Physicochemical Parameters and Nine Heavy Metals in the Euphrates River, Iraq. E-Journal of Chemistry, 7, 685-692.

43. Mohiuddin, K.M., Ogawa, Y., Zakir, H.M., Otomo, K. and Shikazomo, N. (2011) Heavy Metals Contamination in Water and Sediments of an Urban River in a Developing Country. International Journal of Environmental Science and Technology, 8, 723-736.

44. Jumbe, A.S. and Nandini, N. (2009) Heavy Metals Analysis and Sediment Quality Values in Urban Lakes. American Journal of Environmental Sciences, 5, 678-687.

45. Chapman, P.M. (2007) Determining When Contamination Is Pollution—Weight of Evidence Determinations for Sediments and Effluents. Environment International, 33, 492-501.

46. Yisa, J., Jacob, J.O. and Onoyima, C.C. (2012) Assessment of Toxic Levels of Some Heavy Metals in Road Deposited Sediments in Suleja, Nigeria. American Journal of Chemistry, 2, 34-37.

47. Rafiei, B., Bakhtiari, N.M., Hashemi, M. and Khodaei, A.S. (2010) Distribution of Heavy Metals around the Dashkasan Au Mine. International Journal of Environmental Research, 4, 647-654.

48. Charkhabi, A.H., Sakizadeh, M. and Rafiee, G. (2005) Seasonal Fluctuation in Heavy Metal Pollution in Iran's Siar- houd River: A Preliminary Study. Environmental Science and Pollution Research, 12, 264-270.

49. APHA (2005) Standard Methods for the Examination of Water and Wastewater Analysis. APHA-AWWA-WEF, Washington DC.

50. CEBR (2013) Cebr's World Economic League Table. Centre for Economics and Business Research, London.

51. Amarasinghe, U.A., Sharma, B.R., Aloysius, N., Scott, C., Smakhtin, V. and de Fraiture, C. (2004) Spatial Variation in Water Supply and Demand Across River Basins of India. Research Report 83, International Water Management Institute, Colombo.

Chapter 12

Elements in Water, Suspended Particulate Matter and Sediments of The Sava River

Janja Vidmar [1,2] & Tea Zuliani[1] & Petra Novak[1,2] & Ana Drinčić[1,2] & Janez Ščančar[1,2] & Radmila Milačič[1,2]

[1].*Department of Environmental SciencesJožef Stefan InstituteLjubljanaSlovenia*

[2].*Jožef Stefan International Postgraduate SchoolLjubljanaSlovenia*

ABSTRACT

Purpose

River ecosystems are under pressure from several different stressors. Among these, inorganic pollutants contribute to multiple stressor situations and the overall degradation of the ecological status of the aquatic environments. The main sources of pollution include different industrial activities, untreated effluents from municipal waste waters and intensive agriculture. In the present study, water, suspended particulate matter (SPM) and sediments of the Sava River were studied in order to assess the pollution status of this river system.

Materials and methods

Sampling was performed during the first sampling campaign of the EU 7th FW funded GLOBAQUA project in September 2014, at 18 selected sampling sites along the Sava River. In 2014, floods predominated from spring to fall. Water samples were collected to determine the total element concentrations, the dissolved (0.45 μm) fraction and element concentrations in SPM. In order to assure comparative results with other river basins, the fraction below 63 µm was analysed in sediments. The extent of pollution was estimated by determination of the total element concentrations and by the identification of the most hazardous highly mobile element fractions (extraction 0.11 mol L^{-1} acetic acid) and anthropogenic inputs of elements to sediments (normalization to aluminium (Al) concentration). Concentrations of selected elements were determined by inductively coupled plasma mass spectrometry (ICP-MS).

Results and discussion

Since during sampling campaign the water level was extremely high, water samples contained high amounts of SPM (in general between 80 and 100 mg L^{-1}). The data of chemical analysis revealed that concentrations of elements in water, SPM and sediments in general increase along the Sava River from its origin to the confluence with the Danube River. Elevated concentrations of chromium (Cr) and nickel (Ni) in SPM and sediments were observed at industrially exposed sites. Concentrations of Cr and Ni in sediments were up to 320 and 250 mg kg^{-1}, respectively. Nevertheless, these elements were present in sparingly soluble forms and hence did not represent an environmental threat. Phosphorus (P) was found in elevated concentrations (up to 1500 mg kg^{-1}) at regions with intensive agricultural activities and cities with dense population.

Conclusions

With respect to element concentrations, the pollution of the Sava River is similar to other moderately polluted European rivers. The data from the present study are beneficial for the water management authorities and can contribute to sustainable utilization, management and protection of the Sava River water resources.

Keywords: Elements Environmental status Sediments Suspended particulate matter The Sava River Water

1 INTRODUCTION

The quality of groundwater within the Sava River basin, which is an important source of drinking water for the population of the riparian countries, depends strongly on the quality of the Sava River water (International Sava River Basin

Commission 2013). Sediments that represent a source of nutrients for microorganisms have a strong impact on the ecosystem functioning. On the other hand, they may accumulate different contaminants, which can be remobilised into the aqueous phase, causing hazardous effects on the aquatic organisms (Kwok et al. 2014). So, water and sediment quality is of great importance for sustaining the health of benthic life. The sediment quality is essential in the development of management plans when dredging of sediments is necessary for restoring waterways and when accumulated sediments from hydroelectric power plant accumulation basins need to be removed (International Sava River Basin Commission 2002). Moreover, chemical analyses of sediments are frequently applied as they reveal spatial and temporal variation in elemental concentrations (Loring and Rantala 1992; Grosbois et al. 2006; Meybeck et al. 2007; Mohiuddin et al. 2012). Among different pollutants, elements present in elevated concentrations in water and sediments may considerably contribute to multiple stressor conditions of the aquatic environment.

For water analysis, standard methods for the examination of surface waters are recommended (Gottler 2012; ISO 15586 2003; ISO 5667-3 2012). These methods are also regulated in the Water Framework Directive (WFD) 2000/60/EC (Official Journal of the European Communities 2000) and its daughter Directive 2009/90/EC (Official Journal of the European Union 2009), and in environmental objectives for surface waters (European Communities Environmental Objectives 2009). In the EU directives (Official Journal of the European Union2009; European Communities Environmental Objectives 2009) and ISO 6777-3 standard (ISO 5667-3 2012), filtration of whole water sample through a 0.45-μm filter is required for the determination of the dissolved metal fraction. In standard methods for the examination of water and wastewater (Gottler 2012), in surface water analyses, the dissolved metal fraction (filtering through 0.45 μm filter) and total metal fraction, which are defined operationally (nitric acid digestion, nitric acid-sulphuric acid digestion, nitric acid-perchloric acid digestion, nitric acid-perchloric acid-hydrofluoric acid digestion and microwave-assisted digestion), are recommended for regular analysis.

In sediment analysis, it is not possible to estimate the extent of pollution only by the determination of the total element concentration. Namely, the bioavailability and toxicity of element depend primarily on chemical forms in which it is present in a given environmental compartment. By applying fractionation procedures, valuable information on the partitioning of elements between easily and sparingly soluble fractions of sediments can be obtained. Knowing the concentration of a particular element in the easily soluble fraction enables evaluation of its potential harmful effects to the aquatic biota. In sediment analyses, Tessier's fractionation procedure (Tessier et al. 1979) or BCR scheme proposed by the Community Bureau of Reference

(Quevauviller et al. 1997) was commonly applied. The leaching protocol of the first step of these sequential extraction procedures is also used to investigate the easily soluble element fraction, applying aqueous solutions of 1 mol L^{-1} ammonium chloride (Tessier et al.1979) or 0.11 mol L^{-1} acetic acid (Quevauviller et al. 1997) as extracting agents. Data on mobility and potential bioavailability of elements in sediments represent a good basis for the estimation of the environmental burden (Svete et al. 2001; Vasile et al. 2010; Milačič et al. 2010; Dundar et al. 2012) and are useful also in the management of dredged sediments (Šömen Joksič et al. 2005; Alvarez-Guerra et al. 2007; Zuliani et al. 2016). For estimation of the natural and anthropogenic input of elements in sediments, normalization procedures, based on correlations between the concentration of trace elements and the element that is naturally present in the environment investigated, are applied. For this purpose, aluminium (Al) as a major constituent of aluminosilicates and iron (Fe) as a clay mineral indicator element have been the most frequently used (Loring and Rantala 1992; Svete et al. 2001; Kotnik et al. 2003; Milačič et al.2010).

Elements like cadmium (Cd), lead (Pb), copper (Cu), zinc (Zn), chromium (Cr), nickel (Ni) and arsenic (As) are frequently present in sediments due to industrial (Giusti and Taylor 2007; Karadede-Akin and Ünlü 2007; Sakan et al. 2010), municipal (Grosbois et al. 2006) and mining activities (Svete et al. 2001; Bird et al. 2010), whereas phosphorous (P) is accumulated in sediments mainly as a consequence of the use of fertilisers at riparian agricultural areas (House and Denison 2002) or is released from municipal sewage treatment plants (Howell 2010).

For evaluation of the environmental status of sediments, various sediment quality guidelines have been proposed, based mainly on the biological effect approaches (CCME 2001; McCready et al. 2006; Long et al. 2006). MacDonald et al. (2000) developed and evaluated consensus-based sediment quality guidelines for freshwater ecosystems, including a threshold effect concentration (TEC) below which harmful effects are unlikely to be observed, and a probable effect concentration (PEC) above which harmful effects are likely to be observed. As a part of common implementation strategy for the WFD, the European Commission has issued technical guidance for the derivation of sediment quality guidelines (European Communities Technical Report 2010-041 2010). To optimise sediment management alternatives and to improve quality assessment criteria, sediment guidelines are still being developed (Bakke et al. 2010; Kwok et al.2014; Merrington and Sprang 2014).

In support of reducing the pressure of multiple stressors on aquatic ecosystems and for optimal management of the water resources, the European Union granted the 7th FW project GLOBAQUA (Navarro-Ortega et al. 2015) in which the Sava River is one of the river basins investigated.

Elements are among pollutants that contribute to the multiple stressor conditions. Despite the fact that national environmental agencies regularly monitor the water quality of the Sava River, data on the environmental status of the river basin are still missing. Therefore, the aim of the present work was to determine the concentrations of elements in water, suspended particulate matter (SPM) and sediments of the Sava River, collected during a sampling campaign in September 2014.

2 MATERIALS AND METHODS

2.1 Sampling

The Sava River sampling campaign was carried out from September 1 to 10, 2014, on nine water units along the river at automatic gauging stations, providing the data for water level, temperature and discharge. The Lagrangian approach to sample of the same water mass upstream at the input (1) and downstream at the output (2) boundaries of the selected river water unit was applied (Writer et al. 2011). This enabled to relate transformations of dissolved and suspended contaminant compounds with the travel time. The water travel time was determined empirically by a velocity metre. The water units were 10 to 30 km long. Sampling sites, sample codes and sampling site characteristics are provided in Fig. 1 and online measurement parameters (dissolved oxygen, electrical conductivity and pH) in Table S1(Electronic Supplementary Material). The normal average discharge for Litija (Slovenia) is around 50 $m^3 s^{-1}$ and for Sremska Mitrovica (Serbia) around 500 $m^3 s^{-1}$. During the sampling campaign, flood conditions prevailed, which was evident from elevated water discharges: the water discharge in Litija (Slovenia) was 300 $m^3 s^{-1}$ and in Sremska Mitrovica (Serbia) 3300 $m^3 s^{-1}$. Sampling of water was carried under requirements of the WFD (Official Journal of the European Communities 2000; Official Journal of the European Union 2009; European Communities Environmental Objectives 2009) and ISO 5667-3 standard (2012), while that of SPM and sediments is described in the sediment quality guidelines (European Communities Technical Report 2010-041 2010). Since in Jasenovac, Slavonski Brod and Županja (Croatia) the area was flooded, it was not possible to take the sediment samples.

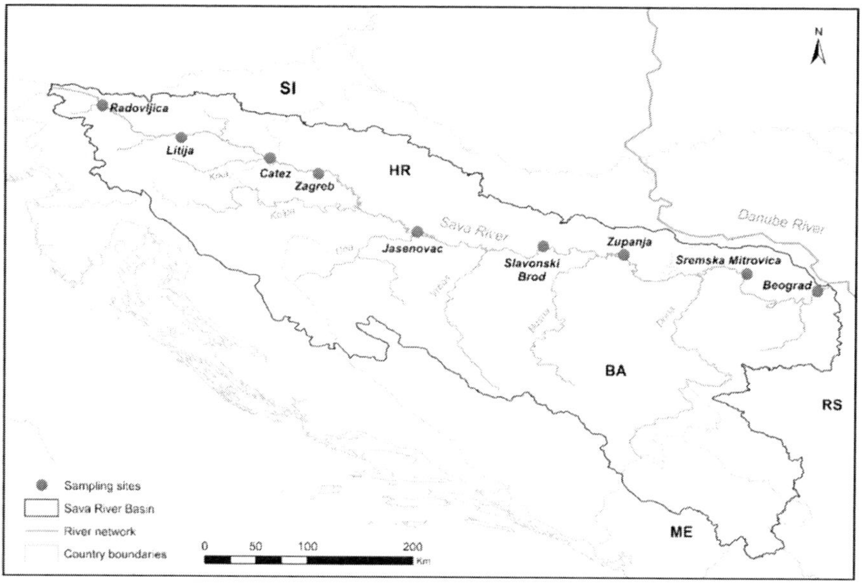

Figure 1. The Sava River sampling locations, sampling codes and main sampling site characteristics (sampling campaign September 2014). Radovljica (SI) RAD1 = 46.340840, 14.165080; RAD2 = 46.292979, 14.260754 (small city at the confluence of the Sava Dolinka and the Sava Bohinjka, both headwaters of the Sava River). Litija (SI) LIT1 = 46.084561, 15.582632; LIT2 = 46.066191, 14.850857 (land use, abandoned mining area). Čatež (SI) CAT1 = 45.896355, 15.627037; CAT2 = 45.860357, 15.692068 (land use). Zagreb (HR) ZAG1 = 45.783014, 16.003184; ZAG2 = 45.770897, 16.069357 (big city, urban pollution). Jasenovac (HR) JAS1 = 45.263968, 16.894432; JAS2 = 45.250861, 16.951914 (navigation, land use). Slavonski Brod (HR) SLB1 = 45.152703, 18, 013,906; SLB2 = 45.126224, 18.084751 (navigation, oil refinery, land use). Županja (HR) ZUP1 = 45.074805, 18.687667; ZUP2 = 5.014880, 18.740313 (navigation, oil refinery, metallic and mining industry, land use). Sremska Mitrovica (RS) SRM1 = 44.973390, 19.596213; SRM2 = 44.913582, 19.752412 (navigation, oil refinery, heavy industry, land use). Beograd (RS) BEO1 = 44.769133, 20.355572; BEO2 = 44.806317, 20.443683 (navigation, big city, urban pollution)

From each sampling site, water was collected into 10-L plastic buckets. An aliquot of 1 L was then taken for SPM analysis and 1 L for the determination of elements in whole water samples and dissolved element fraction. Approximately 2 kg of sediments were collected in 2-L plastic bottles.

2.2 Analytical procedures

Water samples

To determine the soluble metal concentrations, water samples were filtered before analysis (Official Journal of the European Communities 2000; European Communities Environmental Objectives 2009; Gottler 2012). For the determination of the total metal concentrations, 10 mL of whole water samples was transferred to Teflon vessels and 4 mL of supra pure HNO_3, 2 mL of supra pure HF and 1 mL of supra pure HCl added. The Teflon vessels were covered and subjected to microwave-assisted digestion by ramping the temperature to 200 °C in 30 min, holding the same temperature for the next 60 min and cooling the contents in the following 30 min. Supra pure H_3BO_3 (12.5 mL, 4 % aqueous solution) was then added for complexation of fluorides and the contents digested again by ramping the temperature to 200 °C in 15 min, holding the same temperature for the next 30 min and cooling the contents in the following 30 min (Zuliani et al. 2016). Acidification of the sample as recommended in the standard methods for the examination of water and wastewater (Gottler 2012) was also applied to determine the total metal content. For this purpose, 1 mL of supra pure HNO_3 was added to 1 L of the whole water sample. Concentrations of elements in water samples were determined by ICP-MS (Novotnik et al. 2015).

Sediment and SPM samples

For the determination of the total element content in sediments, approximately 0.25 g of samples was added to Teflon vessels and microwave-assisted digestion as described above performed. To determine the total element content in SPM, samples were first filtered (Gottler2012), the filter transferred to Teflon vessels and microwave digestion of the mineral phase as described above applied. Metal concentrations in sediments and SPM were determined by ICP-MS.

Quality control

The accuracy of the analytical procedures used in the present study was checked by the analysis of appropriate certified reference materials. As is evident from Table S2 (Electronic Supplementary Material), good agreement of data (better than ±5 %) was obtained between the determined concentrations of elements and the reported certified values. The expanded uncertainty of analytical procedures applied was better than ±3 % (k = 2).

Determination of the easily soluble metal concentrations in sediments and normalization to Al

Acetic acid (0.11 mol L^{-1}) was used to extract the easily soluble element fraction from sediment (Milačič et al. 2010) and the element content determined by ICP-MS. In order to account for geochemical variations along the Sava River, a normalization of total sediment concentrations by a conservative element Al was applied (Milačič et al. 2010). Significant deviations from the linear relationship were used to differentiate between natural vs. anthropogenic inputs.

3 RESULTS AND DISCUSSION

3.1 Analysis of element concentrations in whole water samples

For the analysis of element concentrations in the whole water samples, microwave-assisted digestion and acidification of samples with HNO_3, as described in Sect. 2.2, were applied and concentrations of elements determined with ICP-MS. These results are presented in Table 1 and Table S3 (Electronic Supplementary Material) and the mass of SPM in the water samples in Fig.2. As evident, great differences in element concentrations are observed between the microwave-assisted digestion of water samples and acidification procedure, since acidification just desorbed the elements from particles, while elements that are associated with the matrix constituents could not be decomposed in this way. Due to this reason, microwave-assisted digestion gave significantly higher results than did acidification with HNO_3, in particular in elements, which are major constituents of the mineral phase of the SPM (e.g. Fe, Al, Ti) (Table1). This is related to high mass of SPM in water samples, ranging from around 50 up to 140 mg L^{-1}. Similar behaviour, but in lesser extent, was observed also for Cr, Ni, Cd, Zn, Cu and As (Table 1), as well as for Se, Sb, Co and Mn (Table S3, Electronic Supplementary Material). These data revealed that in samples with high SPM content, microwave-assisted digestion of water samples should be applied for the determination of the element concentrations in the whole water samples.

Table 1. Concentrations of elements in whole water samples of the Sava River collected in September 2014 determined by ICP-MS after microwave assisted digestion (MW) or acidification with HNO_3

Element digestion		Sampling site								
		RAD1	RAD2	LIT1	LIT2	CAT1	CAT2	ZAG1	ZAG2	JAS1
Fe (μg/L)	MW	1764	2124	1721	859	1047	1005	1242	866	474
	HNO_3	212	346	213	143	170	166	258	202	181
Al (μg/L)	MW	4633	5233	4306	2080	2471	2649	3107	2124	1078
	HNO_3	157	222	166	89	119	148	168	138	90.3
Ti (μg/L)	MW	173	193	148	79.4	92.4	108	137	93.0	51.1
	HNO_3	2.09	3.87	2.61	2.21	2.49	3.30	3.81	5.21	2.54
Cr (μg/L)	MW	4.30	4.94	2.71	1.23	1.55	1.83	3.07	1.60	0.570
	HNO_3	0.59	0.826	0.542	0.399	0.506	0.523	0.689	0.586	0.487
Ni (μg/L)	MW	13.1	8.12	22.6	2.57	4.12	9.00	19.1	32.2	16.6
	HNO_3	0.24	0.612	0.071	<0.05	0.131	0.058	0.512	0.325	0.618
Cd (μg/L)	MW	0.031	0.039	0.014	0.018	0.006	0.020	<0.010	0.007	<0.010
	HNO_3	<0.010	<0.010	<0.010	<0.010	<0.010	<0.010	<0.010	<0.010	<0.010
Pb (μg/L)	MW	1.95	3.73	1.59	2.17	1.00	1.68	1.38	0.910	0.567
	HNO_3	0.942	1.72	1.07	1.63	0.739	0.754	1.12	0.892	0.531
Zn (μg/L)	MW	8.20	6.77	3.65	3.96	1.53	3.71	3.31	1.14	0.640
	HNO_3	1.09	3.39	2.60	2.55	0.98	2.93	2.22	0.87	0.452
Cu (μg/L)	MW	2.53	3.85	2.62	2.29	2.36	1.88	2.36	2.07	1.98
	HNO_3	0.487	1.11	1.24	1.06	1.09	1.27	1.43	1.33	1.22
As (μg/L)	MW	4.85	3.13	2.47	1.67	1.66	1.06	0.823	0.711	0.943
	HNO_3	0.278	0.361	0.389	0.354	0.449	0.409	0.606	0.479	0.649

Element digestion		Sampling site								
		JAS2	SLB1	SLB2	ZUP1	ZUP2	SRM1	SRM2	BEO1	BEO2
Fe (μg/L)		718	1609	1291	2459	2438	3429	2202	1871	3998
		184	277	225	545	360	557	594	441	559
Al (μg/L)		1637	3609	2641	4740	4777	5479	4377	3659	7936
		145	214	181	407	267	370	384	338	358
Ti (μg/L)		89.3	165	120	234	234	290	220	188	445
		4.27	3.57	3.88	5.41	4.60	6.03	8.92	4.76	5.23
Cr (μg/L)		1.29	6.69	5.35	12.2	11.9	13.3	10.5	8.50	20.0
		0.574	1.14	1.22	2.81	2.07	2.96	3.09	2.29	2.78
Ni (μg/L)		17.6	24.8	13.9	22.0	16.2	20.6	29.7	27.2	30.4
		0.572	3.45	3.85	10.8	7.27	10.6	10.2	8.03	10.3
Cd (μg/L)		<0.010	<0.010	<0.010	<0.010	<0.010	<0.010	<0.010	0.028	0.038
		<0.010	<0.010	<0.010	0.005	<0.010	<0.010	<0.010	<0.010	0.023
Pb (μg/L)		0.644	1.04	0.869	1.72	1.70	2.30	1.82	1.98	3.75
		0.462	0.964	0.571	1.20	1.17	1.40	1.57	1.63	2.53
Zn (μg/L)		0.534	0.992	10.7	2.96	3.72	19.4	2.96	5.35	11.3
		0.344	0.610	1.24	1.55	2.12	6.72	1.63	3.22	8.22
Cu (μg/L)		2.00	3.56	3.36	5.85	5.92	6.80	5.88	6.84	9.02
		1.06	1.52	1.61	2.92	2.46	2.09	2.76	3.62	3.02
As (μg/L)		0.637	0.939	0.973	1.56	1.65	2.23	2.08	2.59	3.39
		0.489	0.596	0.708	0.858	0.792	1.53	1.63	1.85	1.68

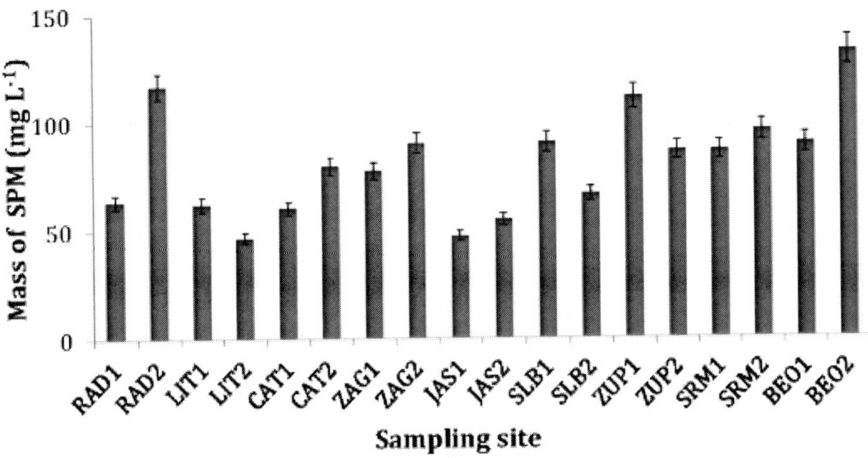

Figure 2. Mass of SPM in water samples of the Sava River collected in September 2014

Cr

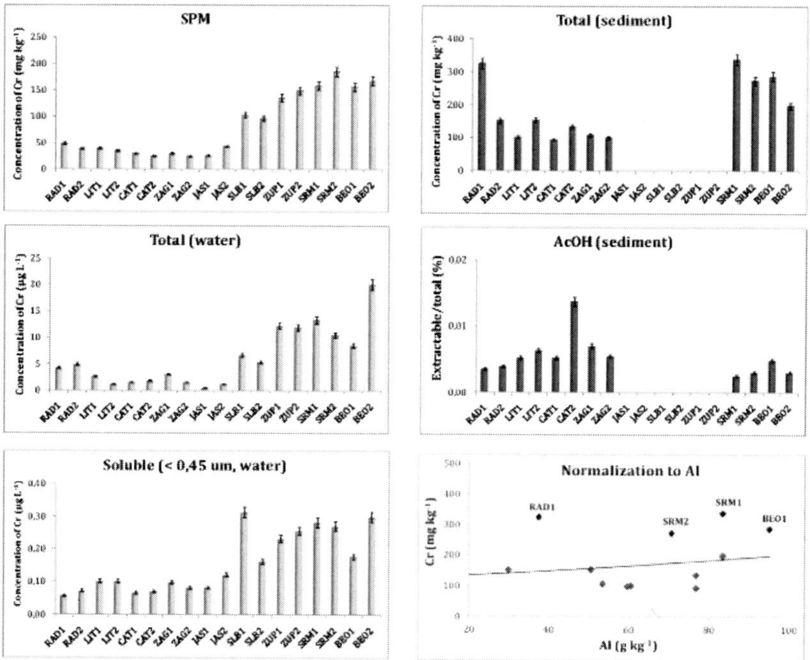
SPM
Concentration of Cr (mg kg⁻¹)
Total (sediment)
Concentration of Cr (mg kg⁻¹)
Total (water)
Concentration of Cr (µg L⁻¹)
AcOH (sediment)
Extractable/total (%)
Soluble (< 0,45 um, water)
Concentration of Cr (µg L⁻¹)
Normalization to Al
Cr (mg kg⁻¹)
Al (g kg⁻¹)
RAD1
SRM2
SRM1
BEO1
RAD1 RAD2 LIT1 LIT2 CAT1 CAT2 ZAG1 ZAG2 JAS1 JAS2 SLB1 SLB2 ZUP1 ZUP2 SRM1 SRM2 BEO1 BEO2

Ni

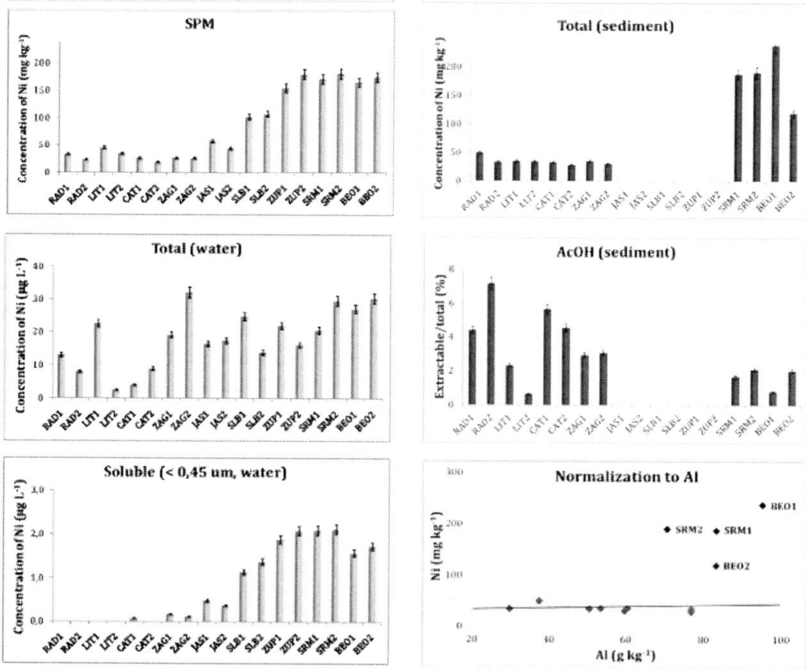
SPM
Concentration of Ni (mg kg⁻¹)
Total (sediment)
Concentration of Ni (mg kg⁻¹)
Total (water)
Concentration of Ni (µg L⁻¹)
AcOH (sediment)
Extractable/total (%)
Soluble (< 0,45 um, water)
Concentration of Ni (µg L⁻¹)
Normalization to Al
Ni (mg kg⁻¹)
Al (g kg⁻¹)
BEO1
SRM2
SRM1
BEO2
RAD1 RAD2 LIT1 LIT2 CAT1 CAT2 ZAG1 ZAG2 JAS1 JAS2 SLB1 SLB2 ZUP1 ZUP2 SRM1 SRM2 BEO1 BEO2

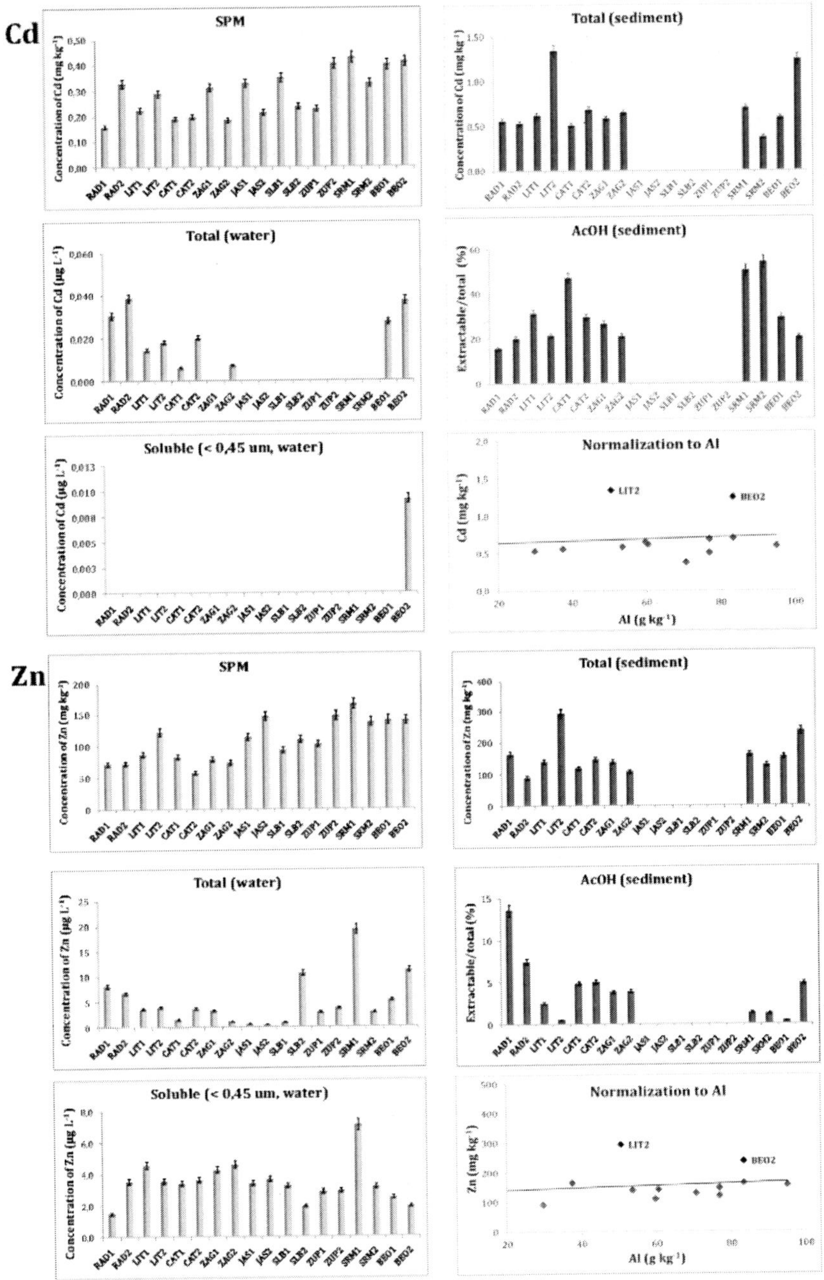

Figure 3. Total element concentrations in water samples, suspended particulate matter and sediments. Data for the easily soluble element fraction in sediments and normalization to Al are also given

3.2 Selected elements in water samples, suspended particulate matter and sediments

Concentrations of 18 selected elements, Cr, Ni, Cd, Zn, P, Pb, As, Cu, Sb, V, Mo, Co, Se, Ba, Mn, Ti, Fe and Al, were determined in the water (whole water samples (microwave-assisted digestion) and dissolved element fraction (<0.45 µm)), SPM and sediment samples (total element concentration after microwave-assisted digestion and the easily soluble element fraction (extraction in 0.11 mol L^{-1} acetic acid)). Normalization of the total element concentrations in sediments vs. total Al concentration was also applied to assess the anthropogenic inputs of elements in sediments. The results of these investigations are for Cr, Ni, Cd and Zn presented in Fig. 3. Data for Al that was used for normalization are presented in Fig. S1 (Electronic Supplementary Material) and for other elements in Fig. S2 (Electronic Supplementary Material).

The data indicate that concentrations of elements in water, SPM and sediments in general increase along the Sava River towards its confluence with the Danube River (Belgrade). Total element concentrations in the whole water samples are significantly higher from the dissolved element fraction (samples filtered through a 0.45-µm filter), which is related to flooding conditions during the sampling campaign and, consequently, to high SPM content in the water samples. From comparison of concentrations of elements in the whole water samples at the input (1) and output (2) boundaries of the selected river water unit (Fig. 3 and Fig. S2 in the Electronic Supplementary Material), there is no evidence on general attenuation of total element concentrations in the whole water samples. It can be therefore concluded that the Lagrangian approach is not applicable to big rivers, such as Sava. Regarding the dissolved element fraction, concentrations of Cd, Pb, Ni, Cu, Cr, As and Zn did not exceed the limit values regulated by the WFD and environmental objectives for surface waters (Official Journal of the European Communities 2000; European Communities Environmental Objectives 2009). Concentrations of soluble Fe, Mn, Zn and Co determined in the present study were similar, dissolved Cd concentrations were lower, while concentrations of Pb, Ni and Cu were found slightly higher (mainly at industrial sites) in comparison to those previously reported for the Sava River by Dragun et al. (2009, 2011).

From data of Fig. 3 and Fig. S2 (Electronic Supplementary Material), it is further evident that concentrations of elements in SPM along the Sava River were lower than total element concentrations in sediments. The grain size of SPM is typically around 30 µm and is usually smaller than in sediments (Bouche et al. 2011). The main constituents of SPM are phyllosilicate minerals and carbonates, which are relatively light and are preferentially transported at the surface of the water column (Bouche et al. 2011). In comparison to clay fraction, which is an important constituent of bed sediments and rich in trace elements, phyllosilicate and carbonate mineral phases contain lower concentrations of trace elements. The other possible reason that concentrations of elements determined in SPM were lower than in sediments is that during flooding events, particles from soils, which may be less contaminated than sediments, were discharged in the river.

The data in Fig. 3 and Fig. S2 (Electronic Supplementary Material) also indicate that the easily soluble element fraction represented in general below 5 % of the total element concentrations. Higher percentages of the soluble element fractions (20–50 %) were observed for Ba, Cd, Mn and Co.

Considering the total element concentrations in sediments, the Sava River is moderately polluted with Cr and Ni. At all sampling sites, Cr concentrations in sediments were higher than the TEC value (43.4 mg kg^{-1}) (MacDonald et al. 2000). Cr concentrations highly exceeded also the PEC value of 111 mg kg^{-1} (MacDonald et al. 2000) at sampling sites RAD1, SRM1 and SRM2 and BEO1 and BEO2, while at sampling sites RAD2, LIT2 and CAT2, Cr concentrations were slightly higher than the PEC value. Similarly as for Cr, Ni concentrations exceeded TEC values (22.7 mg kg^{-1}) at all sampling sites. The PEC value (48.6 mg kg^{-1}) was slightly exceeded at RAD1, while at sampling sites SRM1 and SRM2 and BEO1 and BEO2, total Ni concentrations in sediments highly exceeded the PEC value. Data from normalization to Al indicate on anthropogenic inputs of Cr at sampling sites RAD1, SRM1 and SRM2 and BEO1. For Ni, anthropogenic inputs are evident at sampling sites SRM1 and SRM2 and BEO1 and BEO2. The sampling site RAD1 is impacted by the Acroni metal works Jesenice upstream of the Radovljica city, whereas sediments in Sremska Mitrovica and Beograd (SRM1, SRM2 and BEO1, BEO2) are influenced by the heavy metal and chemical industry from the area. Cr and Ni concentrations at sampling sites influenced by the industrial activities ranged from 180 up to 320 mg kg^{-1} and from 100 to 250 mg kg^{-1}, respectively. These Cr and Ni concentrations are comparable to those determined in sediments of the Sava River in 2005–2006 (Milačič et al. 2010). Vignati et al. (2003) reported similar concentrations for Cr and Ni found in sediments of the Po River, Italy, which were determined during normal and high flow conditions (120 to 230 and 100 to 240 mg kg^{-1}, respectively). Nevertheless, Cr and Ni concentrations in sediments of the Sava River from industrial areas are higher than most of the data reported for the Danube River (around 150 and 100 mg kg^{-1}, respectively) (ICPDR 2008; Sakan et al. 2010; Vignati et al. 2013; Comero et al. 2014). Data from Fig. 3 further revealed extremely low Cr mobility and low mobility of Ni. Namely, below 0.015 % of total Cr and not more than 8 % of total Ni content were extracted with 0.11 mol L^{-1} acetic acid. As these two elements exist mainly in the sparingly soluble forms, it can be expected that Cr and Ni concentrations in sediments at industrially exposed sites do not represent an environmental burden.

Cd concentrations in sediments exceeded the TEC value (0.99 mg kg^{-1}) only at LIT2 and BEO2, but were below the PEC value (4.98 mg kg^{-1}), while Zn concentrations were in general higher than the TEC value (121 mg kg^{-1}), but did not exceed the PEC value (459 mg kg^{-1}). Anthropogenic inputs of Cd and Zn (normalization to Al data) at LIT2 arise from the former mining and metallurgy activities in Litija (Šajn and Gosar 2007) and in BEO2 as a consequence of an untreated sewage system in Belgrade (municipal wastewater treatment system in construction). Concentrations of Cd in sediments of the Sava River are comparable to sampling of sediments in 2005–2006 at the Sava River (Milačič et al. 2010) and to sediments of the Po River, Italy (0.4 to 1.4 mg kg^{-1}) (Vignati et al. 2003). Meybeck et al. (2007) reported slightly higher Cd concentrations for the sediments of the Seine River, France (1 to 2 mg kg^{-1}), and Kohušová et al. (2011) for the sediments of the Bílina River, Czech Republic (around 3 mg kg^{-1}), while considerably higher Cd concentrations were found in sediments of the Danube River (around 2 to 3 mg kg^{-1} Cd) (ICPDR 2008). Nevertheless, recent reports indicate that Cd concentrations in Danube sediments tend to decrease (Sakan et al. 2010; Vignati et al. 2013; Comero et al. 2014). High proportions of the mobile fractions in sediments of the Sava River found for Cd (50 %) (Fig. 3) do not represent an environmental hazard, since the total concentrations of Cd were low. At sampling sites Litija and Beograd, where total Cd content in sediments was around 1.4 mg kg^{-1}, the proportion in the easily soluble fraction was 20 %, so also at these sites the environmental burden from Cd was low. Concentrations of Zn in sediments of the Sava River are similar to those collected in 2005–2006 at the Sava River (Milačič et al. 2010) and are in most cases lower than that reported for the Danube River (ICPDR 2008; Sakan et al. 2010; Vignati et al. 2013; Comero et al. 2014) and River Po, Italy (Vignati et al. 2003) (200 to 500 mg kg^{-1} Zn), as well as for the Seine River, France (Meybeck et al. 2007), and the Bílina River, Czech Republic (Kohušová et al. 2011) (600 to 800 mg kg^{-1} Zn). Due to low mobility of Zn (concentrations in the easily soluble fraction was in general below 5 %), the potential environmental hazard and threat for the aquatic organisms from Zn was appraised to be low.

The highest concentrations of total P in the sediments of the Sava River (Fig. S2) (between 1000 and 1500 mg kg^{-1} of total P) were found mainly due to the use of P-containing fertilisers in agricultural areas (Litija) and also due to impact of untreated municipal sewage from big cities (use of P-containing detergents in household) in big cities (Zagreb, Belgrade). These concentrations are similar to those found in the sampling of the Sava sediments in 2005–2006 (Milačič et al. 2010). Concentrations of Pb, As, Cu,

Sb, V, Mo, Co, Se, Ba, Mn, Ti and Fe in sediments were in general low (see data from Fig. S2, Electronic Supplementary Material).

4 CONCLUSIONS

Data from the sampling campaign performed at the Sava River in September 2014 revealed that the Lagrangian approach of sampling the same water mass at the input and output boundaries of the water unit, to relate transformations of dissolved and suspended contaminant compounds with the travel time, was not applicable to big rivers, such as Sava. It was demonstrated that acidification of samples did not dissolve elements associated with SPM, so microwave-assisted digestion should be applied to decompose trace elements bearing minerals in whole water samples. Soluble element concentrations (<0.45 µm) were significantly lower than concentrations in the whole water samples and were also below the maximal allowable values regulated by the WFD and its daughter directives. Among elements, Cr and Ni were found in moderate concentrations (around 320 and 220 mg kg^{-1}, respectively) in sediments at industrially exposed sites. However, Cr and Ni exist in sparingly soluble forms and as such do not represent an environmental burden. Elevated P concentrations from 1200 to 1500 mg kg^{-1}were found at agricultural areas and big cities. In slightly elevated concentrations, Cd and Zn were also present in sediments at sites impacted by the former mining activities and untreated sewage system from the big cities. However, their portion in the easily soluble fractions was low, so Cd and Zn did not represent an environmental threat. Concentrations of Pb, As, Cu, Sb, V, Mo, Co, Se and Ba in water, SPM and sediments of the Sava River were low. Regarding environmental status, pollution of the Sava River with elements is comparable to other moderately polluted rivers in Europe.

ACKNOWLEDGMENTS

This work has been supported by the European Communities 7th Framework Programme Funding under grant agreement no. 603629-ENV-2013-6.2.1-Globaqua and by the Ministry of Higher Education, Science, and Technology of the Republic of Slovenia (programme group P1-0143).

REFERENCES

1. Alvarez-Guerra M, Viguri JR, Casado-Martínez MC, DelValls TÁ (2007) Sediment quality assessment and dredged material management in Spain: part I, application of sediment quality guidelines in the Bay of Santander. Integr Environ Assess Manag 3:529–538

2. Bakke T, Källqvist T, Ruus A, Breedveld GD, Hylland K (2010) Development of sediment quality criteria in Norway. J Soils Sediments 10:172–178

3. Bird G, Brewer PA, Macklin MG, Nikolova M, Kotsev T, Mollov M, Swain C (2010) Dispersal of contaminant metals in the mining-affected Danube and Maritsa drainage basins, Bulgaria, Eastern Europe. Water Air Soil Pollut 206:105–127

4. Bouche J, Gaillardet J, France-Lanord C, Maurice L, Dutra-Maia P (2011) Grain size control of river suspended sediment geochemistry: clues from Amazon River depth profiles. Geochem Geophys Geosyst 12:1–24

5. CCME - Canadian Council of Ministers of the Environment (2001) Canadian sediment quality guidelines for the protection of aquatic life

6. Dundar MS, Altundag H, Eyupoglu V, Keskin SC, Tutunoglu C (2012) Determination of heavy metals in lower Sakarya river sediments using BCR-sequential extraction procedure. Environ Monit Assess 184:33–41

7. European Communities Environmental Objectives (2009) (Surface waters) regulations S.I. no. 272/2009

8. European Communities Technical Report 2010-041 (2010) Common implementation strategy for the water framework directive (2000/60/EC), guidance document No. 25 on chemical monitoring of sediment and biota under the water framework directive

9. Comero S, Vaccaro S, Locoro G, De Capitani L, Gawlik BM (2014) Characterization of the Danube River sediments using the PMF multivariate approach. Chemosphere 95:329–335

10. Dragun Z, Roje V, Mikac N, Raspor B (2009) Preliminary assessment of total dissolved trace metal concentrations in Sava River water. Environ Monit Assess 159:99–110

11. Dragun Z, Kapetanović D, Raspor B, Teskeredžić E (2011) Water quality of medium size watercourse under baseflow conditions: the case study of river Sutla in Croatia. Ambio 40:391–407

12. Grosbois C, Meybeck M, Horowitz A, Ficht A (2006) The spatial and temporal trends of Cd, Cu, Hg, Pb and Zn in Seine River floodplain deposits (1994-2000). Sci Total Environ 356:22–37

13. Gottler RA (2012) Part 3000 metals. In: Rice EW, Braid RB, Eaton AD, Clesce LS (eds) Standard methods for the examination of water and waste water, 22nd edn. Port City press, Baltimore, pp. 3-1–3-11

14. Giusti I, Taylor A (2007) Natural and antrophogenic contamination of the Fratta-Gorzone River (Veneto, Italy). Environ Monit Assess 134:211–231

15. House WA, Denison FH (2002) Total phosphorus content of river sediments in relationship to calcium, iron and organic matter concentrations. Sci Total Environ 282-283:341–351

16. Howell J-A (2010) The distribution of phosphorus in sediment and water downstream from a sewage treatment works. Biosci Horiz 3:113–123

17. ICPDR (International Commission for the Protection of the Danube River) Joint Danube Survey 2 (2008) Final Scientific report of the International Commission for the protection of the Danube River. Vienna, Austria.https://www.icpdr.org/jds/files/ICPDR_Technical_Report_for_web_low_corrected.pdf. Last Accessed June 2016

18. International Sava River Basin Commission (2002) Protocol on sediment management to the framework agreement on the Sava River Basinhttp://www.savacommission.org/dms/docs/dokumenti/documents_publications/basic_documents/protocols/protokol_nanos.pdf. Last Accessed June 2016

19. International Sava River Basin Commission (2013) Sava River Basin management plan, background paper no.2: groundwater bodies in the Sava River Basin—Annex 1http://www.savacommission.org/dms/docs/dokumenti/srbmp_micro_web/backgroundpapers_final/nnno_2_background_paper_gwbs_in_the_sava_rb.pdf. Last Accessed June 2016

20. ISO 15586 (2003) Water quality—determination of trace elements using atomic absorption spectrometry with graphite furnace

21. ISO 5667-3 (2012) Water quality—sampling—part 3: preservation and handling of water samples

22. Karadede-Akin H, Ünlü E (2007) Heavy metal concentrations in water, sediment, fish and some benthic organisms from Tigris River, Turkey. Environ Monit Assess 131:323–337

23. Kohušová K, Havle L, Vlasák P, Tonika J (2011) A long-term survey of heavy metals and specific organic compounds in biofilms, sediments, and surface water in a heavily affected river in the Czech Republic. Environ Monit Assess 174:555–572

24. Kotnik J, Horvat M, Milačič R, Ščančar J, Fajon V, Kryžanovski A (2003) Heavy metals in the sediments of the Sava River, Slovenia. Geologija 46:263–272

25. Kwok KWH, Batley GE, Wenning RJ, Zhu L, Vangheluwe M, Lee S (2014) Sediment quality guidelines: challenges and opportunities for improving sediment management. Environ Sci Pollut Res Int 21:17–27

26. Long ER, Ingersoll CG, MacDonald DD (2006) Calculation and uses of mean sediment quality guideline quotients: a critical review. Environ Sci Technol 40:1726–1736

27. Loring DH, Rantala RRT (1992) Manual for the geochemical analysis of marine sediments and suspended particulate matter. Earth Sci Rev 32:325–238

28. MacDonald DD, Ingersoll CG, Berger TA (2000) Development and evaluation of consensus-based sediment quality guidelines for freshwater ecosystems. Arch Environ Contam Toxicol 39:20–31

29. McCready S, Birch GF, Long ER, Spyrakis G, Greely CR (2006) An evaluation of Australian sediment quality guidelines. Arch Environ Contam Toxicol 50:306–315

30. Meybeck M, Lestel L, Bonté P, Moilleron R, Colin J-L, Rousselot O, Hervé D, de Pontèves C, Grosbois C, Thevénot DR (2007) Historical perspective of heavy metals contamination (Cd, Cr, Cu, Hg, Pb, Zn) in the Seine River basin (France) following a DPISIR approach (1950-2005). Sci Total Environ 375:204–231

31. Merrington G, Sprang P (2014) Deriving environmental quality standards in European surface waters: when are there too few data? Environ Sci Pollut Res 21:67–76

32. Milačič R, Ščančar J, Murko S, Kocman D, Horvat M (2010) A complex investigation of the extent of pollution in sediments of the Sava River. Part 1. Selected elements. Environ Monit Assess 163:263–275

33. Mohiuddin KM, Otomo K, Ogawa Y, Shikazano N (2012) Seasonal and spatial distribution of trace elements in the water and sediments of the Tsurumi River in Japan. Environ Monit Assess 184:265–279

34. Navarro-Ortega A, Acuña V, Bellin A, Burek P, Cassiani G, Choukr-Allah R, Dolédec S, Elosegi A, Ferrari F, Ginebreda A, Grathwohl P, Jones C, Ker Rault P, Kok K, Koundouri P, Ludwig RP, Milacic R, Muñoz I, Paniconi C, Paunović M, Petrovic M, Sabater S, Skoulikidis NT, Slob A, Teutsch G, Voulvoulis N, Barceló D (2015) Managing the effects of multiple stressors on aquatic ecosystems under water scarcity. Sci Total Environ 503/504:3–9

35. Novotnik B, Zuliani T, Ščančar J, Milačič R (2015) Content of trace elements and chromium speciation in neem powder and tea infusions. J Trace Elem Med Biol 31:98–106

36. Official Journal of the European Communities (2000) Directive 2000/60/EC of the European parliament and of the council establishing a framework for community action in the field of water policy L 327/1

37. Official Journal of the European Union (2009) Commission Directive 2009/90/EC laying down, pursuant to Directive 2000/60/EC of the European Parliament and of the Council, technical specifications for chemical analysis and monitoring of water status. L 201:36–38

38. Quevauviller P, Rauret G, López-Sánchez JF, Rubio R, Ure A, Muntau H (1997) Certification of trace metal extractable contents in a sediment reference material (CRM 601) following a three-step sequential extraction procedure. Sci Total Environ 205:223–234

39. Sakan SM, Đorđević DS, Manojlović DD (2010) Trace elements at traces of environmental pollution in the canal sediments (alluvial formation of the Danube River, Serbia). Environ Monit Assess 167:219–233

40. Svete P, Milačič R, Pihlar B (2001) Partitioning of Zn, Pb, and Cd in river sediments from lead and zinc mining area using the BCR three-step extraction procedure. J Environ Monit 3:586–590

41. Šajn R, Gosar M (2007) Soil pollution in surroundings of Litija as a reflection of mining, metallurgy and natural conditions. Geologija 50:131–145

42. Šömen Joksič A, Katz SA, Horvat M, Milačič R (2005) Comparison of single and sequential extraction procedures for assessing metal

leaching from dredged costal sediments. Water Air Soil Pollut 162:265–283

43. Tessier A, Campbell PGC, Bisson M (1979) Sequential extraction procedure for the speciation of particulated metals. Anal Chem 51:844–851

44. Vasile GD, Nicolau M, Vlădescu L (2010) Zinc speciation in sediments from a polluted river, as an estimate of its bioaccesibility. Environ Monit Assess 160:71–81

45. Vignati D, Pardos M, Diserens J, Ugazio G, Thomas R, Dominik J (2003) Characterisation of bed sediments and suspension of the river Po (Italy) during normal and high flow conditions. Water Res 37:2847–2864

46. Vignati DAL, Secrieru D, Bogatova YI, Dominik J, Céréghino R, Berlinsky NA, Oaie G, Szobotka S, Stanica A (2013) Trace element contamination in the arms of the Danube Delta (Romania/Ukraine): current state of knowledge and future needs. J Environ Manag 125:169–178

47. Writer JH, Keefe SK, Ryan JN, Ferrer I, Thurman ME, Barber LB (2011) Methods for evaluating in-stream attenuation of trace organic compounds. Appl Geochem 26:S344–S345

48. Zuliani T, Mladenovič A, Ščančar J, Milačič R (2016) Chemical characterisation of dredged sediments in relation to their potential use in civil engineering. Environ Monit Assess 188:234

Chapter 13

Sulfur Isotope Fractionation and Sequential Extraction to Assess Metal Contamination on Lake and River Sediments

Lívia Ribeiro de Souza[1] & Kay Knöller[2] & Ana Cláudia Queiroz Ladeira[1]

[1]. *Department of Mineral Processing, Center for the Development of Nuclear TechnologyCDTNBelo HorizonteBrazil*

[2]. *Department of Catchment HydrologyUFZ Helmholtz Centre for Environmental ResearchHalleGermany*

ABSTRACT

Purpose

The present study investigated lake and river sediments affected by metals from an acid mine drainage (AMD) from a former uranium mine. The role of bacterial sulfate reduction in the immobilization of contaminants was evaluated, and the analyses of acid volatile sulfide (AVS) and sequential extraction were performed. Consequently, the potential mobility and bioavailability of contaminants were established.

Materials and methods

Sulfur isotopic fractionation ($\delta^{34}S$), AVS, and sequential extraction procedure were used to assess the sulfate bacterial reduction and the availability of contaminants in the environment at six sampling stations.

Results and discussion

The $\delta^{34}S$ indicated that bacterial reduction is a key process in the natural attenuation of contamination in the Águas Claras reservoir, precipitating metal sulfides. According to the USEPA criteria, adverse biological effects are expected for sample S1 (inside the reservoir) which is likely to be toxic, while for sediment S4 (in the river), the toxicity is uncertain. The other samples were classified as non-toxic, likely because of the decreased solubility of zinc sulfide. A decrease in the concentration of the contaminants downstream of the reservoir was observed. The predominance of U (0.4 %) in the labile fraction and the elevated concentrations of Zn (0.5 %) and Mn (0.7 %) in the sediments inside the reservoir raises concerns regarding the availability of these contaminants in the environment.

Conclusions

The main environmental impact appears to be concentrated in the Águas Claras reservoir, whereas the Antas creek does not seem to be affected by the AMD process. Although the bacterial sulfate reduction is effective in its production of sulfides capable of immobilizing the contaminants, the presence of Zn and U in the labile and reducible fraction is a matter of concern due to its long-term bioavailability. Thus, continuous monitoring of the redox potential of the waters and sediments, mainly in the reservoir, is recommended in order to assess and possibly prevent later dissolution of sequestered contaminants.

Keywords : Acid volatile sulfides Bacterial sulfate reduction Metals Sediments Sulfur isotopes

INTRODUCTION

Acid mine drainage (AMD) is a critical environmental issue caused by chemical and bacterial oxidation of pyrite (FeS_2) and other sulfide bearing minerals exposed to atmospheric conditions during the mining of metal ores and coal (Taylor et al. 1984; Balci et al. 2007). As a result, the generation of acidic waters containing a wide array of toxic metals and sulfur species, mainly sulfate, eventually reaches rivers, groundwaters, and lakes and has adverse effects to both flora and fauna. In particular, dissolved and sedimentary sulfur species and their transformations play an essential role in the hydrochemistry of the waters, since the content of dissolved metals can increase or decrease through interaction with them (Knöller et al. 2004; Sarmiento et al.2009). Sulfate from the AMD can be reduced to sulfide due to bacterial action and can be assessed trough isotopic analyses. In general, the sulfates with lighter sulfur isotopes are more readily metabolized because their bonds are weaker (Seal 2003). As a consequence, a distinct sulfur isotopic signature in sulfide is produced by a depletion in ^{34}S with values as high as 46 ‰ (Rees 1973; Habicht and Canfield 1997) or even depletion up to 75 ‰ as reported by Brunner (2005). Once formed, the sulfide can react with divalent metals to form insoluble sulfides, such as FeS, ZnS, and CuS, decreasing the chemical activity and thus the bioavailability of these contaminants (USEPA 2005). In addition, the metastable iron monosulfides (FeS) can be a precursor of pyrite (FeS_2) that is a fairly stable sulfide which is capable of adsorbing contaminants. Therefore, the assessment of sulfur isotopic fractionation, the quantification of the sulfides, and the determination of the solid phases along with their association with the contaminants are essential for the following reasons: firstly they provide information about distribution, mobility, and bioavailability of the contaminants, and secondly, they indicate the occurrence of natural attenuation processes (Morillo et al. 2008; Sarmiento et al. 2009; Byrne et al. 2010; Charriau et al. 2011).

In Brazil, AMD occurs in a former uranium (U) mine and contains radionuclides and other elements, which are precipitated from the acid water by liming. Due to the inefficiency of the acid water treatment, contaminants overflow from the settling ponds and end up at the Águas Claras reservoir, leading to a concern about their migration to the environment, mainly to the Antas creek (Rodgher et al. 2013). The multiplicity of interactions between the contaminants and the sediments requires a broader approach in order to understand the fixation and/or transport of a wide range of elements, including U. Within this framework, a combination of different techniques was used to elucidate the bio-availability of the most hazardous contaminants present in surficial sediments from the Águas Claras reservoir and Antas creek. Sulfur isotopic fractionation was used to assess the occurrence of the bacterial reduction process. Additionally, the sediment's toxicity was predicted based upon the interactions of toxic metals with the sulfide and organic matter. Finally, the availability of contaminants was assessed by a sequential extraction procedure that indicated the immobilization of contaminants due to sorption, occlusion, and co-precipitation. These data help to infer the mobility and bioavailability of contaminants in the sediments, which can in turn be useful for the decision-making related to remediation of the area.

2 METHODOLOGY

2.1 Study area and sampling

The study area surrounds a former U mine at Poços de Caldas Plateau (Minas Gerais State, Brazil) whose activity was ceased in 1996 leaving behind 44 t of waste rocks that generate AMD. The waste rocks consisted of gibbsite, muscovite, kaolinite, and microcline as well as pyrite and sphalerite. The acid mine water contains mainly sulfate (2400 mg L^{-1}), F (100 mg L^{-1}), U (8 to 12 mg L^{-1}), Mn (up to 170 mg L^{-1}), and Zn (41 mg L^{-1}) and has a pH of ~3.7 (Ladeira and Gonçalves 2007; Rodgher et al. 2013), and it is neutralized with limestone ($CaCO_3$) and lime (CaO) at the water treatment unit. The treated water is released into the Águas Claras reservoir whose waters join the Antas creek. Sediment samples were taken in August 2011 from the reservoir and the Antas creek (Fig. 1). In the Águas Claras reservoir, the sediments were collected using a van Veen grab while in the Antas creek a cylindrical collector was used. Samples were placed in plastic containers and stored at −4 °C before analysis. The sampling locations were defined by the Brazilian Environmental Agency and were labelled S1, S2, S3, S4, S5, and S6. The procedure was repeated in August 2012, and the samples were lyophilized and submitted for isotopic analysis. These samples were taken from the same sampling stations but named S1B, S2B, S3B, S4B, S5B, and S6B.

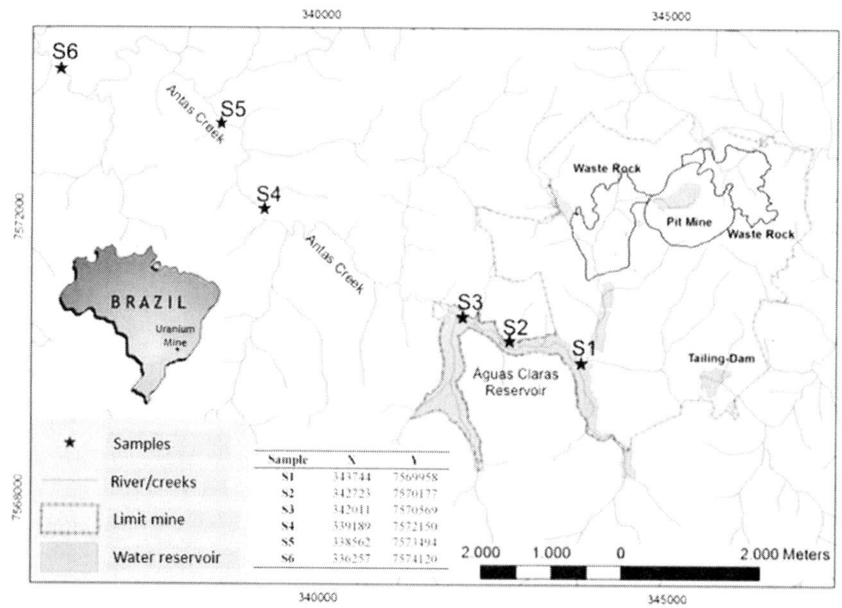

Figure 1. Overview of the study area showing Águas Claras reservoir, the Antas creek, sampling stations (S1, S2, S3, S4, S5, and S6), and the UTM coordinates (southern hemisphere, zone 23)

2.2 Determination of organic carbon, total sulfur, acid volatile sulfide, simultaneously extracted metals, and toxicity of the sediments

The total sulfur and the organic carbon (f_{oc}) were determined using a LECO CS230 (Leco Corporation, St. Joseph, USA). Samples were dried at 40 °C and 2 mL of HCl (4 M) was added to eliminate the inorganic carbon. The samples were dried again and the residual organic carbon determined. The acid volatile sulfide (AVS) and simultaneously extracted metals (SEM) determination was in accordance with the EPA 821-R-91-100 (USEPA 1991) procedures. Wet sediments were weighed and placed in flasks with 100 mL of Milli-Q water, which was deaerated by bubbling N_2. After 20 min, 20 mL of HCl (6 M) was added and the reactor was under continuous flow of nitrogen for the following 2 h, while the volatilized sulfides were trapped in flasks with 0.5 M NaOH. The acid slurry was filtered in a 0.45-μm polyacetate membrane, and the metals in the filtrate (Fe, Mn, Cu, Cd, Pb, Ni, Zn) were determined by flame atomic absorption spectrometry (Varian AA 240FS, Mulgrave, Australia). Detection limits were 0.03 mg L^{-1} for Mn and 0.1 mg L^{-1} for all the other elements. Sulfide was determined by the colorimetric method with methylene blue using a UV–VIS spectrophotometer (Varian Cary 50, Mulgrave, Australia). The calibration curve for the determination of sulfides was obtained with solutions of $Na_2S{\cdot}9H_2O$ in concentrations between 0.2 and 1.2 mg L^{-1}. The percentage of recovery was estimated based on a standard solution of Na_2S, yielding results ranging between 84 and 92 %.

The toxicity of the sediments was estimated by the difference of $\Sigma SEM - AVS$. If the $\Sigma SEM - AVS \leq 0$, the absence of toxicity can be assured; however, predicting which sediments with SEM-AVS >0 will be toxic is less accurate. In this case, the organic carbon was considered as additional binding phase as many dissolved metals readily bind to dissolved organic carbon forming complexes that do not appear to be bioavailable (Di Toro et al. 2005; USEPA 2005); this is a standard procedure to derive concentrations of metal mixtures in sediment, which are protective of the presence of benthic organisms (USEPA 2005). This approach was chosen because it accounts for the varying biological availability of chemicals in different sediments and allows for the incorporation of the appropriate biological effects concentration. The toxicity criteria for the sediments based on AVS, SEM, and organic carbon (f_{oc}) are shown in Table 1.

Table 1. Toxicity criteria for sediments based on AVS, SEM, and f_{oc}. (adapted from USEPA 2005)

Limit values (μmol/g_{OC})	Predicted biological effect
$\Sigma SEM - AVS \leq 0$	Sediment is non-toxic
$\left[\frac{\Sigma SEM-AVS}{f_{oc}}\right] < 130$	Sediment toxicity is not likely to be caused by Cd, Cu, Pb, Ni, Ag, and Zn
$130 < \frac{\Sigma SEM-AVS}{f_{oc}} < 3000$	Sediment toxicity is uncertain due to Cd, Cu, Pb, Ni, Ag, and Zn
$\frac{\Sigma SEM-AVS}{f_{oc}} > 3000$	Sediment toxicity is likely to be caused by Cd, Cu, Pb, Ni, Ag, and Zn

2.3 Isotopic analysis

For the isotopic analysis, three distinct sedimentary sulfur compounds were extracted from the sediments and analyzed separately: (i) the sulfur present in the AVS, (ii) the elemental and pyritic sulfur (chromium reducible sulfur (CRS)), and the (iii) water-soluble sulfate. In order to obtain the AVS and CRS fractions, the freeze-dried sediment samples were reacted with HCl (6 M) under nitrogen gas flow, liberating AVS as H_2S (Fossing and Jorgensen 1989). Subsequently, the remaining slurry was boiled for 2 h after the addition of acidified $CrCl_2$solution, and CRS was carried out in the flow as H_2S. In each step, the produced sulfide was trapped in a zinc acetate solution and the ZnS precipitate was converted to Ag_2S by addition of 0.1 M $AgNO_3$ solution. The water-soluble sulfate fraction was obtained by shaking the freeze-dried sediment with distilled water for 4 h, followed by filtration. The filtrate containing the soluble sulfate was acidified with HCl 6 M to a pH between 2 and 3 and heated for the elimination of CO_2, and then the $BaCl_2$ solution was added leading to the precipitation of sulfate as $BaSO_4$. The stable isotope ratios of sulfur ($^{34}S/^{32}S$) were determined at UFZ—Centre for Environmental Research, Leipzig-Halle, Germany, using continuous-flow isotope ratio mass spectrometry (DeltaS, ThermoFinnigan, Bremen, Germany) with a reproducibility of ±0.4 ‰. The sulfur isotope measurements were reported in a delta notation ($\delta^{34}S$) as part per thousand (‰) deviation relative to the Vienna Canon Diablo Troilite (VCDT) standard. NBS 127 was used as a primary reference material for normalization of the raw data against a standard reference.

2.4 Sequential extraction

The sequential extraction procedure was based on the modified Bureau Communautaire de Références (BCR) (Rauret et al. 1999). The sequential fractions were named labile (E1), reducible (E2), oxidizable (E3), and residual (E4). Among all the constituents of the sediments, the elements Fe, Mn, U, and Zn were selected for investigation. Uranium analyses were carried out using a X-ray spectrometry (Kevex, model Sigma X9050, CA, USA) coupled with a radioactive source of americium (^{241}Am) and a Si(Li) detector with a detection limit of 10 mg L^{-1}. Samples below this detection limit were determined by neutron activation in the Triga Mark I IPR-R1 (General Atomics, San Diego, USA) reactor, and the detection limit was 0.1 mg L^{-1}. The other elements were analyzed by flame AA spectrometry (Varian AA 240FS, Mulgrave, Australia). The sum of the four fractions of U, Fe, Mn and Zn accounted for 82-116 % of the total concentration. Detection limits were 0.13, 0.03 and 0.08 mg L^{-1} for Fe, Mn and Zn, respectively. All reagents used for analysis were ACS grade (Merck), and reagent blanks were determined for each step of analysis.

3 RESULTS AND DISCUSSION

3.1 Sulfur isotopic composition of the sediments

The isotopic ratio ($\delta^{34}S$) is a useful tool to provide information about the occurrence and the extension of sulfur redox processes. In sediments, there are three main inorganic fractions associated with the redox processes: (i) the AVS produced by the precipitation of biogenic hydrogen sulfide with bivalent metal ions, forming the biogenic sulfide; (ii) the CRS is comprised mostly of pyritic sulfur originated either from the aging of the biogenic sulfide and/or from geogenic pyrite, and (iii) the SO_4 generated during the weathering of sulfide bearing rocks and mainly present in AMD. The $\delta^{34}S$ signatures of the three main sulfur compounds (SO_4, AVS and CRS) for each sample are presented in Fig. 2. As shown in Fig. 2, the slight variation of the $\delta^{34}S$ signatures for the SO_4 throughout the reservoir and the creek indicates that this anion derives from the same sulfur source in all sediments. The average $\delta^{34}S$ value for the SO_4 for all samples is −3.03 ‰ and is within the values reported for the pyrite from the Poços de Caldas Plateau that ranges from +1.24 to −3.63 ‰ (Waber et al. 1992). Considering that the SO_4 formed via oxidation of pyrite presents small or negligible sulfur isotope fractionation (Taylor et al. 1984; Balci et al. 2007), it may be concluded that the pyrite oxidation is the main source of sulfate. In the sediments S4B and S6B, no sulfate was detected, probably because the concentrations of dissolved sulfate and Ca^{2+} in the creek are below the saturation limit.

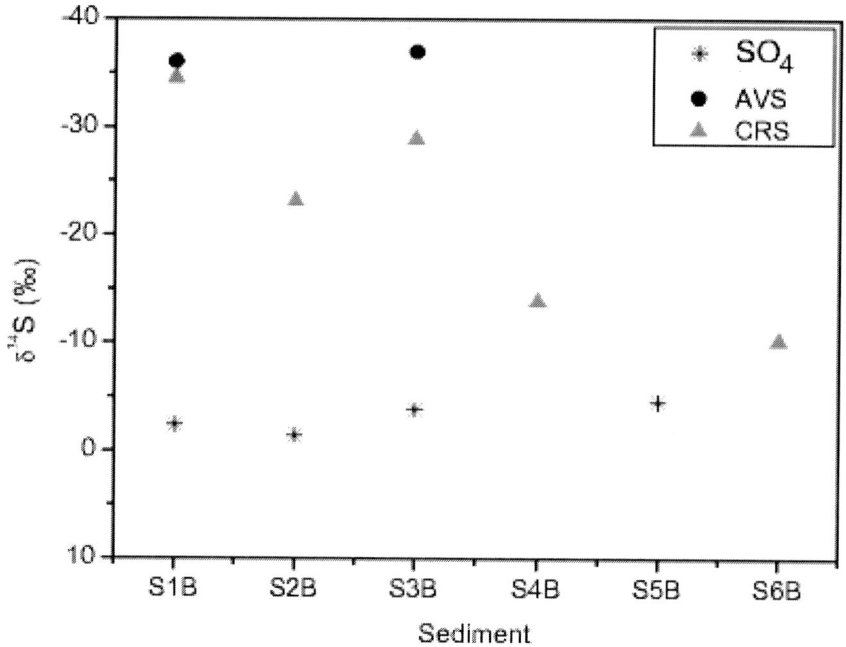

Figure 2. Isotopic signatures of the fractions SO_4, AVS, and CRS for the surface sediments

According to Fig. 2, the AVS isotopic signatures were ~–37 ‰ for the sediments S1B and S3B; the first and last sampling station inside the reservoir. These results are consistent with those reported in the literature where the dissimilatory reduction of sulfate produces sulfide depleted in ^{34}S up to 46 % compared to the initial isotopic composition of SO_4 (Rees 1973; Habicht and Canfield 1997). The sulfate-reducing process and the precipitation of divalent metals with the resulting sulfide is shown in Eqs. (1) and (2), respectively:

$$CH_3COO^- + SO_4^{2-} \rightarrow 2\,HCO_3^- + HS^- \quad (1)$$

$$HS^- + M^{2+} \rightarrow MS + H^+ \quad (2)$$

where the acetate CH_3COO^- represents dissolved organic carbon, although other organic compounds may be found. A group of anaerobic microorganisms, mainly bacteria and archaea (Muyzer and Stams 2008), reduce sulfate (SO_4^{2-}) to sulfide (S^{2-}) using organic electron donors through dissimilatory reduction. The sulfate reduction consumes protons, consequently increasing the pH and yielding sulfides which react with metals (e.g., Fe, Zn, Cu, Cd, Ni, and Pb), leading to the precipitation of insoluble metal sulfides. Most of the sulfate-reducing bacteria are neutrophilic, thriving at a pH between 6 and 8. However, the existence of acidophilic bacteria is also reported, offering the possibility to form insoluble metal sulfides without the need of prior neutralization of the AMD (Sánchez-Andrea et al. 2014).

Commonly, iron is the most abundant divalent metal in the pore water of sediments and susceptible to sulfide precipitation; therefore, iron sulfides can be the predominant phase in AVS (Rickard and Morse 2005). Once the FeS is formed, three main mechanisms have been reported for the formation of neogenic pyrite: (i) a reaction in anoxic conditions (Eq. 3), (ii) a reaction with reduced sulfur or polysulfur (Eq. 4) and (iii) a direct precipitation of dissolved sulfides with iron oxides (Eq. 5) (Lesven et al. 2008):

$$FeS + H_2S \rightarrow FeS_2 + H^+ \quad (3)$$

$$FeS + 1/8\ S_8 \rightarrow FeS_2 \quad (4)$$

$$FeOOH + 3/2HS^- + 1/16S_8 + 3/2H^+ \rightarrow FeS_2 + 2H_2O \quad (5)$$

The $\delta^{34}S$ values for CRS inside the reservoir, from −21 to −33 ‰, are between the AVS and the SO_4 values, which indicate the presence of primary sedimentary pyrite as well as neogenic pyrite formed by the bacterial reduction process. In contrast, the $\delta^{34}S$ value for CRS in the Antas creek were more enriched and closer to the sulfate isotopic value and thus to the $\delta^{34}S$ value of −3.03 ‰ for pyrite in the region. Therefore, inside the Águas Claras reservoir, the bacterial sulfate reduction is a significant process resulting in the precipitation of metallic sulfides and the formation of neogenic pyrite, capable of occluding contaminants in its structure.

3.2 Acid volatile sulfide and simultaneously extracted metals

Once the bacterial reduction of the sulfate was confirmed by isotopic studies, the next step was the assessment of the extension of the reduction process through the quantification of sulfides and their bonded metals (i.e., AVS–SEM) (USEPA 1991). The AVS fraction comprises the free sulfides (S^{2-}), iron sulfides such as FeS (mackinawite) and Fe_3S_4 (greigite) (Rickard and Morse2005), and sulfides from bivalent metals such as Cu, Ni, Zn, Cd, and Pb. The SEM fraction is composed of the metals, normally Cd, Cu, Pb, Ni, Hg, Ag, and Zn, liberated from the sediment during the acidification. In the present work, the concentrations of Ni, Cu, and Cd were below the detection limit

(0.1 mg L^{-1}) and considered not relevant. The results for AVS, metals extracted in the SEM, organic carbon (f_{oc}), and total sulfur (S_{Tot}) are shown in Table 2.

Table 2. Organic carbon (f_{oc}), total sulfur (S_{Tot}), AVS, and metals extracted in the SEM procedure

	f_{oc} (g kg^{-1})	S_{Tot} (g kg^{-1})	AVS (mg kg^{-1})	Fe (g kg^{-1})	Mn (mg kg^{-1})	Zn (mg kg^{-1})	Pb (mg kg^{-1})
S1	5.6	11.6	151	34.5	7184	5845	24
S2	22.2	13.0	2665	42.5	184	1375	35
S3	23.6	9.9	2969	31.7	300	1561	37
S4	10.2	2.0	3	11.9	917	160	23
S5	25.5	4.2	133	21.1	2058	188	23
S6	21.4	3.2	58	17.9	789	101	30

As shown in Table 2, there is a significant variability in the content of AVS in the sediments, ranging from 3 to 2969 mg kg^{-1}, which indicates different environmental conditions that may be unsuitable for the formation and precipitation of AVS phases. For instance, sediments S1 and S4 presented the lowest AVS concentrations, 150 and 3 mg kg^{-1}, respectively, as well as the lowest concentration of organic carbon, 5.584 and 10.214 mg kg^{-1}, respectively, and Eh of 66.4 and 58.8 mV, respectively. For sediment S1, the high content of total sulfur is associated with insoluble phases of sulfate, carried from the water treatment station, which contains ettringite ($Ca_6Al_2(SO_4)_3(OH)_{12}\cdot 26H_2O$) and gypsum ($CaSO_4\cdot 2H_2O$) (Gomes et al. 2012). The other samples inside the reservoir (S2 and S3) present high AVS, ~3000 mg kg^{-1}, and the greatest amount of organic carbon. These results corroborate the bacterial reduction activity suggested by isotopic fractionation and indicate that the sulfide production is prolific, leading to precipitation of metallic ions and reducing the availability of metals. Furthermore, the bacterial reduction could be one of the reasons behind the decreasing sulfate content in the waters downstream of the reservoir, as reported in the Technical Report of the Water Committee (Comissão das Águas 2012). According to this report, the concentration of sulfate in the water at the station S1 was 290 mg L^{-1} while at station S2 the value was 135 mg L^{-1}.

Despite the high concentrations of organic carbon in the sediments S5 and S6, compared to S2 and S3, the content of AVS is much lower, i.e., of 133 and 58 mg kg^{-1}, respectively, which implies that bacterial reduction processes are not pronounced. This fact may be associated with the low content of sulfate in the waters of Antas creek (~35 mg L^{-1}), much lower than the values in the reservoir. In addition, the higher water flow in Antas creek results in an oxic environment, hindering the bacterial sulfate reduction and the formation of metals sulfides. Table 2 also shows the metals, Fe, Mn, Zn, and Pb, extracted in the AVS–SEM procedure after the dissolution of the biogenic metallic sulfides as well as the amorphous and crystalline phases of Fe and Mn. According to Table 1, the toxicity of the sediments that can affect the benthic organisms can be estimated by the difference (ΣSEM – AVS) which is based upon the chemical interactions of metal-sulfide and the equilibrium partitioning theory (Di Toro et al. 2005). When there is an excess of sulfide, the metals precipitate as insoluble phases, such as NiS, ZnS, CdS, PbS, CuS, and Ag_2S, whose preference for precipitation depends on their solubility product (K_{ps}). As a result, the metals are sequestrated from the aqueous phase, what decreases their activity and, consequently, their bioavailability. The results for the assessment of toxicity based upon the difference (ΣSEM – AVS) and the normalized difference ((ΣSEM – AVS)/f_{oc}) for all the sediments are shown in Table 3. According to Table 3, the sediments S2 and S3 (inside the Águas Claras reservoir) and the sediments S5 and S6 (in the river) were classified as non-toxic, since the difference of ΣSEM – AVS were negative. On the other hand, the sediments S1 and S4 presented positive value, i.e., (ΣSEM – AVS) > 0, and were normalized by the fraction of organic carbon (f_{oc}). According to these criteria, the sample S4 presented uncertain toxicity once the value of ΣSEM – AVS)/f_{oc} is between 130 and 3000 µmol g_{OC}^{-1}. The sediment S1 presented (ΣSEM – AVS)/f_{oc} >3000 µmol g_{OC}^{-1}, which implies high risk of adverse biological effects mainly due to the high content of Zn. As will be discussed in the sequential extraction results below, the sample S1 was strongly affected by the overflow of sludge from the water treatment unit.

Table 3. Results of sediment toxicity based upon the difference (ΣSEM – AVS) and the normalized difference (ΣSEM – AVS)/f_{oc})

Sample	AVS (µmol g^{-1})	ΣSEM (µmol g^{-1})	f_{oc} (g_{OC} g^{-1})	(ΣSEM – AVS) (µmol g^{-1})	(ΣSEM – AVS)/f_{oc} (µmol g_{OC}^{-1})
S1	4.70	89.40	0.006	84.70	15,168
S2	83.11	21.20	0.022	–61.91	–2785
S3	92.58	24.05	0.024	–68.53	–2913
S4	0.10	2.56	0.010	2.46	241
S5	4.15	2.98	0.025	–1.17	–46
S6	1.82	1.70	0.021	–0.13	–6

g_{OC} gram of organic carbon

In conclusion, the results show that the soluble sulfate from the reservoir undergoes bacterial reduction producing sulfide. The S^{2-} can precipitate divalent metals, particularly Zn, producing highly insoluble metal sulfides which decrease the availability to the environment. Therefore, despite the high levels of total metals in most sediment, samples S2, S3, S5, and S6 were assessed as non-toxic. On the other hand, the sediments S1 and S4 present low amounts of sulfide; consequently, sample S1 is likely to be toxic, whereas for S4, the biological adverse effects are uncertain. Therefore, it is recommended to monitor the redox potential once its increase can promote the oxidization of sulfides as well as the degradation of organic compounds, accelerating the liberation of the adsorbed/complexed metals (Zhang et al. 2014). Furthermore, an increase in the redox potential can also inhibit the activity of anaerobic sulfate-reducing bacteria and thus hamper the production of sulfides.

3.3 Sequential extraction

The AVS procedure takes into account only the elements Cu, Zn, Pb, Ni, Cd, and Ag in the estimation of the toxicity of the samples. Consequently, other toxic metals present in sediments are not assessed by this method. However, as the study area presents other elements, such as U and Mn, additional methods should be used to identify their association with different mineralogical phases and, consequently, their availability to the environment. Thus, sequential extraction experiments were carried out and four elements were selected to be evaluated: Fe and Mn, because they are considered as a sink for contaminants (Filgueiras et al. 2002), and Zn and U because they are present in high levels in sediments, besides their toxicity. Figure 3shows the sequential extraction results as well as the total metal concentration in the original samples. It is noted that the sample station S1 presents the highest concentration of Mn, Zn, and U (7417, 5285, and 3600 mg $kg-^{1}$, respectively) followed by samples S2 and S3 that are also enriched in U and Zn if compared to the sediments in the Antas creek. This enrichment is caused by part of the precipitate that overflows from the water treatment unit and settles at the bottom of the reservoir.

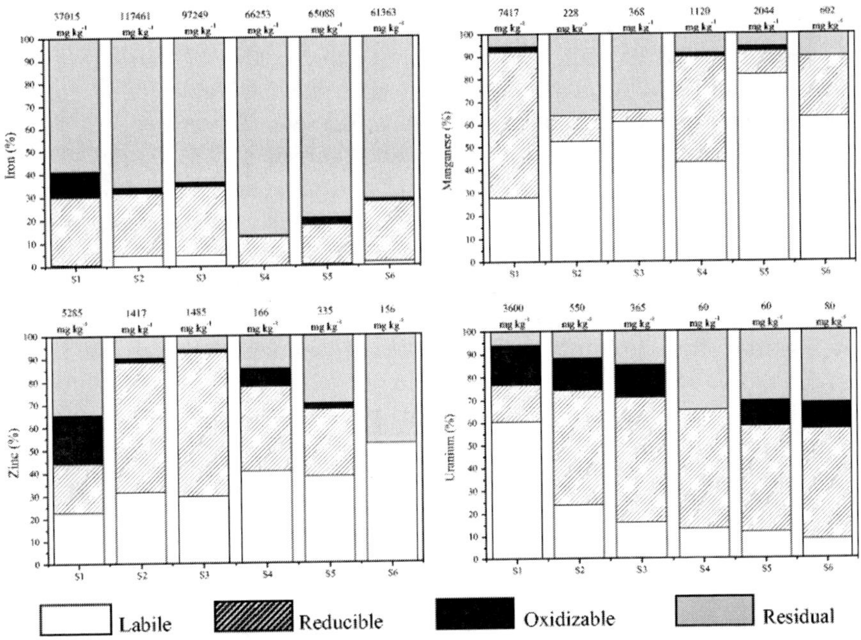

Figure 3. Mn, Zn, U, and Fe partitioning in the sediments, according to mBCR procedure. The total concentration is displayed on the top of each column

Iron As shown in Fig. 3, minor amounts of Fe were associated with the labile and oxidizable fractions. However, it was mainly extracted in the reducible and residual steps which are similar to the results reported in the literature (Morillo et al. 2008; Charriau et al. 2011). The reducible fraction (E2) comprises 12 to 30 % and consists of oxyhydroxides present in highly weathered environments (amorphous Fe). In the sediments, these amorphous phases are important sinks for contaminants, since they have a high capacity of adsorbing other metals in their structure (Filgueiras et al. 2002). The small portion of Fe extracted in the oxidizable fraction is associated with organic matter and sulfides. In all sediments, Fe was preferentially associated with the residual fraction comprising crystalline forms of hematite and goethite, in addition to aluminossilicates containing Fe (II and III) and pyrite.

Manganese

High percentage of Mn (28 to 81 %) was extracted in the labile fraction (E1), most likely due to its association with the carbonate phases formed in the water treatment unit as reported by Gomes et al. (2012). In the reducible fraction E2 (i.e., amorphous Fe oxides and crystalline Mn oxides; Filgueiras et al. 2002), it was extracted 5 to 64 % of the Mn. The highest extraction was for sample S1 which also presented the highest Mn content. The oxidizable fraction (E3) seems to be irrelevant as it represents at maximum of 2 % of the Mn. This is explained by the low content of Mn sulfides as well as the difficulty of Mn to form complexes with organic matter. The Mn present in the residual fraction, which comprises the phases resistant to the weathering and bonded to lithogenic minerals, ranged from 6 to 36 %.

Zinc The mobility of Zn is evidenced by the high levels of extraction in the labile fraction E1 (20 to 50 %) indicating that it could be mainly weakly adsorbed, easily exchangeable, and/or co-precipitated with carbonates (Filgueiras et al. 2002). The labile fraction is also associated with the extraction of Zn precipitated as AVS, since the sample drying may result in the oxidation of the sulfides, increasing the amount of Zn extracted. Since metals in the labile fraction are considered readily and potentially bioavailable, the Zn associated to this fraction is of concern for the samples S1, S2, and S3, which contain high levels of zinc (5285 to 2417 mg kg^{-1}). These results corroborate the findings in the AVS–SEM approach, in which sample S1 was classified as likely to be toxic due to the great amount of Zn which is highly bioavailable. Inside the reservoir, samples S2 and S3 have around 60 % of total Zn extracted in the reducible fraction, mainly in the form of Zn oxide and/or adsorbed in Fe oxyhydroxides since almost 30 % of total Fe is extracted in this phase. The association of Zn in Fe–Mn oxides/hydroxides confirms the known ability of this phase to scavenge Zn from the aqueous phase, and these phases are completely extracted during the AVS–SEM procedure. The sediment S1 was the only one to present significant amount of Zn associated with the oxidizable fraction. Since this sediment presented low contents of organic matter and sulfides, Zn was probably associated with other oxidizable phases such as CRS-Fe. Despite the elevated content of organic matter in the other samples, only 2 % of total Zn was extracted in the oxidizable fraction, indicating it is not significant as a binding phase. The Zn extracted in the residual fraction indicates the presence of sphalerite ZnS, a mineral found in the region, and in part contributes to the acid mine generation. The Zn inside the Águas Claras reservoir is mainly present in sediments due to reactions of precipitation, complexation, and/or adsorption, which enrich the levels of this element in the labile and reducible fractions.

Uranium The sample S1 presented the highest extraction of U in the labile fraction, ~60 %, most likely due to the presence of calcium diuranate (CaU_2O_7) that overflowed from the water treatment unit (Gomes et al. 2012). Minor amounts of U were extracted in the labile fraction in the other sediments, ranging from 2 to 23 %. For these samples, U can be non-specifically adsorbed on the surface of clays or on Fe and Mn oxyhydroxides. The presence of the U in the labile fraction is particularly worrying as its high concentration—3600 mg kg^{-1} in sample S1, for instance—can pose risks to the environment. However, it is interesting to note that the U content in the labile fraction decreases with distance from upstream to downstream the lake, i.e., from sample S1 to S6. In all samples except S1, most of the U is associated with the Mn and Fe oxyhydroxides (step E2), most likely due to the sorption or occlusion of U in hematite (Fe_2O_3) and goethite (Fe_3O_4) (Duff et al. 2002). The lower extraction of U in step E2 for sample S1, ~15 %, is related to the low content of Fe in this sample. The U in the oxidizable fraction E3 is ~15 % and represents the U sorbed in the organic matter (Duff et al. 2002) and/or the U reduced into crystalline oxides (e.g., uraninite) in an effective mechanism of bioremediation. Sample S4, the only one without U in the oxidizable fraction, is also the one which presented the lowest level of total sulfur and organic matter. Significant contents of U in the residual fraction are found in sediments S4, S5, and S6, ~30 %. The U in these sediments may be included in the crystalline iron oxides, silicates, and gibbsite or as uraninite; all of them are typical minerals of the region. On the other hand, the S1, S2, and S3 samples, collected inside the reservoir, showed a low content of U associated with the residual fraction (6 to 14 %), which reinforces the contamination of the lake by U solid phases with great bioavailability.

4 CONCLUSIONS

The results indicate that the Antas creek seems not to be affected by the AMD, while the environmental impact appears to be concentrated in the Águas Claras reservoir and this has been caused by an enrichment of metals in its sediments. Among the samples, point S1 was particularly affected by the discharge of treated effluent as shown by the elevated concentrations of Zn (0.5 %), Mn (0.7 %), and U (0.4 %). In addition, according to the AVS–SEM analyses, the sample S1 is likely to be toxic, probably due to the high content of Zn. All the other samples are unlikely to be toxic or present uncertain toxicity. The isotopic sulfur investigation in the Antas creek showed no evidence of bacterial sulfate reduction, and this is confirmed by the presence of geogenic pyritic typical of the geology of the region. Sequential extraction showed U, Zn, and Mn mostly present in the labile fraction and reducible phase. For this reason, special attention should be given to the sediments inside the reservoir to avoid the spread of the contaminants. Therefore, to prevent later mobilization of the metals it is recommended that the sediments remain in anoxic conditions, thus hindering the re-dissolution of the main scavenging phases.

It is worth mentioning that the present work is a contribution to the remediation of areas contaminated with metals, including radioactive and non-radioactive elements. The full characterization of the bioremediation process, as well as the definition of the chemical availability of the contaminants, could help determine the best remediation plan for such areas, which could be the complete removal of the sediments or the improvement of the bioremediation process. The enhancement of the bacterial sulfate reduction has the advantage of promoting the formation of metal sulfides as well as the reduction of the highly soluble U(VI) to a less soluble U(IV). In addition, the understanding of the natural processes provided by this investigation could lead to possible strategies for artificial immobilization of the metals.

ACKNOWLEDGMENTS

The authors thank CAPES, FAPEMIG, and CNPq—INCT-Acqua, PPM, CT Mineral—for the financial support and the staff of the Stable Isotope Laboratory of UFZ (Leipzig-Halle) for their isotopic analytical assistance. Also, the authors gratefully acknowledge the comments offered by the editor and two anonymous reviewers.

REFERENCES

1. Balci N, Shanks WC, Mayer B, Mandernack KW (2007) Oxygen and sulfur isotope systematics of sulfate produced by bacterial and abiotic oxidation of pyrite. Geochim Cosmochim Acta 71:3796–3811

2. Brunner B (2005) A revised isotope fractionation model for dissimilatory sulfate reduction in sulfate reducing bacteria. Geochim Cosmochim Acta 69:4759–4771

3. Byrne P, Reid I, Wood PJ (2010) Sediment geochemistry of streams draining abandoned lead/zinc mines in central Wales: the Afon Twymyn. J Soils Sediments 10:683–697

4. Charriau A, Lesven L, Gao Y, Leermakers M, Baeyens W, Ouddane B, Billon G (2011) Trace metal behaviour in riverine sediments: role of organic matter and sulfides. Appl Geochemist 26:80–90

5. Comissão das Águas (2012) Avaliação da qualidade das águas e sedimentos das microbacias do Ribeirão das Antas e do Ribeirão de Caldas no Planalto de Poços de Caldas - Relatório Técnico da Comissão das Águas. Available at: pocosdecaldas.mg.leg.br/index.php?download = 52. Accessed 25 Sep 2015

6. Di Toro DM, McGrath JA, Hansen DJ, Berry WJ, Paquin PR, Mathew R, Wu KB, Santore RC (2005) Predicting sediment metal toxicity using a sediment biotic ligand model: methodology and initial application. Environ Toxicol Chem 24:2410–24127

7. Duff MC, Coughlin JU, Hunter DB (2002) Uranium co-precipitation with iron oxide minerals. Geochim Cosmochim Acta 66:3533–3547

8. Filgueiras AV, Lavilla I, Bendicho C (2002) Chemical sequential extraction for metal partitioning in environmental solid samples. J Environ Monit 4:823–857

9. Fossing H, Jorgensen B (1989) Measurement of bacterial sulfate reduction in sediments: evaluation of a single-step chromium reduction method. Biogeochemistry 8:205–222

10. Gomes AFS, Lopez DL, Ladeira ACQ (2012) Characterization and assessment of chemical modifications of metal-bearing sludges arising from unsuitable disposal. J Hazard Mater 199–200:418–425

11. Habicht KS, Canfield DE (1997) Sulfur isotope fractionation during bacterial sulfate reduction in organic-rich sediments. Geochim Cosmochim Acta 61:5351–5361

12. Knöller K, Fauville A, Mayer B, Strauch G, Friese K, Veizer J (2004) Sulfur cycling in an acid mining lake and its vicinity in Lusatia, Germany. Chem Geol 204:303–323

13. Ladeira ACQ, Gonçalves CR (2007) Influence of anionic species on uranium separation from acid mine water using strong base resins. J Hazard Mater 148:499–504

14. Lesven L, Gao Y, Billon G, Leermakers M, Ouddane B, Fischer JC, Baeyens W (2008) Early diagenetic processes aspects controlling the mobility of dissolved trace metals in three riverine sediment columns. Sci Total Environ 407:447–459

15. Morillo J, Usero J, Rojas R (2008) Fractionation of metals and As in sediments from a biosphere reserve (Odiel salt marshes) affected by acidic mine drainage. Environ Monit Assess 139:329–337

16. Muyzer G, Stams AJM (2008) The ecology and biotechnology of sulphate-reducing bacteria. Nat Rev Microbiol 6:441–454

17. Rauret G, López-Sánchez JF, Sahuquillo A, Rubio R, Davidson C, Ure AM, Quevauviller P (1999) Improvement of the BCR three step sequential extraction procedure prior to the certification of new sediment and soil reference materials. J Environ Monit 1:57–61

18. Rees CE (1973) A steady-state model for sulphur isotope fractionation in bacterial reduction processes. Geochim Cosmochim Acta 37:1141–1162

19. Rickard D, Morse JW (2005) Acid volatile sulfide (AVS). Mar Chem 97:141–197

20. Rodgher S, de Azevedo H, Ferrari CR, Roque CV, Ronqui LB, de Campos MB, Nascimento MRL (2013) Evaluation of surface water quality in aquatic bodies under the influence of uranium mining (MG, Brazil). Environ Monit Assess 185:2395–2406

21. Sánchez-Andrea I, Sanz JL, Bijmans MFM, Stams AJM (2014) Sulfate reduction at low pH to remediate acid mine drainage. J Hazard Mater 269:98–109

22. Sarmiento AM, Olías M, Nieto JM, Cánovas CR, Delgado J (2009) Natural attenuation processes in two water reservoirs receiving acid mine drainage. Sci Total Environ 407:2051–2062

23. Seal RR (2003) Stable-isotope geochemistry of mine waters and related solids. In: Jambor JL, Blowes DW, Ritchie AIM (eds) Environmental aspects of mine wastes, Miner. Soc. Canada, Short Course Series 31., pp 303–334

24. Taylor BE, Wheeler MC, Nordstrom DK (1984) Isotope composition of sulphate in acid mine drainage as measure of bacterial oxidation. Nature 308:538–541

25. USEPA (1991) Draft analytical method for determination of acid volatile sulfide and selected simultaneously extractable metals in sediment. EPA-821-R-91-100. US Environmental Protection Agency, Washington, DC

26. USEPA (2005) Procedures for the derivation of equilibrium partitioning sediment benchmarks (ESBs) for the protection of benthic organisms: metal mixtures (cadmium, copper, lead, nickel, silver and zinc). EPA-600-R-02-011. US Environmental Protection Agency, Washington, DC

27. Waber N, Schorscher HD, Peters T (1992) Hydrothermal and supergene uranium mineralization at the Osamu Utsumi mine, Poços de Caldas, Minas Gerais, Brazil. J Geochem Explor 45:53–112

28. Zhang C, Yu Z, Zeng G, Jiang M, Yang Z, Cui F, Hu L (2014) Effects of sediment geochemical properties on heavy metal bioavailability. Environ Int 73:270–281

Index